W9-BIT-177

ADVANCES IN CHEMICAL PHYSICS

VOLUME XXXV

EDITORIAL BOARD

Advances in
CHEMICAL PHYSICS

EDITED BY

I. PRIGOGINE

University of Brussels,
Brussels, Belgium
and University of Texas
Austin, Texas

AND

STUART A. RICE

Department of Chemistry
and
The James Franck Institute
The University of Chicago
Chicago, Illinois

VOLUME XXXV

AN INTERSCIENCE ® PUBLICATION

JOHN WILEY AND SONS

NEW YORK • LONDON • SYDNEY • TORONTO

An Interscience® Publication.

Library of Congess Catalog Card Number: 58-9935
ISBN 0-471-69937-3)

Printed in the United States of America

10 9 8 7 6 5 4 3 2 1

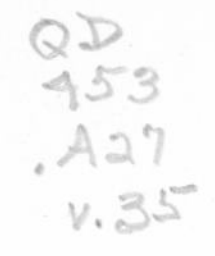

CONTRIBUTORS TO VOLUME XXXV

P. W. ATKINS, Physical Chemistry Laboratory, Oxford, England

R. BYRON BIRD, Chemical Engineering Department and Rheology Research Center, University of Wisconsin, Madison, Wisconsin

C. F. CURTISS, Theoretical Chemistry Institute, University of Wisconsin, Madison, Wisconsin

DAVID R. DION, Department of Chemistry, State University of New York at Stony Brook, Stony Brook, New York

G. T. EVANS, Physical Chemistry Laboratory, Oxford, England

M. J. HAGGERTY, Center for Statistical Mechanics and Thermodynamics, University of Texas, Austin, Texas

OLE HASSAGER, Haslevej 61, DK 3700 Rønne, Denmark

JOSEPH O. HIRSCHFELDER, Theoretical Chemistry Institute, University of Wisconsin, Madison, Wisconsin

G. SEVERNE, Fakulteit van de Wetenschappen, Vrije Universiteit Brussel, Brussels, Belgium

P. J. STEPHENS, Department of Chemistry, University of Southern California, Los Angeles, California

INTRODUCTION

In the last decades chemical physics has attracted an ever-increasing amount of interest. The variety of problems, such as those of chemical kinetics, molecular physics, molecular spectroscopy, transport processes, thermodynamics, the study of the state of matter, and the variety of experimental methods used, makes the great development of this field understandable. But the consequence of this breadth of subject matter has been the scattering of the relevant literature in a great number of publications.

Despite this variety and the implicit difficulty of exactly defining the topic of chemical physics, there are a certain number of basic problems that concern the properties of individual molecules and atoms as well as the behavior of statistical ensembles of molecules and atoms. This new series is devoted to this group of problems which are characteristic of modern chemical physics.

As a consequence of the enormous growth in the amount of information to be transmitted, the original papers, as published in the leading scientific journals, have of necessity been made as short as is compatible with a minimum of scientific clarity. They have, therefore, become increasingly difficult to follow for anyone who is not an expert in this specific field. In order to alleviate this situation, numerous publications have recently appeared which are devoted to review articles and which contain a more or less critical survey of the literature in a specific field.

An alternative way to improve the situation, however, is to ask an expert to write a comprehensive article in which he explains his view on a subject freely and without limitation of space. The emphasis in this case would be on the personal ideas of the author. This is the approach that has been attempted in this new series. We hope that as a consequence of this approach, the series may become especially stimulating for a new research.

Finally, we hope that the style of this series will develop into something more personal and less academic than what has become the standard scientific style. Such a hope, however, is not likely to be completely realized until a certain degree of maturity has been attained—a process which normally requires a few years.

At present, we intend to publish one volume a year, and occasionally several volumes, but this schedule may be revised in the future.

In order to proceed to a more effective coverage of the different aspects of chemical physics, it has seemed appropriate to form an editorial board. I want to express to them my thanks for their cooperation.

I. PRIGOGINE

CONTENTS

THEORIES OF CHEMICALLY INDUCED ELECTRON SPIN POLARIZATION

P. W. ATKINS and G. T. EVANS

Physical Chemistry Laboratory,
South Parks Road,
Oxford, England

CONTENTS

I. INTRODUCTION

In this review we are concerned with the theory of chemically induced electron spin polarization (CIDEP) and not with the experimental observations.[1] The main concentration of effort since the phenomenon was observed has been related to the radical-pair mechanism, which has been so successful for the description of the nuclear analog CIDNP. Recently, the emphasis has changed since evidence has been found that another process, the triplet mechanism, may dominate the other under some circumstances. We shall describe both theories here.

The general scheme preceding the detection of a doublet radical formed by photolysis is summarized in Fig. 1.

The primary process is the absorption of a photon and excitation of a ground-state substrate molecule into some excited singlet S_1. The next step (if the substrate is an aromatic carbonyl) is a rapid intersystem crossing into an excited triplet state T_1. This might be $n\pi^*$ or $\pi\pi^*$ depending on both the nature of the carbonyl and the nature of the solvent. The triplet may then abstract an electron or a hydrogen atom from the solvent, or if that is inert, from another substrate molecule after a diffusion-controlled migration. The transfer of e^- or $H\cdot$ leads to the formation of a pair of doublet radicals in close proximity. These diffuse apart, may reencounter

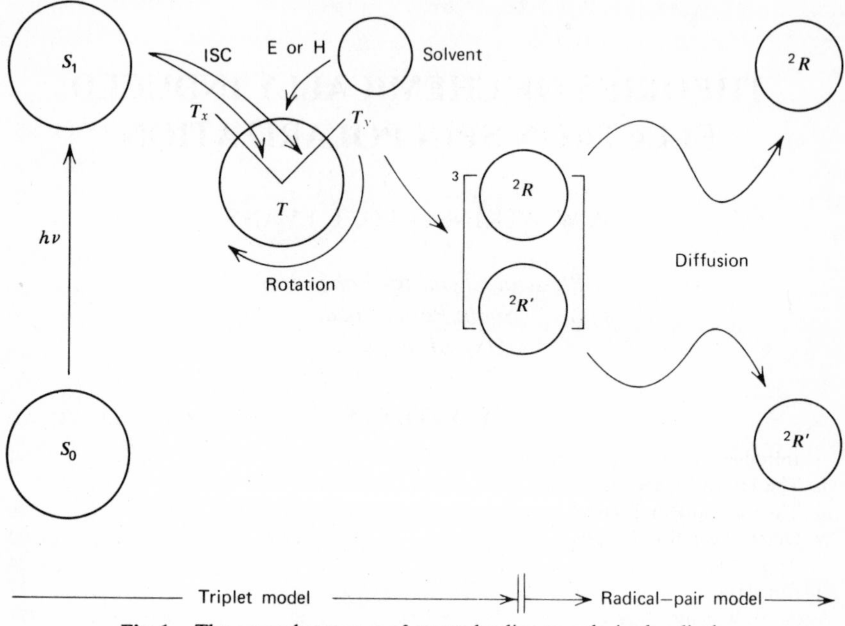

Fig. 1. The general sequence of events leading to polarized radicals.

and depart again, and at least one member of the pair (in favorable cases both) is observed by the e.s.r. spectrometer.

In fluid solution two steps in the above process may be responsible for the generation of polarization in the radicals. The first lies in the intersystem crossing (ISC) step, for the generation of polarization in solids in this step is well known[2]. This mechanism, which, as we shall see, involves the electron or hydrogen atom transfer step, will be called the triplet mechanism (TM) and the sequence of events that contribute to it is so labeled in Fig. 1. Polarization may also occur in the sequence of events following electron or hydrogen atom transfer. As the radicals migrate along their diffusional trajectories, the magnetic perturbations present in the two radicals and their Heisenberg exchange interaction may induce an evolution in their spin states that ultimately appears as a spin polarization. This set of processes constitutes the radical-pair mechanism (RPM) and, until recently, was believed to be the only process that made a significant contribution to the polarization.

II. THE TRIPLET MODEL

Intersystem crossing from a singlet to a triplet state occurs in accord with selection rules relating to the molecular frame. The reason for this is well known, and the group-theoretical arguments run as follows. In the molecular point group the three states of the triplet (labeled T_x, T_y, and T_z according to the direction of the spin vector in the molecular frame) span irreducible representations $\Gamma(T_q)$, $q=x,y,z$. If the orbital state is of symmetry $\Gamma(\mathrm{Or})$ the spin–orbital symmetry of each of the spin states $\Gamma(\mathrm{Or}) \times \Gamma(T_q)$. The perturbation responsible for the intersystem crossing is the spin–orbit interaction, which, being a scalar in the joint orbital and spin space, transforms as the totally symmetrical irreducible representation $\Gamma(A_1)$. It follows that an initial singlet state of symmetry $\Gamma(S)$, which in the joint orbital and spin space spans $\Gamma(S)\times\Gamma(A_i)=\Gamma(S)$ can be connected only to those spin states of the triplet for which $\Gamma(T_q)\times\Gamma(\mathrm{Or})$ contains $\Gamma(S)$. Therefore crossing into one or more of the T_q-states might be forbidden by the symmetry of the molecule. In fact some failure of selection rules may normally be expected because of vibronic effects or departures from the assumed point–group symmetry, and so in general ISC will generate population in all three sublevels. Nevertheless, one may anticipate, and what is more to the point, actually observe[3] that the selection rules operate with sufficient discrimination to lead to different rates of populating the spin levels.

There are, however, two differences between the solid-state and the fluid solution situations, and it is the latter that concerns us. Both differences stem from the rotation of the triplet in the fluid. In the first place we have seen that the selection rules governing ISC relate to the molecular frame whereas we observe the polarization in the fixed laboratory frame (relative to the direction of the applied field). At first sight it would appear that the selection rules would lead to an even distribution of spin projections in the laboratory frame. In the second place we might anticipate that if for some reason a laboratory frame polarization were generated, then the very fast spin relaxation of the rotating triplet molecule would obliterate the polarization extremely rapidly.

These reflections are responsible for the late introduction of the triplet mechanism. The first problem was treated for a static, random distribution of triplets by Atkins and McLauchlan[4] and independently, by Wong, Hutchinson, and Wan.[5] Both sets of authors arrived at the conclusion that the differential ISC rates in the molecular frame could lead to polarization in the laboratory frame. Wong et al., however, made the further significant and essential contribution[11,5] that the triplet relaxation time entered into the theory in competition with the rate of atom (or electron) transfer from the solvent. If the rate of the latter were comparable to the relaxation time, then the doublets might be formed before the triplet spin system had fully relaxed, and therefore the initial polarization would not be lost. (The doublets have spin–lattice relaxation times orders of magnitude longer than the triplet.) Both sets of authors treated the system as a static, random distribution of triplets, but both were aware that the true triplet model would have to be treated as a dynamical problem in which the molecules were allowed to rotate. A solution to the dynamical problem was presented by Atkins and Evans,[6] and the following account is a description of their approach.

Before describing the calculation of the polarization by the TM, it might be helpful to give a simple explanation of why the molecular selection rules can give a laboratory polarization.[4] Let us suppose that the zero-field splitting (ZFS) parameter D is positive, which is likely in many carbonyl compounds,[7] so that T_z lies lowest. Furthermore, let us suppose that the rate of crossing into T_x greatly exceeds the rate of ISC into T_y and T_z (this too has been found for a wide variety of aromatic carbonyls[7]). At high magnetic fields the eigenstates of the system are predominantly quantized in the laboratory frame, and T_{+1} lies highest. T_{+1} may be expressed as a linear combination of the T_q, and the combination varies as the molecule rotates. Nevertheless, since T_x and T_y are the higher energy zero-field states, they always contribute more to T_{+1} than T_z contributes (on simple perturbation arguments). Therefore the upper high-field levels share the higher population of T_x, and the lower high-field levels, which are closer to the relatively empty T_z level, will be relatively empty. Consequently, polarization will appear to the static observer of the rotating triplet, for the order of the levels T_q is independent of orientation.

This was the argument made quantitative by Atkins and McLauchlan[4] and Wong et al.[5] Their conclusions were that the polarization of the emergent triplets should be of the order of D/B, where B is the applied field: When D is zero no level lies closest to T_{+1} and so the polarization is absent; and so it is when B is enormous, for then the differences of amplitudes of the T_q in the T_{+1} linear combination are insignificant. Wong et al.[5,8] deduced the expression

$$P_{+1/2}-P_{-1/2}\sim\left(\frac{4}{15}\right)\left(\frac{D}{B}\right)(P_x+P_y-2P_z) \tag{2.1}$$

for the polarization of a doublet radical emerging from a triplet in which the ISC probabilities are P_x, P_y, P_z for the sublevels T_x, T_y, T_z. If $P_x \gg P_y, P_z$, as Cheng and Hirota[7] observe, the enhancement of the signal is

$$\gamma\sim\frac{1}{2}\left(\frac{P_{+1/2}-P_{-1/2}}{P_{+1/2}+P_{-1/2}}\right)\left(\frac{2kT}{g\mu_B B}\right)\sim-\frac{2}{5}\left(\frac{D}{B}\right)\left(\frac{2kT}{g\mu_B B}\right) \tag{2.2}$$

Since $D\sim 1$ kG, $B\sim 3$ kG, and at room temperature the Boltzmann difference of populations ($g\mu_B B/2k_T$) is about 7.5×10^{-4}, an enhancement of $\gamma\sim-156$ is predicted by the theory.[8] Such a massive enhancement is admirable, but we must remember that it is based on an unphysical static model and will certainly be diminished by molecular tumbling.

The density matrix for the spin of the triplet molecule evolves under two influences. First, there is the *magnetic evolution* of the spin states. The contributions to the Hamiltonian responsible for this are the Zeeman term ($\mathcal{H}_z$) and the ZFS interaction $\mathcal{H}_D$. For simplicity we suppose that the g-factor is isotropic; $\mathcal{H}_D$, however, represents an interaction that depends on the spin orientation in the molecular frame, and therefore if the spin is expressed in the laboratory frame, the interaction is dependent on the molecular orientation. Second, there is the *chemical evolution* of the density matrix which arises because the triplet is formed as the initial singlet decays, and because the triplet itself decays as chemical processes generate from it the doublet radicals. Each of these processes may be represented by the addition of a source (for the formation step) or a sink (for the decay). Since the rate of forming the states of the triplet is different for each zero-field state, the rate constant for the ISC formation step is a matrix whose principal values are the rates of populating the different spin states, and whose principal axes are also the principal axes of the ZFS interaction **D**. It is conceivable that the rate of quenching also depends on the spin level of the triplet; but for this there is as yet no evidence, and so it is assumed that each level is quenched with the same rate constant k_T.

The magnetic evolution and the chemical processes lead to the following overall expression for the evolution of the triplet's density matrix:

$$\dot{\rho}(t)=i\rho(t)*\{\mathcal{H}_z+\mathcal{H}_D(\Omega)\}-k_T\rho(t)+\mathbf{k}P_s(\Omega,t) \qquad (2.3)$$

The singlet population $P_s(\Omega,t)$ has been written as a function of Ω to accommodate a suggestion by Adrian[9] that the use of polarized light might lead to a polarization that varied with the azimuth of the electric field vector. We shall return to this point later.

The manipulation of (2.1) can be greatly simplified by recognizing that the matrix **k** can be written as an operator in the spin space of the system.[6] Thus the elements of the rate matrix are obtained by taking matrix elements of the operator

$$\mathbf{k}=\frac{1}{6}kS^2-\mathbf{S}\cdot\mathbf{K}\cdot\mathbf{S}=\frac{1}{6}kS^2+\mathcal{H}_K(\Omega) \qquad (2.4)$$

between the spin states of the triplet. In this expression **K** is the kinetic anisotropy matrix which measures the deviation of **k** from equal rates into all three spin states. **K** is a traceless matrix diagonal in the principal axes of

D; and as it may be expressed in a second-rank irreducible spherical tensor, it has the same transformation properties as **D**. This means that when Eq. 2.1 is iterated, ensemble averaging will not eliminate cross-terms between $\mathcal{H}_D(\Omega)$ and $\mathcal{H}_K(\Omega)$. It is through this behavior that the effect of the rotation of the molecule on the relaxation (arising from terms quadratic in $\mathcal{H}_D$) and the ISC anisotropy (arising from terms bilinear in $\mathcal{H}_D$ and $\mathcal{H}_K$) makes its appearance in a systematic and consistent way.

The formulation of the problem in the present manner has the further advantage that Eq. 2.1 need not be solved as it stands, for the polarization of the emergent doublets is proportional to a Laplace transform of the density matrix; therefore Eq. 2.4 may be solved algebraically once it has been developed a little further. To this end we introduce an interaction picture based on the transformations

$$\sigma(t)=\exp(k_T t)\exp(i\mathcal{H}_0 t^*)\rho(t)$$
$$\mathcal{H}_A(\Omega_t)=\exp(i\mathcal{H}_0 t^*)\mathcal{H}_A(\Omega) \qquad (A=D \text{ or } K) \tag{2.5}$$

This leads to an equation that, after one iteration and ensemble averaging, gives

$$\begin{aligned}\langle\dot{\sigma}(t)\rangle=&\left\langle\left\{\frac{1}{6}kS^2+\mathcal{H}_K(\Omega_t)\right\}P_s(\Omega,t)\right\rangle\exp(k_T t)\\ &-\int_0^t dt'\langle\{\sigma(t')*\mathcal{H}_D(\Omega_{t'})\}*\mathcal{H}_D(\Omega_t)\rangle\\ &+i\int_0^t dt'\langle P_s(\Omega,t')\mathcal{H}_K(\Omega_{t'})*\mathcal{H}_D(\Omega_t)\rangle\exp(k_T t')\end{aligned} \tag{2.6}$$

This equation of motion is an exact consequence of (2.3) and involves no approximations that are not inherent to that calculation.

The spin polarization of a doublet radical emerging from the triplet depends on the population difference of the T_{+1} and T_{-1} states, and the time after intersystem crossing at which the doublet was formed:

$$S_{Z1}=\frac{1}{2}k_T\int_0^\infty dt\, f(t)\exp(-k_T t) \qquad f(t)=\sigma_{11}(t)-\sigma_{-1-1}(t) \tag{2.7}$$

The population difference obeys the equation

$$\left(\frac{d}{dt}+\omega_1\right)f(t)=\omega_1 f_0(t)+i\int_0^t dt'\,\mathrm{tr}\langle P_s(\Omega,t')S_z\mathcal{H}_K(\Omega_{t'})*\mathcal{H}_D(\Omega_t)\rangle\exp(k_T t') \tag{2.8}$$

which is derived on the basis of a Redfield approximation on the second term in (2.6), which is responsible for the spin relaxation of the triplet. $f_0(t)$ is defined elsewhere,[6] and $\omega_1 = 1/T_1$, the triplet spin–lattice relaxation time.

The population of the excited singlet arises on account of the photochemical excitation from the ground state and is proportional to $(\boldsymbol{\mu}\cdot\mathbf{E})^2$, where $\boldsymbol{\mu}$ is the transition moment and $\mathbf{E}$ the electric field of the incident light. In the original calculation[6] the anisotropy of $(\boldsymbol{\mu}\cdot\mathbf{E})^2$ was ignored, but its effect has been analyzed by Adrian[9] who concludes that incident light of different polarizations can influence the extent of the polarization. Equation 2.8 is sufficiently general to incorporate Adrian's suggestion, and we develop it on the basis of plane polarized light in which the $\mathbf{E}$ vector makes an angle χ to the laboratory magnetic field. (For convenience we regard χ as the colatitude in a set of Eulerian angles Ω_{EB}.) In the laboratory frame the singlet population obeys

$$\begin{aligned} P_s(\Omega,t) &= \exp\{-(k+k_s)t\}\{1+\exp(D\nabla^2 t)\alpha(\Omega)\} \\ &= \exp\{-(k+k_s)t\}P_s(\Omega_t) \end{aligned} \tag{2.9}$$

where k is the total ISC rate (the trace of $\mathbf{k}$), k_s is the quenching rate of the excited singlet, and $\alpha(\Omega)$ is the correction due to the anisotropy of the singlet formation rate:

$$\alpha(\Omega) = \sqrt{6}\, \mathfrak{D}_{00}^{(2)}(\Omega_{EB}) \sum_q T_q^{(2)} \mathfrak{D}_{-q,0}^{(2)}(\Omega) \tag{2.10}$$

with

$$T_q^{(2)} = \sum_{pp'} \frac{(11pp'|2q)\,\mu_p \mu_{p'}}{\boldsymbol{\mu}\cdot\boldsymbol{\mu}}$$

Equations 2.6 through 2.10 yield the result

$$\langle S_{Z1}\rangle = \frac{1}{2}\left(\frac{k}{k+k_s}\right)\left\{ S_Z^\circ + \frac{(S_Z^* - S_Z^\circ)k_T T_1}{1+k_T T_1}\right\} \tag{2.11}$$

S_Z° is the triplet thermal equilibrium magnetization, and S_Z^* is given by

$$\begin{aligned} S_Z^* = i(k+k_s)\int_0^\infty dt \int_0^t dt' \operatorname{tr}\{S_Z\langle P_s(\Omega,t')\mathcal{H}_K(\Omega_{t'})_* \mathcal{H}_D(\Omega_t)\rangle\} \\ \times \exp(-k_T(t-t')) \end{aligned} \tag{2.12}$$

S_Z^* can be decomposed into two parts, one arising from unpolarized light (S_{ZU}^*) and the other (S_{ZP}^*) from the anisotropy of the excitation when plane polarized light is used. (In practice, with an external light source, the formation is unlikely to be strictly isotropic because the electric vector is zero in the propagation direction, apart from contributions from reflections within the cavity.)

$$S_{ZU}^* = 2\omega_0\{DK[4j(2\omega_0\tau_2) + j(\omega\tau_2)] + 3EI[4j(2\omega_0\tau_2') + j(\omega_0\tau_2')]\} \tag{2.13}$$

$$S_{ZP}^* = \frac{4}{7}\omega_0(k+k_s)\mathfrak{D}_{00}^{(2)}(\Omega_{EB})\left\{\left(\frac{M\tau_2}{1+(k+k_s)\tau_2}\right)[DKh(\omega_0\tau_2) - 3EIh(\omega_0\tau_2')]\right.$$
$$\left. - \left(\frac{3N\tau_2'}{1+(k+k_s)\tau_2'}\right)[DIh(\omega_0\tau_2) + EKh(\omega_0\tau_2')]\right\} \tag{2.14}$$

where we have used the following definitions:

$$j(\omega\tau) = \frac{(2\tau^2/15)}{(1+\omega^2\tau^2)} \qquad h(x) = 8j(2x) - j(x)$$

$$\tau_2 = \frac{1}{6D_\perp} \qquad \tau_2' = \frac{1}{(2D_\perp + 4D_\parallel)}$$

$$kK = \frac{1}{2}(k_{xx} + k_{yy}) - k_{zz} \qquad kI = \frac{1}{2}(k_{yy} - k_{xx})$$

$$\mu^2 M = \frac{1}{2}(\mu_x\mu_x + \mu_y\mu_y) - \mu_z\mu_z \qquad \mu^2 N = \frac{1}{2}(\mu_y\mu_y - \mu_x\mu_x)$$

$$\omega_1 = T_1^{-1} = \frac{D^2}{\tau_2}[4j(2\omega_0\tau_2) + j(\omega_0\tau_2)] + \frac{3E^2}{\tau_2'}[4j(2\omega_0\tau_2') + j(\omega_0\tau_2')]$$

In deriving the foregoing expressions we have made use of the realistic limit $\omega_0 \gg k_T$, which is normally true for X-band e.s.r. If $k_T \gg \omega_0$, the calculation predicts $S_Z \sim k_T^{-2}$. This dependence on the triplet lifetime arises because the TM polarization originates from the cross-correlation of $\mathcal{H}_D$ and $\mathcal{H}_K$. For times less than the inverse of the Larmor frequency the interaction picture $\mathcal{H}_D$ and $\mathcal{H}_K$ commute and therefore do not give polarization—the triplet must live long enough so that the cross-correlation has time to develop.

If the triplet-quenching process is diffusion-controlled, the concentration of the quencher enters explicitly into (2.10) because we may write $k_T = k_q[Q]$. If also $k \gg k_s$, taking the thermal magnetization of the doublet as $S^{\circ}_{Z1} = (3/8)S^{\circ}_Z$, where S°_Z is the triplet thermal magnetization, enables us to write

$$4(3\gamma_D - 1)^{-1} = \gamma_T^{-1} + (\gamma_T k_q[Q]T_1)^{-1} \tag{2.15}$$

where

$$\gamma_D = \frac{(S_{Z1} - S^{\circ}_{Z1})}{S^{\circ}_{Z1}} = \frac{(S_{Z1} + |S^{\circ}_{Z}|)}{S^{\circ}_{Z1}}$$

$$\gamma_T = \frac{(S_Z - S^{\circ}_Z)}{S^{\circ}_Z}$$

The enhancement factor γ_D is a measure of the polarization: For emission ($S_{Z1} > 0$) it is the observed signal height plus the Boltzmann signal height divided by the latter.

This expression has been tested experimentally by Atkins et al.,[1u] who observed the initial polarization of duroquinone in fluid solution in the presence of a variety of amines. The triplet-quenching rate k_T was varied by controlling the amine concentration,[9] in good agreement with an equation in the form of (2.15). In fact their equation had $(\gamma_D - 1)^{-1}$ on the left because the doublet's thermal magnetization was not distinguished from the triplet's. Reanalyzing their data on the basis of (2.15) gives an initial triplet polarization of $\gamma_T = -24.4$ and -26.8 for duroquinone in triethylamine and *sec*-butylamine, respectively.

The determination of the dependence of the initial polarization on the quencher concentration has an important application for the determination of *triplet* spin relaxation times in fluid solutions.[1u] If γ_T is known from the intercept, since k_q may be determined absolutely by nanosecond flash photolysis techniques, T_1 may be determined from the gradient of the $(3\gamma_D - 1)^{-1}$ versus $[A]^{-1}$ plot. This has been done in a number of cases[1u,10] and duroquinone triplet relaxation times of 16 and 4 nsec for duroquinone in the same two amines may be calculated on the basis of (2.16). Further experimental work on this system[10a] has indicated that $\omega_0^2\tau_2^2 \gg 1$ (i.e., that the slow-motion limit applies), in which case nanosecond triplet T_1's are plausible.

The general form of (2.15) has been established, but there remains the question of whether the absolute polarizations are correctly predicted by the theory. It has been shown elsewhere[6] that the TM is significant only if

$\omega_0\tau_2 \gg 1$, and we restrict attention to this slow-motion limit. Both S^*_{ZU} and S^*_{ZP} simplify:

$$S^*_{ZU} \sim \left(\frac{8}{15\omega_0}\right)(DK+3EI) \tag{2.16}$$

$$S^*_{ZP} \sim \left(\frac{4}{105}\right)\left(\frac{k+k_s}{\omega_0}\right)(3\cos^2\chi-1) \times \left\{ \frac{M\tau_2[DK-3EI]}{1+(k+k_s)\tau_2} - \frac{3N\tau_2'[DI+EK]}{1+(k+k_s)\tau_2'} \right\} \tag{2.17}$$

Equation 2.16 is the result of Wong et al.[5] generalized to include E and I values. Equation 2.17 is the result of Adrian[9] generalized to an anisotropically rotating triplet with E, I, and N nonzero. The physical content of (2.17) has been discussed by Adrian.[9] It follows from it that S^*_{ZP} will vanish if the light is isotropic in the cavity, which means more than using unpolarized light. When the light is unpolarized, but with no component in one direction perpendicular to the field, $(3\cos^2\chi - 1)$ should be replaced by $1/2$. The contribution of S^*_{ZP} also vanishes if the transition moments are isotropic ($M=N=0$), when molecular rotation is fast compared with the singlet's ISC lifetime ($(k+k_s)\tau_2 \ll 1$), and if χ is set at the magic angle. The size of S^*_{ZP} is of the order of 10 to 20% of S^*_{ZU}, and very recently two experiments showing the effect have been reported,[10c,d] although in some cases the magnitude is slightly larger than theory indicates.

Numerical estimates of the polarization that might be expected from an axially symmetric carbonyl molecule with $D/\omega_0 \sim 1/3$ show that in the slow-motion limit

$$\frac{S_{Z1}}{S^\circ_{Z1}} \sim -237 K k_T T_1 \tag{2.18}$$

This suggests that the TM is significant when $k_T T_1 > 0.01$ and K is about unity (dominant population of one substate).

III. THE RADICAL-PAIR THEORY

The radical-pair theory of CIDE(N)P was put forward first by Kaptein and Oosterhoff,[11] although the connection between CIDNP and CIDEP was first recognized by Fischer and Bargon.[12] The mechanism depends on the existence of a pair of radicals produced by some homolysis or electron or atom transfer, on the presence of magnetic interactions within the radicals, and on an exchange interaction between them. If the spins of the radicals experience different local magnetic fields (by virtue of unequal g-values, hyperfine interactions, or conceivably spin rotational interactions) they precess at different rates and about different directions. These perturbations, in combination with the ordering of levels arising from the exchange and Zeeman interactions, constitute a spin-sorting process, and the final result is a spin polarized ensemble of radicals. The singlet state of the radical pair lies closest to the $M_s = 0$ state of the triplet at all but the smallest interradical separations and high magnetic fields. The magnetic perturbations therefore bring about $S - T_0$ mixing with the greatest efficiency. When $S - T_0$ mixing occurs the polarization arises from a relative rephasing of the individual spins of the pair, and the total numbers of α and β spins are unchanged: Spins of one projection are applied predominantly to one radical and spins of the other projection are applied to the other. If the perturbations are independent of the nuclear spin states, this results in one radical having an excess of α-spins and the other radical having a compensating excess of β-spins. If nuclear spin states play a significant role, different hyperfine levels will carry different polarizations, but the overall polarization of both radicals will be zero.

If mixing between the singlet and the T_{-1} level is important (when the exchange interaction is negative) the overall number of α- and β-spins may be changed. Significant mixing of T_{-1} and S eliminates β-spins from the system (if S is initially empty), and both radicals will show an excess of α-spins. The perturbations play a different role in $S - T_{-1}$ mixing because they have to dissipate angular momentum: One route is dissipation into the nuclear spins *via* terms of the form $S_+ I_-$, another is into the rotation of the radical itself through the spin–rotation coupling,[4, 13, 14] and a third is into the magnet through the anisotropy of a g-value.[1g]

These remarks have been turned into quantitative theories in a variety of different ways. In terms of the radical-pair spin density matrix expressed in a coupled-electron basis the Z-component of the electronic magnetization of fragment 1 is given by

$$\begin{aligned} S_{Z1}(t) &= \operatorname{tr} \sigma(t) S_{Z1} \\ &= \operatorname{Re} \sigma(t)_{ST_0} + \frac{1}{2}\{\sigma(t)_{T_1T_1} - \sigma(t)_{T_{-1}T_{-1}}\} \end{aligned} \tag{3.1}$$

For isolated radicals in low magnetic fields $\sigma_{T_1T_1}$ and $\sigma_{T_{-1}T_{-1}}$ do not depart from their equilibrium values unless the triplet mechanism is operative, or unless there is significant $S - T_{-1}$ mixing. We shall disregard both effects for the present and consider the polarization to be given by the first term: The fact that $S_{Z1}(t)$ depends on an off-diagonal element of the matrix reflects the remark that $S-T_0$ mixing depends on a rephasing of the spins. Put another way, the approximation embodied in (3.1) is equivalent to ignoring the TM and retaining only the secular terms in the spin Hamiltonian:

$$\mathcal{H}_\mu = g_\mu \mu_B B S_{Z\mu} + \sum_\nu \mathbf{S}_\mu \cdot \mathbf{A}(\mu\nu) \cdot \mathbf{I}_\nu$$

$$\sim \left\{ g_\mu \mu_B B + \sum_\nu A(\mu\nu) I_{Z\nu} \right\} S_{Z\mu} \qquad (\mu = 1, 2) \tag{3.2}$$

where $A(\mu\nu)$ is the hyperfine interaction between electron μ and nucleus ν. Equation 3.1 suggests that to a good approximation the TM polarization can simply be added to the RPM polarization (the TM modifies the diagonal elements of σ). The failure of the secular approximation, which will permit an $S-T_{-1}$ process to modify the diagonal elements, will be mentioned again later.

The Hamiltonian in (3.2) must be augmented by a term representing intermolecular electron exchange:

$$\mathcal{H}_J = -J\left\{\tfrac{1}{2} + 2\mathbf{S}_1 \cdot \mathbf{S}_2\right\} \tag{3.3}$$

J is some function of the interradical separation. The total Hamiltonian is then

$$\mathcal{H} = \mathcal{H}_1 + \mathcal{H}_2 + \mathcal{H}_J = \mathcal{H}_0 + \mathcal{H}_J \tag{3.4}$$

As an approximation to the true situation, Kaptein and Oosterhoff restricted their considerations to the range of intermolecular separations where the exchange interaction was of the same order of magnitude as the magnetic perturbations.[11] Furthermore, they simplified the problem by regarding the exchange interaction as a constant in the range of interest. Integration of the density matrix equation of motion away from the initial condition $\sigma_{ij}(0) = \sigma_i \delta_{ij}$ is then a trivial matter, and their expression for the magnetization is

$$S_{Z1}(t) = \left\{ \left(\frac{2V_n J}{V_n^2 + J^2} \right) \sin^2 \left(V_n^2 + J^2 \right)^{1/2} t \right\} (\sigma_S - \sigma_T) \tag{3.5}$$

where

$$V_n = \langle S, \{I,m\} | \mathcal{H}_0 | T_0, \{I,m\} \rangle$$
$$= \frac{1}{2}(g_1 - g_2)\mu_B B + \frac{1}{2}\sum_\nu A(1,\mu)m_\nu - \frac{1}{2}\sum_\nu A(2,\nu)m_\nu \qquad (3.6)$$

The subscript of V_n denotes the collection of nuclear spin quantum numbers. If the collisional dimer described by the constraint that $J(r) \sim A$ and $(g_1 - g_2)\mu_B B$ has an exponential distribution of lifetimes with mean value τ_{CKO} (Closs,[15] and Kaptein and Oosterhoff[11]) the time average of $S_{Z1}(t)$ gives the polarization of the radical emerging from the pair with a particular nuclear spin configuration as

$$S_{Z1}^{(n)} = \frac{1}{\tau_{\text{CKO}}} \int_0^\infty dt\, S_{Z1}(t) \exp\left(\frac{-t}{\tau_{\text{CKO}}}\right)$$
$$= \frac{(\sigma_S - \sigma_T) V_n \tau_{\text{CKO}} (2J\tau_{\text{CKO}})}{1 + 4(J^2 + V_n^2)\tau_{\text{CKO}}^2} \qquad (3.7)$$

The lifetime τ_{CKO} is ill defined, and its magnitude has been the subject of much speculation[16]: In the original theory a value of 10^{-9} sec was chosen.

Despite the drastic simplifications of the model and this presentation, the form of (3.7) has been reproduced using more sophisticated theories. It is important to note that the dependence of the sign of the polarization on the initial conditions (whether escape occurs from a singlet or triplet), the relative g-factors of the pair, and the sign of the exchange interaction is embodied in (3.7), and that these are unchanged in the more sophisticated versions. In a typical situation, for example, the pair comes from a triplet precursor ($\sigma_S - \sigma_T < 0$), and the singlet of the pair lies beneath the triplet ($J < 0$); then $S_{Z1}^{(n)}$ carries the sign of V_n, which carries the sign of $g_1 - g_2$. Therefore, if radical 1 has the larger g-value, $S_{Z1}^{(n)}$ is positive, corresponding to a predominance of α-spins, which in turn corresponds to an emissive situation. Conversely, the radical with the lower g-value should appear in enhanced absorption. These rules have been verified.[1]

The first improvement of the CKO theory was due to Adrian.[17] He treated the diffusional character of the relative motion of the members of the radical pair in a more satisfactory fashion, and built into his theory the short-range nature of $J(r)$. Adrian concentrated on the evolution of the state of a single radical pair whose history was essentially the following. A radical pair is born by photochemical or thermal excitation; the members of the pair execute a random walk which might or might not be terminated at a time t_1 by a spin selective chemical reaction that occurs on reencounter. Even if chemical reaction does not occur on this reencounter, there is at least a spin–exchange interaction. The pair of reencountered radicals dissociates at t_2 and additional state evolution of the escaping radicals is supposed to be unimportant.

The equation of motion of the density matrix separates into two parts:

$$\dot{\sigma}(t)=i[\sigma, \mathcal{H}_0], \qquad 0 \leqslant t \leqslant t_1 \tag{3.8a}$$

$$\dot{\sigma}(t)=i[\sigma, \mathcal{H}_J], \qquad t_1 < t \leqslant t_2 \tag{3.8b}$$

Note that Eq. 3.8*a*, which describes the state evolution during the trajectory corresponding to initial departure up to reencounter, is independent of $\mathcal{H}_J$ because $J(r)$ is supposed to have a very short range, far shorter than typical distances attained during the diffusion. In the collision, which occurs in the interval between t_1 and t_2, the exchange interaction dominates the magnetic perturbations (in the sense that level separations are made so large that the perturbations are impotent), and this is reflected by the omission of $\mathcal{H}_0$ from (3.8*b*).

At the time t_2 the polarization is determined by the real part of $\sigma_{ST_0}(t_2)$:

$$S_{Z1}(t_2)=(\sigma_S-\sigma_T)\sin(2V_n t_1)\sin(2J[t_2-t_1]) \tag{3.9}$$

Observe that the exchange interaction is essential because the generation of polarization depends on the relative disposition of the levels. This also shows that a collision is essential in this model, since $S_{Z1}(t_2)$ disappears when $t_2=t_1$. (In general it is not necessary for a pair to achieve a collisional separation to foster CIDEP. Since $J(r)$ has a finite range, the pair may experience exchange at noncollisional separations. The artificial separation of the pair trajectory into $J=0$ and $J \gg V_n$ regions of space overemphasizes the necessity of a collision.) Adrian's theory disregards the polarization generated by species that do not reencounter, but in high-viscosity media this might be a poor approximation[18] for significant polarization might be generated if the diffusional separation is slow. The interval t_2-t_1 corresponds to the duration of the collision, and this will be written τ_A. The time t_1 was removed by integration over the duration of the random walk. Although Adrian parametrized this in terms of the number of steps on the random walk (using Noyes' theory of radical reencounter[19]) it is equivalent to integrating Eq. 3.9 using the distribution

$$P(t) \propto t^{-3/2} \exp \frac{-\tau_A}{t} \tag{3.10}$$

The integrated result gives

$$S_{Z1}=(\sigma_S-\sigma_T)(V_n\tau_A)^{1/2}\operatorname{sgn}(JV_n)\sin(2|J|\tau_A) \tag{3.11}$$

Freed[17b] has suggested that the quantity $\sin(2|J|\tau_A)$ should be averaged over a distribution of collisional lifetimes. For an exponential distribution with average lifetime τ_s, this amounts to the replacement

$$\sin(2|J|\tau_A)\rightarrow\frac{2|J|\tau_s}{(1+4J^2\tau_s^2)}$$

In a subsequent paper,[17c] Adrian presented an alternative scheme of implementing the radical-pair dynamics by noting that $J(r)$ had implicit time-dependence. This arises because $J(r)$ is a function of the relative radical separation, and this separation is modulated by the translational Brownian motion. Adrian transformed the Liouville equation

$$\dot{\sigma}=i[\sigma,\mathcal{H}_0+\mathcal{H}_J] \tag{3.12}$$

into the interaction picture using the transformations

$$\sigma^I(t)=\exp(i\mathcal{H}_0t^*)\sigma(t)$$

$$\mathcal{H}_J^I(t)=\exp(i\mathcal{H}_0t^*)\mathcal{H}_J \tag{3.13}$$

$\mathcal{H}_J^I$ was treated as a perturbation, and the density matrix equation solved by one iteration: Because of the size of the perturbation, this truncation to first-order is really very dubious. This procedure led to

$$S_{Z1}(t)=2(\sigma_S-\sigma_T)\int_0^t dt'\,J(r[t'])\sin(2V_nt') \tag{3.14}$$

The explicit time dependence of $J(r)$ was projected using

$$J(t)=\int_{r_A}^{\infty}dr\,4\pi r^2J(r)P(r;t;r_0) \tag{3.15}$$

with $P(r;t;r_0)$ the solution of the translational diffusion equation with a reflecting wall at $r=0$ and an isotropic initial distribution:

$$P(r,t=0)=\left(\frac{1}{4\pi r_0^2}\right)\delta(r-r_0). \tag{3.16}$$

The lower limit in (3.15) corresponds to the distance at which $J(r_A)\tau_A \sim 0(\pi/2)$. Adrian argued that trajectories attaining separations less than r_A diminish the polarization by spin–exchange, and their contribution should therefore be neglected. When a long-time approximation for $J(t)$ is made, and the integral in (3.14) is evaluated, the polarization is found to be

$$S_{Z1} = (3\pi\sqrt{3})(\sigma_S - \sigma_T)\left(1 - \frac{r_0}{r_A}\right)\left\{\frac{\lambda r_A + 2}{(\lambda\sigma)^3}\right\}(V_n\tau_c)^{1/2} \tag{3.17}$$

In this expression $1/\lambda$ is the range of the exchange interaction [i.e., $J(r) = J_0 \exp(-\lambda r)$], τ_c is the translational correlation time, and σ a collisional diameter.

Although Eq. 3.17 does predict CIDEP enhancements, it does so by ignoring dissipative terms (e.g., those embodied in the discussion involving r_A), and the whole development is based on first-order perturbation theory with the exchange interaction as the perturbation. A perplexing feature of the result is the dependence of S_{Z1} on $1-(r_0/r_A)$: It is difficult to see the physical reason for the disappearance of S_{Z1} when $r_0 = r_A$.

Atkins and Moore[20] considered the parametrization of the exchange energy by expressing it in a linearly or an exponentially decreasing function of time and incorporating this explicit time dependence into the Liouville equation for the density matrix. The exponential time-dependence is physically more plausible, and we confine our attention to it. The aim of the calculation was to find the polarization of a radical where $S - T_0$ mixing was induced by a constant perturbation, and the S and T_0 levels converged exponentially with a time constant τ_J. Using $J = J_0 \exp(-t/\tau_J)$ the polarization can be found exactly without restriction on the rate of convergence of the levels. The result is given by the somewhat daunting expression

$$S_{Z1} = \frac{\pi}{2}(\sigma_S - \sigma_T)\operatorname{sech}(\pi V_n \tau_J)\operatorname{Im}\left\{J_{1/2 - iV_n\tau_J}(J_0\tau_J)J_{1/2 + iV_n\tau_J}(J_0\tau_J)\right\} \tag{3.18}$$

The $J_\eta(z)$ are Bessel functions. This result may be simplified by limiting it to first-order terms in $V_n\tau_J$ (but leaving unrestricted the order of $J_0\tau_J$). Then

$$S_{Z1} = (\sigma_S - \sigma_T)V_n\tau_J Si(2J_0\tau_J) \tag{3.19}$$

where $Si(z)$ is a sine integral, which is tabulated.[21] [This result is correct to $0(V_n^2\tau_J^2)$.] An even simpler form can be obtained by averaging this expression over a distribution of initial interactions assumed to follow

$$P(J_0)=\left(\frac{J_0}{\bar{J}}\right)\exp\left(-\frac{J_0}{\bar{J}}\right) \tag{3.20}$$

(this peaks in the vicinity of $\bar{J}$ but has tails at high and low values of J_0, and is not grossly unrealistic). Then

$$S_{Z1}=(\sigma_S-\sigma_T)V_n\tau_J\left\{\frac{2\bar{J}\tau_J}{1+4\bar{J}^2\tau_J{}^2}+\arctan\left(2\bar{J}\tau_J\right)\right\} \tag{3.21}$$

This is a good and usable approximation to the exact result in (3.18).

The last expression for S_{Z1} has two contributions. The first bracketted term in (3.21) is simply the CKO term with τ_{CKO} replaced by τ_J. The second term, the arctangent, is quite different because even in the limit $\bar{J}\tau\rightarrow\infty$ (i.e., very strong exchange) spin polarization is predicted in contradiction to the other theories (except one, see the following discussion): This is connected with the exponential character of the exchange dynamics.

Evans, Fleming, and Lawler[22] have produced a pseudopotential model for CIDEP (and CIDNP) by approximating the spatial characteristics of spin selective reaction and Heisenberg exchange by a δ-function term added to a stochastic Liouville equation for the radical-pair density matrix. Their equation of motion is

$$\dot{\sigma}(\mathbf{r},t)=L(\mathbf{r})\sigma(\mathbf{r},t)+i\sigma*\{\mathcal{H}_0+\mathcal{H}_J(\mathbf{r})\}-\frac{1}{2}K(\mathbf{r})[\sigma,0_S]_+ \\ -k_S\sigma+R\sigma \tag{3.22}$$

where $L(\mathbf{r})=D\nabla^2$ and D is the pair diffusion constant, and the first term takes care of the diffusional dynamics. The second term represents the magnetic and exchange evolution. The third term is the spin selective sink, which eliminates singlet-phased radical pairs [O_S is the projection operator $\frac{1}{2}(2-S^2)$ which selects singlet states, and K is a rate density]. The fourth term is a scavenging term with rate constant k_S. The fifth term is the spin relaxation contribution, and $\mathbf{R}$ is the Redfield tetradic, which is constrained to contain only secular terms in the coupled representation.

The stochastic Liouville equation was solved exactly using

$$K(r)=\left(\frac{K}{4\pi a^2}\right)\delta(r-a)$$

$$J(r)=\left(\frac{J}{4\pi a^2}\right)\delta(r-a) \tag{3.23}$$

the boundary conditions

$$r^2\left\{\frac{\partial\sigma(r,t)}{\partial r}\right\}_{r=0}=\left\{\frac{r^2\partial\sigma(r,t)}{\partial r}\right\}_{r=\infty}=0 \tag{3.24a}$$

and the initial condition

$$\sigma_{ij}(r,0)=\frac{\sigma_i\delta_{ij}\delta(r-r_0)}{4\pi r_0^{\,2}} \tag{3.24b}$$

S_{Z1} was calculated by averaging over the ensemble and integrating the time variable. [Identical results are obtained if the time integration is omitted but the $t\to\infty$ limit taken of the solution of (3.22) with the omission of the quenching sink $k_S\sigma$.] The strength of the pseudopotentials that appear in (3.23) are obtained by matching their volume integrals to the volume integrals of more realistic exponential functions of distance. In the limit of $V_n\tau_a\ll 1$, with $\tau_a=a^2/D$, the translational correlation time for displacements of magnitude a (which is approximately a bond length in the theory of Evans et al., but more realistically is probably a parameter of the size of a molecular radius), the authors obtained

$$S_{Z1}=\frac{(a/r_0)\,\mathrm{sgn}(V_n)\lambda_J(1-|\lambda_J|)(1-\lambda_K)^2([1-\lambda_K]\sigma_S-\sigma_T)(V_n\tau_a)^{1/2}}{\left[(1-\lambda_K)^2(1-\lambda_K/2)+4\lambda_J^2(1-\lambda_K)^2\right]\left[1-a\lambda_K\sigma_S/r_0\right]} \tag{3.25}$$

with

$$\lambda_J=\frac{\alpha J_0\tau_a}{1+|\alpha J_0\tau_a|} \tag{3.26}$$

and an analogous expression for λ_K with K_0 replacing J_0. α is an anisotropy and geometrical factor defined in the original paper. [Equation

3.25 differs slightly from the original by the insertion into the denominator of an extra factor of 4.]

This theory is a hybrid of several theories. It contains the square-root frequency dependence in accord with Adrian's theory and, in so doing, reflects the diffusional character of the mixing process. It has the advantage of predicting that CIDEP will occur from random reencounters of reactive free radicals, and does so without resorting to the heuristic arguments used by Adrian.[17b]

For unreactive radicals produced with definite phasing (i.e., produced from singlet or triplet precursors) the equation simplifies to

$$S_{Z1}=(\sigma_S-\sigma_T)\,\mathrm{sgn}(V_n)\left(\frac{a}{2r_0}\right)\left\{\frac{2\alpha J_0\tau_a}{1+4\alpha^2J_0^2\tau_a^2}\right\}(V_n\tau_a)^{1/2} \qquad (3.27)$$

The J-dependence of this equation is consistent with the CKO formalism as expressed by (3.7). It is also consistent with Adrian's view that strong collisions (strong in the sense that $J_0\tau_a \gg 1$ will randomize and dissipate spin polarization.) It is also part of the result obtained in the exponential departure calculation of Atkins and Moore[20] (Eq. 3.21). The pseudopotential model disagrees with the exponential departure model in so far as it predicts no polarization for collisions in which J rises to a large value. Thus, despite the apparently plausible representation of short-range interactions by appropriately weighted δ-functions, it seems that a significant portion of the CIDEP enhancement may have been lost. (We say *may* have been lost: It is not clear what role the stochastic element of the formulation plays in modifying the exact exponential solutions.)

In a series of papers, Pedersen and Freed[23] have considered various models of the spatial dependence of the exchange interaction in a stochastic Liouville equation for the spin density matrix. The simplest model for exchange that they investigated was a contact model. The numerically computed results were found to fit the form

$$S_{Z1}=(\sigma_S-\sigma_T)\,\mathrm{sgn}(V_nJ_0)f(V_n,D)\left\{\frac{2J_0\tau_1}{1+4J_0^2\tau_1^2}\right\} \qquad (3.28)$$

with τ_1 being the lifetime of the contact dimer. For a fixed integration volume of (200 Å)^3 and using $V_n \sim 10^8$/sec, it was found that $f(V_n, D)$ behaved linearly in V_n for $D \sim 10^{-3}$ cm^2/sec, as $V_n^{0.48}$ for $D \sim 10^{-5}$ cm^2/sec, and as $V_n^{0.2}$ for $D \sim 10^{-7}$ cm^2/sec. The conclusion relating to $D \sim 10^{-3}$ cm^2/sec was found to be volume dependent, and increasing the integration volume to (900 Å)^3 led to a $V_n^{0.48}$ dependence. This behavior is not inconsistent with the remarks of Evans et al.,[18b] who anticipated a power series in $(V_n\tau_a)^{1/2}$ and believed that the square root alone was inadequate.

In a more significant calculation, Pedersen and Freed[23b] demonstrated that by increasing the range of $J(r)$, asymptotic polarizations persist even in the strong exchange limit. This is a significant departure from the contact and pseudopotential models, but in agreement with the result of Atkins and Moore.[20] An optimum value for the polarization is given by $J(\max)\sim D/d^2$, d being the distance of closest approach. On either side of the polarization maximum, the behavior may be summarized by

$$S_{Z1}\sim h\left(\frac{V_n d^2}{D}\right)\left(\frac{J_0 d^2}{D}\right)\left(\frac{r_{ex}}{d}\right) \qquad J_0<J(\max)$$

$$S_{Z1}\sim g\left(\frac{V_n d^2}{D}\right)\left(\frac{r_{ex}}{d}\right) \qquad \text{asymptotic} \tag{3.29}$$

where g and h are functions of $V_n d^2/D$ and r_{ex} is defined by $J(r+r_{ex})\sim 10^{-5}J(r)$. Unfortunately neither the quadratic nor the linear dependence on the viscosity suggested by (3.29) accommodates the rather sparse experimental observations.[1v] The asymptotic polarization, which increases linearly with range, is an interesting result: It is significant that an r_{ex} of 2 Å can still give rise to asymptotic polarization, and so even apparently negligible range parameters can have a substantial effect.

In their third paper[23c] Pedersen and Freed have employed the Smoluchowski diffusion operator to represent the effects of interradical valence and Coulomb forces on the diffusion dynamics. The diffusion operator in the stochastic Liouville equation for the pair density matrix is now given by

$$L(r)=D\nabla\cdot\left\{\nabla+\frac{1}{kT}\left[\nabla U(r)\right]\right\}$$

$U(r)$ is the potential energy of interaction of the radicals and has as its source in the exchange and Coulomb forces. Since this interaction may be spin dependent, the above replacement represents a spin dependent diffusion. The authors chose different models for the spatial dependence of $J(r)$ and incorporated them into both the Liouville term and the diffusive motion term. Pedersen and Freed permitted reaction to occur from only the pure singlet state and found that S_{Z1} is only weakly dependent on the recombination rate,[23d] presumably because recombination occurs at separations such that $(hJ(r)/kT)>1$, whereas polarization arises for regions in which $(hJ(r)/kT)<1$. Attractive Coulomb interactions between the members of a radical pair were shown by Pedersen and Freed either to increase or to decrease S_{Z1} depending on the magnitude of J_0. Increasing the range of the Coulomb interaction decreases S_{Z1} in the strong exchange limit ($J\tau>1$), but increases S_{Z1} for weak exchange.

One note of caution is worth mentioning in the context of this work, and that arises from the mutual incompatibility of high-frequency, short-wavelength exchange and reaction in a low-frequency, long-wavelength diffusion equation. Because of this inconsistency, the results of Pedersen and Freed should be viewed as illustrative rather than definitive.

In a further refinement of the diffusional dynamics of the radical pair theory, Deutch[24a] has noted that the appropriate diffusion operator is

$$L(r)=\nabla\cdot\{D\mathbf{1}-2kT\mathbf{T}(\mathbf{r})\}\cdot\{\nabla+(kT)^{-1}\nabla U(r)\}$$

with $\mathbf{T}(\mathbf{r})$ the Oseen tensor. For separations large compared with the particle dimensions $\mathbf{T}(\mathbf{r})$ may be approximated by

$$\mathbf{T}(\mathbf{r})\sim(8\pi\eta r)^{-1}\{\mathbf{1}+r^{-2}\mathbf{rr}\}$$

The physical interpretation of the Oseen tensor is that it accounts for the effect of particle 1 establishing a vortex flow that accelerates particle 2 towards particle 1. The overall effect is a reduction in the effective pair diffusion constant. So far no theory, crude or elaborate, has taken this hydrodynamic correction into account.

The thread of similarity running through the expressions for S_{Z1} calculated using various radical-pair theories is striking. It represents the rather insensitive character of CIDEP to the subtleties of different theories. This insensitivity has been underlined by Deutch,[24b] who points out that, aside from the reaffirmation of the Noyes' diffusional models, CIDEP (and CIDNP) conveys little information about the microdynamics of fluids. Perhaps the main value of CIDEP lies in the mechanistic problems it is able to elucidate, and the measurement of doublet and triplet electron spin relaxation times that it permits. We shall make some remarks on these aspects in Section V.

There is very little experimental evidence on the viscosity dependence of the polarization, and only a little more on the frequency dependence. Adrian[17a] has demonstrated a $V_n^{1/2}$ dependence in qualitative accord with the CIDEP intensities of the $CH_3\dot{C}HCO_2^-$ and $(CH_3)_2\dot{C}CO_2^-$ radical anions.[1f] This indicates the basic correctness of the Noyes model for the dynamics of radical pairs, but sheds doubt on the accuracy on the quantitative character of the CKO and Atkins and Moore model. The

$(\eta/T)^{1/2}$-dependence has been observed for the benzophenone ketyl radical,[1v] and it was argued that this supported Adrian's model. Dobbs[10] has pointed out that an $(\eta/T)^{1/2}$ dependence is also predicted by the TM (because the kinetics have been observed to have a square-root viscosity dependence), and so the situation remains obscure.

IV. OTHER THEORETICAL MODELS

In this section we mention some of the other approaches that have been made in an attempt to explain various aspects of CIDEP. Unlike CIDNP, which collects its *D* from early but discredited explanations in terms of dynamic polarization processes based on the Overhauser effect, no published theories have been based on the generation of electron polarization by cross relaxation. (The recent confirmation of the TM, and the accompanying large electron spin polarization, suggests that an Overhauser mechanism for CIDNP might be effective in some cases. Lawler and Halfon[25a] have invoked the Overhauser mechanism to explain the anomalies in the zero-field biradical CIDNP of cyclo-octanone, and in recent work Adrain[25b] has used it to explain ^{19}F polarization.)

Some early theories were based on models which assumed wholly adiabatic character in the level-crossing schemes. Fischer[4] considered the time-evolution of a two-level system on a purely adiabatic trajectory, and Glarum and Marshall[1h] considered a complicated evolution involving two adiabatic transformations. Of course, neither author supposed that the processes were adiabatic: They intended to obtain upper limits of the polarization. An account of Glarum and Marshall's calculation, which related to the polarization of electrons ejected from Rb^- in ethereal solution, has been given elsewhere.[4]

Hutchinson et al. have proposed a mechanism for the generation of electron polarization in atoms[27a] and have extended it to account for polarization in molecules.[27b] The crux of the theory is that the selection rules governing transitions in atoms should be expressed in terms of the total angular momentum. Since this depends on the coupling of the nuclear spin into the orbital and electron spin states, it follows that the decay of an excited atom should be at least weakly dependent on the nuclear spin state. The mechanism seems to require strong interactions between the various angular momenta. In their second paper the group exploits the consequences of $\mathbf{I}+\mathbf{S}$ coupling in alkyl radicals. For alkyl radicals, the ratio of the hyperfine splitting and the magnetic field is about 10^{-3} for a 10 kG field, and consequently, the amount of selection invoked by this scheme is about 10^{-6}, but this is smaller than the nuclear Boltzmann factor in that field. In light of the recent work by Pedersen and Freed,[22b] which has shown that an exponentially varying exchange interaction gives the correct order of magnitude of the CIDEP enhancements observed by Fessenden and Schuler,[1a] it does not appear necessary to invoke the scheme. Furthermore, like the TM this theory predicts that a polarization should be present immediately after the photolytic flash, but recent reports by Verma and Fessenden[1t] and by Pedersen[28] have demonstrated that the long-time tail in hydrogen atom CIDEP arises from spin selective bimolecular recombination, which depletes the ensemble of singlet states.

Another theory that has fallen out of fashion (indeed it hardly ever fell into it) is the $S-T_{-1}$ version of the RPM. This was put forward by Atkins et al.[1g,14] and has been speculated about by Adrian.[13] The only difference between the $S-T_{-1}$ RPM and the $S-T_0$ RPM is the selection of the levels that are involved in the mixing: It assumes that the radicals attain a separation such that the electron exchange interaction is approximately equal to the Zeeman energy (so that S lies close to T_{-1}). This model depends on perturbations being able to transform β-spins into α-spins (if the T_{-1} state is occupied), and so nonsecular perturbations are necessary. A variety of perturbations possess components transverse to the Z-direction. The hyperfine interaction is the simplest to consider, but it will give hyperfine-dependent polarizations unless the nuclear relaxation rate is very fast (e.g., by virtue of it being part of a radical pair[1g]). The g-value of the radicals can be an agent for the perturbation if it is not isotropic.[1g] This arises because an anisotropic g-value can give components of magnetic field perpendicular to the applied field if that field does not lie along one of its principal axes: This field component interacts through $S_{\pm}$, and so can induce $S-T_{-1}$ mixing. The effect disappears when the molecule is rotating rapidly, but does not require a difference of g-values for its existence: Two identical but anisotropic radicals emerging from the pair can be polarized so long as the S and T_{-1} levels are close for a sufficiently long time, and as long as the radicals do not rotate too rapidly as they separate. A third perturbation that could cause $S-T_{-1}$ mixing is the spin–rotation interaction. There is no evidence that it does operate, and no compelling reason to suppose that it does, apart from the fact that it seems to be necessary in order to account for the spin–lattice relaxation times of a number of organic molecules in fluid solution,[29] as well as inorganic complexes and small inorganic radicals.[30] Spin–rotation processes have the double difficulty in the present context that their correlation time must not be so short as to eliminate their power, and yet the level crossing must be sufficiently slow for there to be time for their operation. If evidence did arise that spin–rotation effects were operative, the present theories[13,14] could be made more sophisticated; but the effort does not yet seem worthwhile.

The possibility of $S-T_{-1}$ mixing should not be dismissed as wholly unreasonable. It had been thought that the formation of exciplexes might so retard translational separation of the radical–ion pair that S and T_{-1} levels would be degenerate for a sufficient period. The evidence for this, however, has been obscured by the realization that the TM operates in just those situations in which exciplex function is expected, and there seems to be no reason to suppose that the exciplex retardation effect has been observed. (Of course, since the RPM operates after the TM, it may be that the observed enhancements represent the effect of a tandem process with the RPM stage enhanced by exciplex retardation.) $S-T_{-1}$ mixing is known to occur in biradicals[31] where the spins are constrained to small separations by the geometry of the species: The effect of magnetic fields and chain lengths on biradical CIDNP has been explained in terms of $S-T_{-1}$ mixing.[32] Unfortunately there are no reports of electron polarization in biradicals. In a pertinent report hydrogen atom CIDEP in acidic glasses at $-110°C$ has been observed.[1aa] The low-field hydrogen e.s.r. line was emissive and the high field line absorptive, with the intensity of emission exceeding that of absorption. Since the nonsecular components of the hyperfine interaction link the S electron–nuclear spin state with the initially occupied $T_{-1}\alpha$ state, the $S-T_{-1}$ mechanism would preferentially populate the low-field e.s.r. line as is observed. Hydrogen atom diffusion in the glassy phase is expected to be highly retarded, and this is precisely the regime where $S-T_{-1}$ processes are plausible.

V. APPLICATIONS

The generation of electron spin polarization leads to a spin system with a nonequilibrium population distribution. Furthermore, the time-scale for the development of the polarization (of the order of 10^{-8} sec) is short compared with the electronic spin–lattice relaxation time (of the order of 10^{-6} to 10^{-4} sec). This aspect of CIDEP has not been overlooked, and has been put to use in doublet T_1 determinations[1r] and in transient nutation experiments.[1x] When the TM is operative, the technique can also be used to measure triplet spin–lattice relaxation times in fluid solution, and a few measurements have been made.[1y,1o]. Another application is to mechanistic studies, and a particularly fruitful procedure is to study CIDEP in systems showing CIDNP.[1v] In this section we shall consider only the more physical applications and concentrate on the measurement of relaxation times.

A. Doublet T_1 Measurements

A general review of T_1-measurements on electrons has been given by Hyde[33] who, like Venkataraman et al.[29] and Salikhov et al.,[34] has also discussed pulse experiments. The CIDEP determinations of T_1 is by no means as widely applicable as these pulse experiments because it depends on the ability to induce polarization chemically, it depends on this production process being much faster than the T_1, and it requires the radicals to have a sufficiently long life so that the initial polarization is not dissipated too quickly. Nevertheless, the method can be applied, and a few measurements have been reported.

In the experiments performed by the Oxford group,[1r] the decay of both the absorption and dispersion signals were monitored, and the observed time constants for the exponential decay were extrapolated to zero microwave power to eliminate the contribution of the stimulated transitions. The Bloch equations were solved for the x- and y-components of the magnetization using the constraints that m_x and m_y were zero at zero field (transverse electron polarization has not yet been observed, or at least it has not yet been recognized) but with m_z variable. The time-dependent m_x and m_y expressions were differentiated with respect to frequency in order to mimic the effect of modulation, but this procedure is only approximate and a further analysis of this point seems desirable. When $T_1 \gg T_2$ the solutions are[1r]

$$\left(\frac{\partial m_x}{\partial \omega}\right)_{\text{resonance}} = m_{z,\infty} b T_2^2 \left\{ \left[z_0 - (1 + b^2 T_1 T_2)^{-1} \right] \exp\left[-(T_1^{-1} + b^2 T_2) t \right] + \left[1 + b^2 T_1 T_2 \right]^{-1} \right\} \tag{5.1a}$$

$$\left(\frac{\partial m_y}{\partial \omega}\right) = \left[\frac{2 b m_{z,\infty} \delta\omega T_2^3}{(1 + \delta\omega^2 T_2^2)^2}\right] \left\{ \left[z_0 - \left(\frac{1 + \delta\omega^2 T_2^2}{1 + \delta\omega^2 T_2^2 + b^2 T_1 T_2} \right)^2 \right] \times \exp\left[-\left(T_1^{-1} + \frac{b^2 T_2}{1 + \delta\omega^2 T_2^2} \right) t \right] + \left(\frac{1 + \delta\omega^2 T_2^2}{1 + \delta\omega^2 T_2^2 + b^2 T_1 T_2} \right) \right\} \tag{5.1b}$$

where $\delta\omega$ is the offset from the resonance frequency, $z_0 m_{z,\infty}$ is the initial magnetization expressed as a multiple of the equilibrium magnetization $m_{z,\infty}$, and $b = \gamma B_1$, where B_1 is the microwave magnetic field. The former of these expressions gives the time dependence of the dispersion signal at resonance, and the latter gives the time dependence of the first-derivative absorption signal (which has zero amplitude at resonance, which is why the more general expression is required). Inspection of these expressions gives the following effective time constants:

$$\left(T_{1\,\mathrm{eff}}^{-1}\right)_{\mathrm{abs}} = T_1^{-1} + \frac{b^2 T_2}{1 + \delta\omega^2 T_2^2}$$

$$\left(T_{1\,\mathrm{eff}}^{-1}\right)_{\mathrm{disp}} = T_1^{-1} + b^2 T_2 \tag{5.2}$$

Experiments on the ketyl radical from photolyzed benzophenone in liquid paraffins confirmed the exponential behavior and the time constants obeyed the power dependence required by (5.2). In the viscosity range studied (0.3 to 30 P) the measured T_1's were in the range 10^{-6} to 10^{-4} sec. No hyperfine dependence was studied because the spectra were overmodulated in order to enhance sensitivity.

The Bloch equations for CIDEP have also been examined by Pedersen. He extended the aforementioned analysis by incorporating the rate of formation of the radicals (this had also been done by Moore but remains unpublished[35]). The calculation shows that the time dependence of the signal intensity reflects the kinetics of the system in a way that enables the initial and recombination polarizations to be distinguished. Where his approach may be criticized is in his neglect of the effect of the chemical dissipation of the x- and y-components of the magnetization. Since, however, the analysis is concerned in the main with the limit $T_1 = T_2 \ll k_1^{-1}$, where k_1 is the first-order loss of doublets, the shortcomings in the treatment are not serious and the results are valid. Pedersen's treatment accommodates the kinetics observed by Livingston and Zeldes[1e] and Verma and Fessenden[1f] in a plausible manner.

B. Triplet T_1 Measurements

The measurement of electronic T_1's of triplet states in fluids by direct spin–resonance detection is made very difficult by their shortness and their correspondingly broad resonance lines (too broad in fact to be detected). As demonstrated in Section II, the triplet T_1's may in fact be measured by relating them to the observed polarization if the polarization is due to the TM, and if the chemical lifetime of the triplet lies in a range responsive to the addition of triplet quenchers. This point has been recognized by Atkins et al.[1x] Reinterpretation of their results by Dobbs[10a] on the basis of (2.16), which takes into account the doublet equilibrium population, gives values of the duroquinone triplet T_1 ranging from 2.7 nsec (in 0.5 cP methanol) to 17 nsec (in 57 cP cyclohexanol). The corresponding doublet T_1's of the duroquinone radical anion are 2 and 18 μsec respectively.

VI. CONCLUSION

The present state of the theory underlying RP and TM CIDEP seems adequate to account for current experimental observations.[100] In fact, it is a difficult problem to devise experiments to differentiate between the subtleties of the existing theories.

RP and TM CIDEP are similar phenomena in the sense that the coupling between the spin and lattice degrees of freedom does not vanish when the coupling is averaged over the initial ensemble; that is, in RP the quantity $\langle J(r)\rangle$ does not vanish, and in TM $\langle H_D(\Omega)H_K(\Omega)\rangle$ does not vanish. Although the stochastic Liouville method is not hampered by this shortcoming, there is no equivalent Redfield-like theory that can treat the TM to infinite order in H_D or the RP model to infinite order in J. In this concluding note, we wish to suggest that even in the cases where CIDEP does not probe the dynamics of fluids in a sensitive way, it may still be useful for structuring theories of quantum statistical mechanics in which the usual Redfield assumptions break down.

Acknowledgments

We thank the Science Research Council for Research Assistantship (for GTE). We are grateful to our colleagues for helpful discussions, and in particular acknowledge extensive comments from Dr. A. J. Dobbs. We are also grateful to Dr. F. J. Adrian, Professor J. H. Freed, and Professor R. W. Fessenden for sending preprints of their work.

References

1*a*. R. W. Fessenden and R. H. Schuler, *J. Chem. Phys.*, **39**, 2147 (1963).
1*b*. R. Livingston and H. Zeldes, *J. Amer. Chem. Soc.*, **88**, 4333 (1966).
1*c*. B. Smaller, J. R. Remko, and E. C. Avery, *J. Chem. Phys.*, **48**, 5174 (1968).
1*d*. P. J. Krusic and J. K. Kochi, *J. Amer. Chem. Soc.*, **90**, 7155 (1968).
1*e*. R. Livingston and H. Zeldes, *J. Chem. Phys.*, **53**, 1406 (1970).
1*f*. H. Paul and H. Fischer, *Z. Naturforsch.*, **25a**, 443 (1970).
1*g*. P. W. Atkins, I. C. Buchanan, R. C. Gurd, K. A. McLauchlan, and A. F. Simpson, *Chem. Comm.*, 513 (1970).
1*h*. S. H. Glarum and J. H. Marshall, *J. Chem. Phys.*, **52**, 5555 (1970).
1*i*. E. Eiben and R. W. Fessenden, *J. Phys. Chem.*, **75**, 1186 (1071).
1*j*. P. Neta, R. W. Fessenden, and R. H. Schuler, *J. Phys. Chem.*, **75**, 1654 (1971).
1*k*. B. Smaller, E. C. Avery, and J. R. Remko, *J. Chem. Phys.*, **55**, 2414 (1971).
1*l*. S. K. Wong and J. K. S. Wan, *J. Amer. Chem. Soc.*, **94**, 7197 (1972).
1*m*. S. K. Wong and J. K. S. Wan, *J. Chem. Phys.*, **59**, 3859 (1973).
1*n*. S. K. Wong , D. A. Hutchinson, and J. K. S. Wan, *J. Amer. Chem. Soc.*, **95**, 622 (1973).
1*o*. R. Livingston and H. Zeldes, *J. Magn. Resonance*, **9**, 331 (1973).
1*p*. P. W. Atkins, K. A. McLauchlan, and P. W. Percival, *Chem. Comm.*, 121 (1973).
1*q*. P. W. Atkins, A. J. Dobbs, and K. A. McLauchlan, *Chem. Phys. Lett.*, **22**, 209 (1973).
1*r*. P. W. Atkins, K. A. McLauchlan, and P. W. Percival, *Mol. Phys.*, **25**, 281 (1973).

1*s*. R. W. Fessenden, *J. Chem. Phys.*, **58**, 2489 (1973).
1*t*. N. C. Verma and R. W. Fessenden, *J. Chem. Phys.*, **58**, 2501 (1973).
1*u*. P. W. Atkins, J. M. Frimston, P. G. Frith, R. C. Gurd, and K. A. McLauchlan *J. Chem. Soc. Faraday Trans. II*, **69**, 1542 (1973).
1*v*. P. W. Atkins. J. K. Duggan, K. A. McLauchlan, and P. W. Percival, *Chem. Phys. Lett.*, **24**, 565 (1974).
1*x*. P. W. Atkins, A. J. Dobbs, and K. A. McLauchlan, *Chem. Phys. Lett.*, **25**, 105 (1974).
1*y*. P. W. Atkins, A. J. Dobbs, G. T. Evans, K. A. McLauchlan, and P. W. Percival, *Mol. Phys.*, **27**, 769 (1974).
1*z*. A. D. Trifunac and E. C. Avery, *Chem. Phys. Lett.*, **27**, 141 (1974).
1*aa*. H. Shiraishi, H. Kadoi, Y. Katsumura, Y. Tabata and K. Oshima, *J. Phys. Chem.*, **78**, 1336 (1974).
1*bb*. A review of experimental results has recently been published by J. K. S. Wan, S. K. Wong, and D. A. Hutchinson, *Acct. Chem. Res.*, **7**, 58 (1974).
2. M. S. de Groot, I. A. M. Hesselmann, and J. H. van der Waals, *Mol. Phys.*, **12**, 259 (1967).
3. See, for example,
3*a*. M. A. El-Sayed, *Acct. Chem. Res.*, **1**, 8 (1967).
3*b*. S. P. McGlynn, T. Azumi and T. Kinoshita, *Molecular Spectroscopy of the Triplet State*, Prentice-Hall, Englewood Cliffs, N. J., 1959.
3*c*. D. A. Antheunis, J. Schmidt and J. H. van der Waals, *Mol. Phys.*, **27** 1521 (1974).
4. P. W. Atkins and K. A. McLauchlan, in *Chemically Induced Magnetic Polarization*, edited by A. R. Lepley and G. L. Closs, Interscience, New York, 1973.
5. S. K. Wong, D. A. Hutchinson, and J. K. S. Wan, *J. Chem. Phys.*, **58**, 985 (1973).
6*a*. P. W. Atkins and G. T. Evans, *Chem. Phys. Lett.*, **25**, 108 (1974).
6*b*. P. W. Atkins and G. T. Evans, *Mol. Phys.*, **27**, 1633 (1974).
7. T. H. Cheng and N. Hirota, *Chem. Phys. Lett.*, **14**, 415 (1972).
8. Independent calculations have yielded a population difference one half of the result quoted in (2.2).
9. F. J. Adrian, *J. Chem. Phys.*, **61**, 4875 (1974).
10*a*. A. J. Dobbs, *Mol. Phys.*, **30**, 1073 (1975).
10*b*. P. W. Atkins, A. J. Dobbs, and K. A. McLauchlan, *Chem. Phys. Lett.*, **29**, 616 (1974).
10*c*. A. J. Dobbs and K. A. McLauchlan, *Chem. Phys. Lett.* **30**, 257, (1975).
10*d*. B. B. Adeleke, K. Y. Chuo, and J. K. S. Wan, *J*. Chem. Phys., **62**, 3822 (1975).
11. R. Kaptein and L. J. Oosterhoff, *Chem. Phys. Lett.*, **4**, 195 (1969).
12. H. Fischer and J. Bargon, *Acct. Chem. Res.*, **2**, 110 (1969).
13. P. W. Atkins, R. C. Gurd, K. A. McLauchlan, and A. F. Simpson, *Chem. Phys. Lett.*, **8**, 55 (1971).
14. F. J. Adrian, *Chem. Phys. Lett.*, **10**, 70 (1971).
15. G. L. Closs, *J. Amer. Chem. Soc.*, **91**, 4552 (1969).
16. See, for example,
16*a*. H. Fischer and M. Lehnig, *J. Phys. Chem.*, **75**, 3410 (1971).
16*b*. J. G. Garst, R. H. Cox, J. T. Barbas, R. D. Roberts, J. I. Morris, and R. C. Morris, *J. Amer. Chem. Soc.*, **92**, 5761 (1970).
17*a*. F. J. Adrian, *J. Chem. Phys.*, **54**, 3918 (1971).
17*b*. J. H. Freed, *Ann. Rev. Phys. Chem.* **23**, 265 (1972).
17*c*. F. J. Adrian, *J. Chem. Phys.*, **57**, 5107 (1972).
18. P. W. Atkins, *Chem. Phys. Lett.*, **18**, 290 (1973).
19. R. M. Noyes, *J. Chem. Phys.*, **22**, 1349 (1954).
20. P. W. Atkins and E. A. Moore, *Mol. Phys.*, **25**, 825 (1973).

21. M. Abramowitz and I. A. Stegun, *Handbook of Mathematical Functions*, Dover, New York, 1965.
22. G. T. Evans, P. D. Fleming, and R. G. Lawler, *J. Chem. Phys.*, **58**, 2071 (1973).
23*a*. J. B. Pedersen and J. H. Freed, *J. Chem. Phys.*, **57**, 1004 (1972).
23*b*. J. B. Pedersen and J. H. Freed, *J. Chem. Phys.*, **58**, 2746 (1973).
23*c*. J. B. Pedersen and J. H. Freed, *J. Chem. Phys.*, **59**, 2869 (1973).
23*d*. The result that S_{Z1} is weakly dependent upon the recombination event follows since CIDEP arises from off-diagonal elements and the off diagonal elements were not reacted. If one employs the model of Evans et al.,[22] and recalculates S_{Z1} subject to the constraint that only pure singlets react, one obtains

$$S_{Z1}=(\sigma_S(1-\lambda_K)-\sigma_T)\,\mathrm{sgn}(V_n)\left(\frac{a}{2r_0}\right)\left\{\frac{2\alpha J_0\tau_a}{1+(2\alpha J_0\tau_a)^2}\right\}[V_n\tau_a]^{1/2}$$

Thus, even if $J(r)$ and $K(r)$ have identical range parameters, S_{Z1} depends on K only through the spin selected initial conditions.
24*a*. J. M. Deutch, *J. Chem. Phys.*, **59**, 2762 (1973).
24*b*. J. M. Deutch, in *Transport Phenomena—1973*, (Brown University Seminar), edited by J. Kestin, Amer. Inst. Phys., New York, 1974, p. 17.
25*a*. R. G. Lawler and M. Halfon, private communication (1973).
25*b*. F. J. Adrian, *Chem. Phys. Lett.*, **26**, 437 (1974).
26. H. Fischer, *Chem. Phys. Lett.*, **4**, 611 (1970).
27*a*. D. A. Hutchinson, S. K. Wong, J. P. Colpa, and J. K. S. Wan, *J. Chem. Phys.*, **57**, 3308 (1972).
27*b*. S. K. Wong, D. A. Hutchinson, and J. K. S. Wan, *J. Chem. Phys.*, **60**, 2987 (1974).
28. J. B. Pedersen, *J. Chem. Phys.*, **59**, 2656 (1973).
29. S. K. Rengan, M. P. Khakhar, B. S. Prabhananda, and B. Venkataraman, *J. Pure Appl. Chem.*, **32**, 287 (1972).
30. P. W. Atkins, in *Electron Spin Relaxation in Liquids* edited by L. J. Muus and P. W. Atkins, (1972), Plenum Press p. 279.
31*a*. G. L. Closs and C. E. Doubleday, *J. Amer. Chem. Soc.*, **95**, 2736 (1973).
31*b*. R. Kaptein, H. R. Hill, and R. Freeman, *Chem. Phys. Lett.*, **26**, 104 (1974).
32. P. W. Atkins and G. T. Evans, *Chem. Phys. Lett.*, **24**, 44 (1974).
33. J. S. Hyde, *Ann. Rev. Phys. Chem.*, **25** (1973).
34. A. D. Milov, K. M. Salikhov, and Yu. D. Tvetsko, *Chem. Phys. Lett.*, **8**, 523 (1971).
35. E. A. Moore, D. Phil. Thesis, Oxford (1973).

KINETIC THEORY AND RHEOLOGY OF MACROMOLECULAR SOLUTIONS

C. F. CURTISS

Theoretical Chemistry Institute,
University of Wisconsin,
Madison, Wisconsin

R. BYRON BIRD and OLE HASSAGER*

Chemical Engineering Department and
Rheology Research Center,
University of Wisconsin,
Madison, Wisconsin

CONTENTS

*Present address: Instituttet for Kemiteknik, Danmarks tekniske Højskole, Lyngby, Denmark.

I. INTRODUCTION

We consider here the fluid dynamical phenomena occurring in a system that is made up of several chemical species $\alpha = 0, 1, 2, 3, \ldots, A$, some or all of which may be macromolecules. This might be a solution of one polymeric species in a single solvent, a solution of several polymeric species, or a polymer melt. It is known from elementary continuum mechanical arguments that, if the fluid is in an isothermal condition and if there are no chemical reactions or external forces, the statements of conservation of mass for the various species and the statement of conservation of momentum lead to the following equations[1]:

$$\frac{\partial}{\partial t}\rho_\alpha = -\left(\frac{\partial}{\partial \mathbf{r}}\cdot\rho_\alpha\mathbf{u}\right) - \left(\frac{\partial}{\partial \mathbf{r}}\cdot\mathbf{j}_\alpha\right), \qquad \alpha = 0, 1, 2, \ldots, A \tag{1.1}$$

$$\frac{\partial}{\partial t}\rho\mathbf{u} = -\left(\frac{\partial}{\partial \mathbf{r}}\cdot\rho\mathbf{u}\mathbf{u}\right) - \left(\frac{\partial}{\partial \mathbf{r}}\cdot\mathsf{p}\right) \tag{1.2}$$

Here ρ is the fluid density, ρ_α is the mass concentration of species α (grams of α per cubic centimeter), and $\mathbf{u}$ is the mass-average velocity of the fluid defined by $\rho\mathbf{u} = \Sigma_\alpha \rho_\alpha \mathbf{u}_\alpha$ where $\mathbf{u}_\alpha$ is the velocity of species α; the symbol $\mathbf{j}_\alpha$ is the mass flux of species α with respect to a coordinate system moving with the mass-average velocity, and p is the momentum flux. The $A+1$ equations of continuity can be added together to give the overall equation of continuity for the fluid mixture,

$$\frac{\partial \rho}{\partial t} = -\left(\frac{\partial}{\partial \mathbf{r}}\cdot\rho\mathbf{u}\right) \tag{1.3}$$

In order to solve fluid dynamical and diffusion problems with (1.1) and (1.2), it is necessary to have expressions for the fluxes p and $\mathbf{j}_\alpha$, as well as boundary and initial conditions. Our purpose is to show how nonequilibrium statistical mechanics can be used to obtain an expression for p.

The first formal theory dealing with the mechanical properties of dilute macromolecular solutions was that of Kramers[2] who generalized earlier work of Hermans[3] and Kuhn.[4] He showed that for steady-state potential flows a kinetic theory can be developed that enables one to write down immediately the phase-space distribution function; the latter can then be used to obtain the stresses in the flowing solution by making use of an expression for the stress tensor that Kramers derived by elementary kinetic theory arguments. Kramers set up this theory to describe a particular macromolecular model ("pearl necklace model") consisting of a series of beads connected by rigid rods, with universal joints; the beads undergo Stokes' drag as they move through the solvent, which in turn is viewed as a continuum. Kramers used generalized coordinates so that systems with constraints (such as models containing rigid rods or fixed bond angles) could be described.

Kramers' approach was extended slightly by Bird, Johnson, and Curtiss,[5] who considered the macromolecular model of beads joined by springs. For such models generalized coordinates are not needed, but interparticle (i.e., "inter-bead") potentials have to be included to describe the springs. Bird, Johnson, and Curtiss showed how the zero-shear-rate viscosity of the bead–rod model (Kramers' "pearl necklace model") could be obtained from that of the bead–spring model (with Fraenkel springs[6]), by freezing out the degrees of freedom associated with the springs. It was discovered subsequently by Hassager[7] that when an analogous spring-to-rod limiting process is performed for second-order terms, that is, those associated with the zero-shear-rate normal stresses, an inconsistency arises. This observation emphasizes the fact that such freezing-out of degrees of freedom may lead to results which are different from those obtained by the introduction of constraints, e.g., rigid rods.

The second major advance in the formulation of kinetic theories for macromolecular solutions was the Kirkwood–Riseman[8] theory, which was developed further by Kirkwood and his students. Kirkwood and Riseman used a configuration-space formalism, introducing an explicit expression for the Brownian motion force. They retained Kramers' use of generalized coordinates and focused their attention on linear chain macromolecular models that had constraints imposed by the constancy of the bond angles of the carbon backbone of a macromolecule. They further refined the theory by including hydrodynamic interaction with the Oseen tensor approximation, and in subsequent calculations introduced the further simplification of equilibrium-averaging the hydrodynamic interaction. Their applications were all to shearing flows, and they used Kramers' expression for the shear stress, even though his expression had been specifically derived for potential flow. It should also be pointed out that the equation of motion used in the configuration-space theories is difficult to interpret since it contains a mixture of purely dynamical quantities (e.g., velocities of beads) and statistical quantities (e.g., orientational distribution function).

Kotaka,[9] Giesekus,[10] and Prager[11] all calculated normal stresses for rigid rod-like molecules in shear flow, using a slight generalization of the Kirkwood formalism, which had dealt with shear stresses only.

Zimm's[12] and Williams'[13] calculations for shearing flows of suspensions of bead–spring models were based directly on the Kirkwood–Riseman theory; since the model they considered had no constraints, the theory could be simplified by using Cartesian coordinates. Zimm and Williams also used Kramers' expression for the stress tensor. Zimm's model is an extension of Rouse's[14] in that hydrodynamic interaction is included approximately. Lodge and Wu[15] showed how a constitutive equation could be obtained for the models of Rouse and Zimm. These theories have been subjected to extensive experimental testing in the linear viscoelastic range.[16]

Giesekus[17] showed how the Kramers' stress tensor could be put into an alternative form that involves one of Oldroyd's[18] convected derivatives. This form, usually more useful for computations, is applicable to all kinds of flows and is also helpful for making a connection with the continuum theories of rheology. Later Hassager and Bird[19] showed how the Giesekus expression could be extended slightly to include equilibrium-averaged hydrodynamic interaction.

A third major development is the work of Fixman,[20] who avoided the equilibrium-averaging in the hydrodynamic interaction, and who also considered more concentrated solutions where polymer–polymer interactions have to be included in the expression for the stress tensor. Fixman's theory has been used for calculations by Pyun and Fixman[21] and by Williams.[22] Fixman has set forth his theory in a configuration-space formalism.

The aforementioned theories are compared and contrasted in Table I. It can be seen that a more general formulation of the nonequilibrium statistical mechanics of polymer solutions is needed, if only to provide a common starting point for the discussion of the theories presently available. It is also necessary to have a more general formulation if one wishes to extend any of the present theories, since in the latter it is not always clear what assumptions are implied or to what extent a model is self-consistent. It was pointed out in 1969 by Curtiss[23] that a phase-space formalism for polymer solutions, patterned after the Irving–Kirkwood[24] theory for monatomic liquids, could provide the more general formulation that is needed. In this article the phase-space kinetic theory for macromolecular fluids is developed, and then it is shown how the "diffusion equation" for the configuration-space distribution function and the expression for the momentum-flux tensor can be simplified to obtain the equations given by previous investigators.

TABLE I. A Comparison of Various Macromolecular Theories[a]

	Type of theory (1)	Type of flow (2)	Coordinates used (3)	Model used (4)	Stress tensor used (5)	Brownian motion (6)	Hydro dynamic inter action (7)	Polymer–polymer inter action (8)
Kramers[2]	Phase	Pot	Gen	Constr	Kr	Not spec	No	No
Bird–Johnson–Curtiss[5]	Phase	Pot	Cart	Poten	Kr	Not spec	No	No
Kirkwood–Riseman[8]	Config	Shear	Gen	Constr Poten	Kr	Ch	Avg Oseen	No
Zimm[12]	Config	Shear	Cart	Poten	Kr	Ch	Avg Oseen	No
Rouse[14]	Config	Shear	Cart	Poten	*	Ch	No	No
Giesekus[17]	Config	Arb	Cart	Poten	Kr	Ch	No	No
Hassager–Bird[19]	Config	Arb	Cart	Poten	Kr	Ch	Avg Oseen	No
Fixman[20]	Config	Arb	Cart	Poten	C.S.	Ch	Oseen	Incl
Present work	Phase	Arb	Gen	Constr Poten	P.S.	Deriv	No	Incl

[a] * Used energy-dissipation arguments. (1) Phase space theory (Phase); configuration space theory (Config). (2) Potential flow only (Pot); shearing flows (Shear); arbitrary (Arb). (3) Generalized (Gen); Cartesian (Cart). (4) Model with constraints (Constr); model with interparticle potential (Poten). (5) Kramers' expression (Kr); new derivation using configuration space (C.S.) or phase space (P.S.) arguments. (6) No expression specified (Not spec); used relation attributed to Chandrasekhar (Ch); derived in the theory (Deriv). (7) Not included (No); equilibrium-avg Oseen (Avg Oseen); Oseen tensor (Oseen). (8) Not included (No); included (Incl).

Sections I through IV contain the mathematical and dynamical background needed in the subsequent development; this includes the definitions of coordinates, velocities, forces, and constraints that are used in the description of macromolecules. In Section V the statistical ideas are introduced; here the Liouville equation and ensemble averages are discussed. It is then shown how to obtain the equations of change from the Liouville equation, and thereby derive a formal statistical expression for the momentum-flux tensor. In Sections VI, VII, and VIII lower-order phase-space distribution functions are defined and their time evolution examined; also the momentum-flux tensor is rewritten in terms of these functions.

In Section IX we develop the kinetic equations in configuration space including the equation of continuity for the configuration-space distribution function and the equations of motion for the center of mass and the internal degrees of freedom. After introducing some assumptions in Section X, the equations of continuity and motion are combined in Section XI to give an approximate "diffusion equation" for the configurational distribution function.

Sections XII through XV deal with various expressions for the momentum-flux tensor in terms of the configurational distribution function, including those of Kramers, Kirkwood, Kotaka, Fixman, and Giesekus. In addition a proof is given of the symmetry of the momentum-flux tensor; this proof, however, is apparently possible only after a number of assumptions have been introduced into the theory.

Sections XVI and XVII give some applications of the macromolecular kinetic theory and rheology. First, it is shown how the Giesekus form of the momentum-flux tensor can be used to develop general expressions for the second-order fluid constants for arbitrary bead–spring–rod models. Then a development is given of the high-frequency limit of the dynamic viscosity for macromolecular models of arbitrary complexity.

To assist the reader in keeping track of the assumptions introduced in the theory, all assumptions are listed in Section X. In subsequent sections letters next to the equation numbers indicate which assumptions have been used to obtain the equations. In addition conventions on indices are adopted in Section II, and a chart is provided in Fig. 1 to assist in clarifying the development.

II. DEFINITIONS, NOMENCLATURE, AND COORDINATES

We consider as the system a mixture of several types of molecules, each of which may have internal structure. This could be, for example, a solvent with several dissolved macromolecular species. For the ith molecule of species α (hereinafter referred to as "molecule αi"), we use the following notation to specify the location of a system point in the phase space:

$\mathbf{r}_{\alpha i}$ Position of the center of mass (in laboratory-fixed coordinate system)

$\mathbf{p}_{\alpha i}$ Momentum of the center of mass

$Q_{\alpha i s}$ Generalized coordinates (in center-of-mass system) describing the orientation and internal configuration

$P_{\alpha i s}$ Generalized momenta conjugate to the $Q_{\alpha i s}$

where $s = 1, 2, \ldots, d_\alpha$, and d_α is the number of internal degrees of freedom of species α. Molecules of species α of mass m_α are presumed to be composed

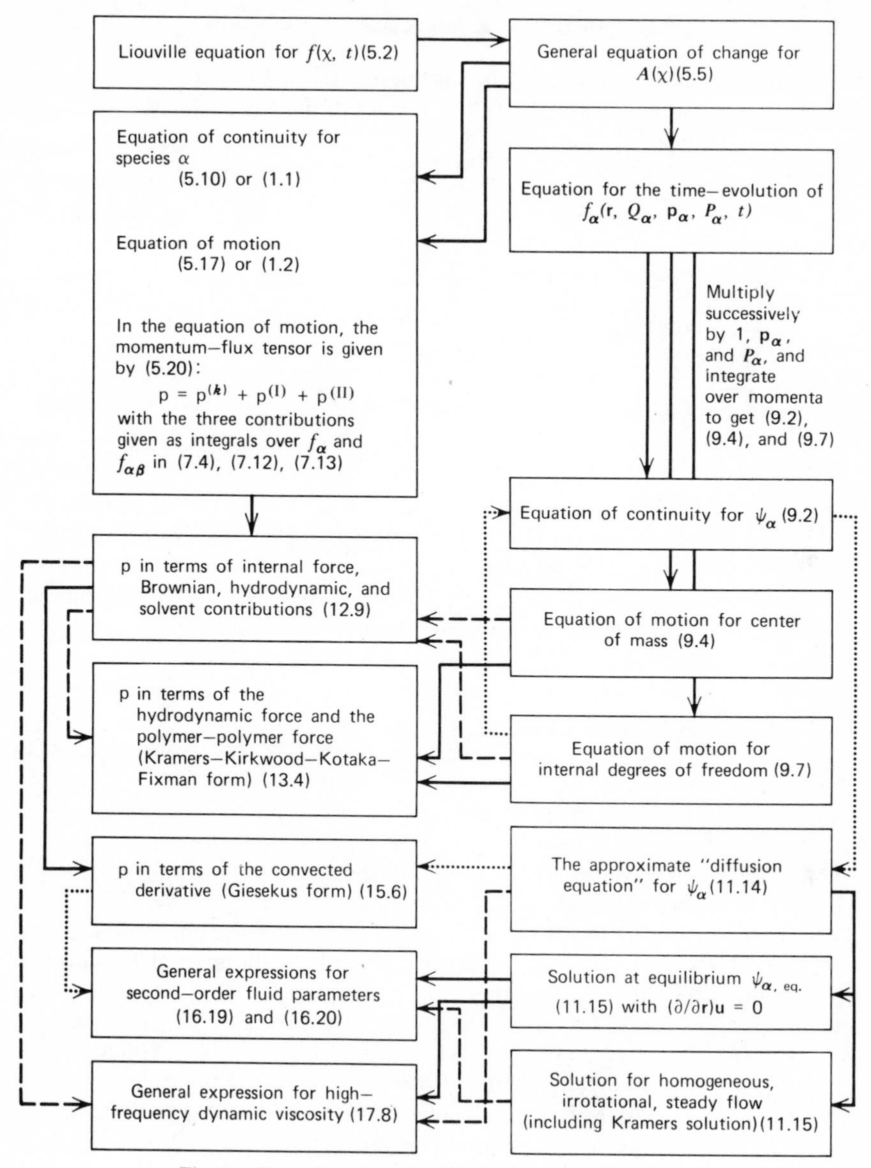

Fig. 1. Chart describing the kinetic theory developments.

of N_α mass points (these could be atomic nuclei or the "beads" in various bead–spring–rod models). The following notation is used for the νth mass point of molecule αi:

$m_{\alpha\nu}$	Mass
$\mathbf{r}_{\alpha i\nu}$	Position (in laboratory-fixed coordinate system)
$\mathbf{R}_{\alpha i\nu}=\mathbf{r}_{\alpha i\nu}-\mathbf{r}_{\alpha i}$	Position (relative to the center of mass $\mathbf{r}_{\alpha i}$ of the molecule)
$\tilde{\mathbf{R}}_{\alpha i\nu}=\mathbf{r}_{\alpha i\nu}-\tilde{\mathbf{r}}_{\alpha i}$	Position (relative to $\tilde{\mathbf{r}}_{\alpha i\nu}$)
$\mathbf{V}_{\alpha i\nu}$	Velocity (relative to the center of mass of the molecule)

where $\nu=1,2,3,\ldots,N_\alpha$ and $d_\alpha \leqslant 3N_\alpha-3$. Note that

$$\mathbf{r}_{\alpha i}=\frac{\Sigma_\nu m_{\alpha\nu}\mathbf{r}_{\alpha i\nu}}{\Sigma_\nu m_{\alpha\nu}}=\frac{\Sigma_\nu m_{\alpha\nu}\mathbf{r}_{\alpha i\nu}}{m_\alpha} \tag{2.1}$$

$$\tilde{\mathbf{r}}_{\alpha i}=\frac{\Sigma_\nu \zeta_{\alpha\nu}\mathbf{r}_{\alpha i\nu}}{\Sigma_\nu \zeta_{\alpha\nu}}=\frac{\Sigma_\nu \zeta_{\alpha\nu}\mathbf{r}_{\alpha i\nu}}{\zeta_\alpha} \tag{2.2}$$

and

$$\tilde{\mathbf{R}}_{\alpha i\nu}-\mathbf{R}_{\alpha i\nu}=-\zeta_\alpha^{-1}\Sigma_\mu \zeta_{\alpha\mu}\mathbf{R}_{\alpha i\mu} \tag{2.3}$$

The $\zeta_{\alpha\nu}$ may be any weighting factors, and $\zeta_\alpha=\Sigma_\nu\zeta_{\alpha\nu}$. Later in (10.6), the $\zeta_{\alpha\nu}$ are identified as friction coefficients, and then $\tilde{\mathbf{r}}_{\alpha i}$ is the "center of resistance" of the molecule.

In what follows we shall, wherever possible, adhere to the following conventions for the use of indices:

Indices	Used For	Range
$\alpha,\beta,\ldots$	Labeling chemical species	$0,1,2,\ldots,A$
$i,j,\ldots$	Identifying molecules of a particular chemical species	$1,2,\ldots$
$\nu,\mu,\eta,\ldots$	Enumerating the mass points ("beads") within a molecule of species α	$1,2,\ldots,N_\alpha$
s,t,u,v,w	Labeling generalized coordinates	$1,2,\ldots,d_\alpha$
m,n,p,q,r	Indices for Cartesian components	$1,2,3$ only
$a,b,c,\ldots$	Enumerating the links in a linear chain	$1,2,3,\ldots,N_\alpha-1$

In this way limits on sums, restrictions on free indices, and orders of matrices will not need to be given explicitly.

In applying macromolecular kinetic theories it is necessary to use some kind of structural model to represent the macromolecule. A few sample models are shown in Fig. 2. It will be noted that many of these models involve some kind of "constraints," such as fixed bond lengths or fixed bond angles. The kinetic theory developed here is capable of describing models with or without constraints; when no constraints are involved, many of the results simplify considerably.

We may now introduce several quantities that depend solely on the coordinates defined previously. First, we define two kinds of basis vectors:

$$\mathbf{b}_{\nu s}^{(\alpha i)} = m_{\alpha\nu}^{1/2} \frac{\partial}{\partial Q_{\alpha i s}} \mathbf{R}_{\alpha i \nu} \tag{2.4}$$

$$\tilde{\mathbf{b}}_{\nu s}^{(\alpha i)} = \zeta_{\alpha\nu}^{1/2} \frac{\partial}{\partial Q_{\alpha i s}} \tilde{\mathbf{R}}_{\alpha i \nu} \tag{2.5}$$

Use of (2.3) enables us to obtain

$$\tilde{\mathbf{b}}_{\nu s}^{(\alpha i)} = \left(\frac{\zeta_{\alpha\nu}}{m_{\alpha\nu}} \right)^{1/2} \mathbf{b}_{\nu s}^{(\alpha i)} - \zeta_{\alpha\nu}^{1/2} \Sigma_\mu \frac{\zeta_{\alpha\mu}}{\zeta_\alpha} \frac{\mathbf{b}_{\mu s}^{(\alpha i)}}{m_{\alpha\mu}^{1/2}} \tag{2.6}$$

Furthermore, using Eqs. 2.3 and 2.6, as well as Eqs. 2.11 and 2.12, one may verify that

$$\begin{aligned} \Sigma_\nu \zeta_{\alpha\nu}^{1/2} \tilde{\mathbf{b}}_{\nu}^{(\alpha i)} \mathbf{R}_{\alpha i \nu} &= \Sigma_\nu m_{\alpha\nu}^{-1/2} \zeta_{\alpha\nu} \mathbf{b}_{\nu t}^{(\alpha i)} \tilde{\mathbf{R}}_{\alpha i \nu} \\ &= \Sigma_\nu \zeta_{\alpha\nu}^{1/2} \tilde{\mathbf{b}}_{\nu t}^{(\alpha i)} \tilde{\mathbf{R}}_{\alpha i \nu} \end{aligned} \tag{2.7}$$

The following useful relations may be obtained from the foregoing definitions:

$$\Sigma_\nu m_{\alpha\nu} = m_\alpha \tag{2.8}$$

$$\Sigma_\nu m_{\alpha\nu} \mathbf{R}_{\alpha i \nu} = 0 \tag{2.9}$$

$$\Sigma_\nu \zeta_{\alpha\nu} \tilde{\mathbf{R}}_{\alpha i \nu} = 0 \tag{2.10}$$

$$\Sigma_\nu m_{\alpha\nu}^{1/2} \mathbf{b}_{\nu s}^{(\alpha i)} = 0 \tag{2.11}$$

$$\Sigma_\nu \zeta_{\alpha\nu}^{1/2} \tilde{\mathbf{b}}_{\nu s}^{(\alpha i)} = 0 \tag{2.12}$$

Equation 2.9 follows from the fact that the $\mathbf{R}_{\alpha i \nu}$ are referred to the centers of mass, and Eq. 2.11 comes from (2.4) and (2.9).

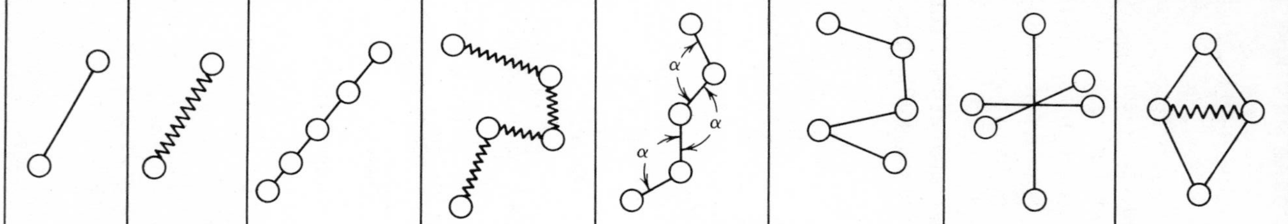

	Rigid dumbbell	Elastic dumbbell	Rigid rod with five beads	Rouse–Zimm model with flexible joints	Kirkwood–Riseman chain with fixed bond angles and fixed bond lengths	Kramers pearl–necklace model with flexible joints and fixed bond lengths	Rigid tridumbbell model with fixed angles	Plane elastic rhombus with flexible joints
$3N_\alpha - 3$	3	3	12	12	12	12	15	9
d_α	2	3	2	12	5	8	3	4

Fig. 2. Several macromolecular models; d_α, number of internal degrees of freedom; N_α, number of mass points in model.

Next, using the basis vectors defined in (2.4) and (2.5), we can form the corresponding covariant metric tensor components for the d_α-dimensional configuration spaces thus:

$$g_{st}^{(\alpha i)} = \Sigma_\nu \mathbf{b}_{\nu s}^{(\alpha i)} \cdot \mathbf{b}_{\nu t}^{(\alpha i)} \tag{2.13}$$

$$\tilde{g}_{st}^{(\alpha i)} = \Sigma_\nu \tilde{\mathbf{b}}_{\nu s}^{(\alpha i)} \cdot \tilde{\mathbf{b}}_{\nu t}^{(\alpha i)} \tag{2.14}$$

We note further that

$$g_{st}^{(\alpha i)} = \Sigma_\nu \frac{m_{\alpha\nu}}{\zeta_{\alpha\nu}} \tilde{\mathbf{b}}_{\nu s}^{(\alpha i)} \cdot \tilde{\mathbf{b}}_{\nu t}^{(\alpha i)} - m_\alpha \left(\Sigma_\nu \frac{\zeta_{\alpha\nu}}{\zeta_\alpha} \frac{\mathbf{b}_{\nu s}}{m_{\alpha\nu}^{1/2}} \right) \cdot \left(\Sigma_\mu \frac{\zeta_{\alpha\mu}}{\zeta_\alpha} \frac{\mathbf{b}_{\mu t}}{m_{\alpha\mu}^{1/2}} \right) \tag{2.15}$$

$$\tilde{g}_{st}^{(\alpha i)} = \Sigma_\nu \frac{\zeta_{\alpha\nu}}{m_{\alpha\nu}} \mathbf{b}_{\nu s}^{(\alpha i)} \cdot \mathbf{b}_{\nu t}^{(\alpha i)} - \frac{1}{\zeta_\alpha} \left(\Sigma_\nu \frac{\zeta_{\alpha\nu}}{m_{\alpha\nu}^{1/2}} \mathbf{b}_{\nu s} \right) \cdot \left(\Sigma_\mu \frac{\zeta_{\alpha\mu}}{m_{\alpha\mu}^{1/2}} \mathbf{b}_{\mu t} \right) \tag{2.16}$$

When $\zeta_{\alpha\nu} = C_\alpha m_{\alpha\nu}$, where C_α is a constant, $\tilde{\mathbf{R}}_{\alpha\nu} = \mathbf{R}_{\alpha\nu}$ and

$$\mathbf{b}_{\nu s}^{(\alpha i)} = C_\alpha^{-1/2} \tilde{\mathbf{b}}_{\nu s}^{(\alpha i)} \tag{2.17}$$

$$g_{st}^{(\alpha i)} = C_\alpha^{-1} \tilde{g}_{st}^{(\alpha i)} \tag{2.18}$$

and thus only one set of basis vectors and metric tensor components is needed.

For the contravariant metric tensor components we use the corresponding capital letters, so that

$$\Sigma_t G_{st}^{(\alpha i)} g_{tu}^{(\alpha i)} = \delta_{su} \tag{2.19}$$

$$\Sigma_t \tilde{G}_{st}^{(\alpha i)} \tilde{g}_{tu}^{(\alpha i)} = \delta_{su} \tag{2.20}$$

From the foregoing definitions, several relations can be derived that will be used later:

$$\frac{\partial}{\partial Q_{\alpha it}} \mathbf{b}_{\nu s}^{(\alpha i)} = \frac{\partial}{\partial Q_{\alpha is}} \mathbf{b}_{\nu t}^{(\alpha i)} \tag{2.21}$$

$$\Sigma_\nu \left(\frac{\partial}{\partial Q_{\alpha is}} \mathbf{b}_{\nu t}^{(\alpha i)} \right) \cdot \mathbf{b}_{\nu u}^{(\alpha i)} = \frac{1}{2} \left(\frac{\partial}{\partial Q_{\alpha is}} g_{tu}^{(\alpha i)} + \frac{\partial}{\partial Q_{\alpha it}} g_{us}^{(\alpha i)} - \frac{\partial}{\partial Q_{\alpha iu}} g_{st}^{(\alpha i)} \right) \tag{2.22}$$

$$\Sigma_{st} g_{st}^{(\alpha i)} \frac{\partial}{\partial Q_{\alpha iu}} G_{st}^{(\alpha i)} = - \frac{\partial}{\partial Q_{\alpha iu}} \ln g_{\alpha i} \tag{2.23}$$

where $g_{\alpha i}$ is the determinant of the $g_{st}^{(\alpha i)}$. We note that the right-hand side of (2.22) is in fact the Christoffel symbol $[st,u]$ for the d_α-dimensional space (see, for example, Ref. 25, p. 141, Eq. (4)). Equations 2.4 through 2.23 will be needed at various points in the subsequent theory.

In later sections we shall have occasion to use many of the symbols mentioned previously without the index i; this will be particularly true when, for example, $Q_{\alpha s}$ is used as one of the arguments in a distribution function. All of the aforementioned relations will apply equally well without the index i. In the discussion immediately following, the index i is omitted for brevity.

III. EULER ANGLES AND INTERNAL COORDINATES

More needs to be said about the generalized coordinates $Q_{\alpha s}$. These coordinates may be chosen in many ways, and for much of the subsequent discussion the choice will be left arbitrary. In the developments in Sections XIV and XVI, however, it becomes particularly useful to require that the first three generalized coordinates $Q_{\alpha 1}$, $Q_{\alpha 2}$, $Q_{\alpha 3}$ be the Euler angles that give the orientation of the macromolecular model in space. We shall use the Euler angles α, β, γ as defined on p. 908 of Ref. 1.

In order to specify the locations of the mass points in the macromolecule, it will be useful to work with two sets of Cartesian coordinate axes: A "space-fixed" set of axes with unit vectors $\boldsymbol{\delta}_n$ $(n=1, 2, 3)$ in the three coordinate directions, and a "body-fixed" set of axes, embedded in the macromolecule, with unit vectors $\check{\boldsymbol{\delta}}_m$ $(m=1, 2, 3)$ in the three coordinate directions. The body-fixed axes are embedded in the macromolecule in an arbitrary fashion. The orientation of the body-fixed frame with respect to the space-fixed frame is then given by the Euler angles α, β, γ. Specifically the relations between the two sets of unit vectors is given by

$$\check{\boldsymbol{\delta}}_m = \Sigma_n S_{mn} \boldsymbol{\delta}_n \tag{3.1}$$

in which the S_{mn} are the elements of the rotation matrix:

$$[S_{mn}] = \begin{bmatrix} cC¢ - s\$ & sC¢ + c\$ & -S¢ \\ -cC\$ - s¢ & -sC\$ + c¢ & S\$ \\ cS & sS & C \end{bmatrix} \tag{3.2}$$

in which s, S, \$ are $\sin\alpha$, $\sin\beta$, $\sin\gamma$, and c, C, ¢ are $\cos\alpha$, $\cos\beta$, $\cos\gamma$. The Euler angles vary over the following ranges: α, 0 to 2π; β, 0 to π; γ, 0 to 2π. It is useful to note that $\Sigma_p S_{mp} S_{np} = \delta_{mn}$ and $\Sigma_p S_{pm} S_{pn} = \delta_{mn}$.

The vectors locating the mass points with respect to the center of mass $\mathbf{r}_\alpha$

or the center of resistance $\tilde{\mathbf{r}}_\alpha$ are

$$\mathbf{R}_{\alpha\nu} = \Sigma_m R_{\alpha\nu m} \check{\boldsymbol{\delta}}_m = \Sigma_{mn} R_{\alpha\nu m} S_{mn} \boldsymbol{\delta}_n \tag{3.3}$$

$$\tilde{\mathbf{R}}_{\alpha\nu} = \Sigma_m \tilde{R}_{\alpha\nu m} \check{\boldsymbol{\delta}}_m = \Sigma_{mn} \tilde{R}_{\alpha\nu m} S_{mn} \boldsymbol{\delta}_n \tag{3.4}$$

where the $R_{\alpha\nu m}$ and $\tilde{R}_{\alpha\nu m}$ are the components of $\mathbf{R}_{\alpha\nu}$ and $\tilde{\mathbf{R}}_{\alpha\nu}$ in the body-fixed coordinate system; these quantities are functions of the coordinates $Q_{\alpha s}$ $(s \geqslant 4)$. The S_{mn} are functions of the $Q_{\alpha s}$ $(s \leqslant 3)$. The basis vectors $\mathbf{b}_{\nu s}^{(\alpha)}$ of (2.4) and $\tilde{\mathbf{b}}_{\nu s}^{(\alpha)}$ of (2.5) then become

$$\mathbf{b}_{\nu s}^{(\alpha)} = \begin{cases} m_{\alpha\nu}^{1/2} \Sigma_{mn} R_{\alpha\nu m} \left(\dfrac{\partial S_{mn}}{\partial Q_{\alpha s}} \right) \boldsymbol{\delta}_n & (s \leqslant 3) \\ m_{\alpha\nu}^{1/2} \Sigma_{mn} \left(\dfrac{\partial R_{\alpha\nu m}}{\partial Q_{\alpha s}} \right) S_{mn} \boldsymbol{\delta}_n & (s \geqslant 4) \end{cases} \tag{3.5}$$

$$\tilde{\mathbf{b}}_{\nu s}^{(\alpha)} = \begin{cases} \zeta_{\alpha\nu}^{1/2} \Sigma_{mn} \tilde{R}_{\alpha\nu m} \left(\dfrac{\partial S_{mn}}{\partial Q_{\alpha s}} \right) \boldsymbol{\delta}_n & (s \leqslant 3) \\ \zeta_{\alpha\nu}^{1/2} \Sigma_{mn} \left(\dfrac{\partial \tilde{R}_{\alpha\nu m}}{\partial Q_{\alpha s}} \right) S_{mn} \boldsymbol{\delta}_n & (s \geqslant 4) \end{cases} \tag{3.6}$$

Alternatively, the vectors $\mathbf{b}_{\nu s}^{(\alpha)}$ and $\tilde{\mathbf{b}}_{\nu s}^{(\alpha)}$ may be referred to the body-fixed coordinate system thus:

$$\mathbf{b}_{\nu s}^{(\alpha)} = \Sigma_m b_{\nu s m}^{(\alpha)} \check{\boldsymbol{\delta}}_m = \Sigma_{mn} b_{\nu s m}^{(\alpha)} S_{mn} \boldsymbol{\delta}_n \tag{3.7}$$

$$\tilde{\mathbf{b}}_{\nu s}^{(\alpha)} = \Sigma_m \tilde{b}_{\nu s m}^{(\alpha)} \check{\boldsymbol{\delta}}_m = \Sigma_{mn} \tilde{b}_{\nu s m}^{(\alpha)} S_{mn} \boldsymbol{\delta}_n \tag{3.8}$$

where

$$b_{\nu s m}^{(\alpha)} = \begin{cases} m_{\alpha\nu}^{1/2} \Sigma_{np} R_{\alpha\nu p} S_{mn} \left(\dfrac{\partial S_{pn}}{\partial Q_{\alpha s}} \right) & (s \leqslant 3) \\ m_{\alpha\nu}^{1/2} \left(\dfrac{\partial R_{\alpha\nu m}}{\partial Q_{\alpha s}} \right) & (s \geqslant 4) \end{cases} \tag{3.9}$$

$$\tilde{b}_{\nu s m}^{(\alpha)} = \begin{cases} \zeta_{\alpha\nu}^{1/2} \Sigma_{np} \tilde{R}_{\alpha\nu p} S_{mn} \left(\dfrac{\partial S_{pn}}{\partial Q_{\alpha s}} \right) & (s \leqslant 3) \\ \zeta_{\alpha\nu}^{1/2} \left(\dfrac{\partial \tilde{R}_{\alpha\nu m}}{\partial Q_{\alpha s}} \right) & (s \geqslant 4) \end{cases} \tag{3.10}$$

are the components in the body-fixed system.

In order to factor the metric tensor into parts associated with rotation and the internal motions, it is convenient to define the following matrix:

$$\Lambda_{ts} = \begin{cases} \frac{1}{2}\Sigma_{mnp}\epsilon_{tmn}S_{np}\left(\frac{\partial S_{mp}}{\partial Q_{\alpha s}}\right) & (s,t \leqslant 3) \\ \delta_{ts} & (s,t \geqslant 4) \\ 0 & \text{otherwise} \end{cases} \tag{3.11}$$

in which ϵ_{tmn} is the usual permutation symbol ($\epsilon_{tmn} = +1$ if tmn is a cyclic permutation of 123; $\epsilon_{tmn} = -1$ if tmn is an anticyclic permutation; $\epsilon_{tmn} = 0$ otherwise). For s, $t \leqslant 3$ Eq. 3.11 can be inverted to give

$$\frac{\partial S_{mn}}{\partial Q_{\alpha s}} = \Sigma_{pt}\epsilon_{mpt}\Lambda_{ts}S_{pn} \tag{3.12}$$

The Λ_{ts} and Λ_{ts}^{-1} are the elements of the reciprocal matrices:

$$[\Lambda_{ts}] = \left[\begin{array}{ccc|cccc} -S\text{¢} & \$ & 0 & & & & \\ S\$ & \text{¢} & 0 & & 0 & & \\ C & 0 & 1 & & & & \\ \hline & & & 1 & 0 & 0 & \cdots \\ & & & 0 & 1 & 0 & \cdots \\ & 0 & & 0 & 0 & 1 & \cdots \\ & & & \vdots & \vdots & \vdots & \end{array}\right] \tag{3.13}$$

$$[\Lambda_{ts}^{-1}] = \left[\begin{array}{ccc|cccc} -S^{-1}\text{¢} & S^{-1}\$ & 0 & & & & \\ \$ & \text{¢} & 0 & & 0 & & \\ S^{-1}C\text{¢} & -S^{-1}C\$ & 1 & & & & \\ \hline & & & 1 & 0 & 0 & \cdots \\ & & & 0 & 1 & 0 & \cdots \\ & 0 & & 0 & 0 & 1 & \cdots \\ & & & \vdots & \vdots & \vdots & \end{array}\right] \tag{3.14}$$

Note also that $\det[\Lambda_{ts}] = -S = -\sin\beta$. Now Eqs. 3.9 and 3.10 can be rewritten as

$$b_{\nu sm}^{(\alpha)} = \Sigma_t \Lambda_{ts} b_{(\alpha)}^{\nu tm} \tag{3.15}$$

$$\tilde{b}_{\nu sm}^{(\alpha)} = \Sigma_t \Lambda_{ts} \tilde{b}_{(\alpha)}^{\nu tm} \tag{3.16}$$

where

$$b_{(\alpha)}^{\nu tm} = \begin{cases} m_{\alpha\nu}^{1/2} \, \Sigma_n R_{\alpha\nu n} \epsilon_{nmt} & (t \leqslant 3) \\ m_{\alpha\nu}^{1/2} \left(\dfrac{\partial R_{\alpha\nu m}}{\partial Q_{\alpha t}} \right) & (t \geqslant 4) \end{cases} \tag{3.17}$$

$$\tilde{b}_{(\alpha)}^{\nu tm} = \begin{cases} \zeta_{\alpha\nu}^{1/2} \, \Sigma_n \tilde{R}_{\alpha\nu n} \epsilon_{nmt} & (t \leqslant 3) \\ \zeta_{\alpha\nu}^{1/2} \left(\dfrac{\partial \tilde{R}_{\alpha\nu m}}{\partial Q_{\alpha t}} \right) & (t \geqslant 4) \end{cases} \tag{3.18}$$

Note that the $b_{(\alpha)}^{\nu tm}$ and $\tilde{b}_{(\alpha)}^{\nu tm}$ are easy to compute when the positions of all mass points are known in the embedded coordinate frame.

Finally, combining Eqs. 3.7 and 3.15 (or Eqs. 3.8 and 3.16) gives

$$\mathbf{b}_{\nu s}^{(\alpha)} = \Sigma_{tmn} \Lambda_{ts} b_{(\alpha)}^{\nu tm} S_{mn} \boldsymbol{\delta}_n \tag{3.19}$$

$$\tilde{\mathbf{b}}_{\nu s}^{(\alpha)} = \Sigma_{tmn} \Lambda_{ts} \tilde{b}_{(\alpha)}^{\nu tm} S_{mn} \boldsymbol{\delta}_n \tag{3.20}$$

Note that the $b_{(\alpha)}^{\nu tm}$ and the $\tilde{b}_{(\alpha)}^{\nu tm}$ contain the $Q_{\alpha s}$ $(s \geqslant 4)$; whereas, Λ_{ts} and S_{mn} contain only the Euler angles.

Next we turn our attention to the components of the covariant and contravariant metric tensors. From the definitions in (2.13) and (2.14) and the foregoing results in (3.19) and (3.20) it follows that

$$g_{st}^{(\alpha)} = \Sigma_{uv} \Lambda_{us} \Gamma_{uv}^{(\alpha)} \Lambda_{vt} \tag{3.21}$$

$$\tilde{g}_{st}^{(\alpha)} = \Sigma_{uv} \Lambda_{us} \tilde{\Gamma}_{uv}^{(\alpha)} \Lambda_{vt} \tag{3.22}$$

in which

$$\Gamma_{uv}^{(\alpha)} = \Sigma_{\nu m} b_{(\alpha)}^{\nu um} b_{(\alpha)}^{\nu vm} \tag{3.23}$$

$$\tilde{\Gamma}_{uv}^{(\alpha)} = \Sigma_{\nu m} \tilde{b}_{(\alpha)}^{\nu um} \tilde{b}_{(\alpha)}^{\nu vm} \tag{3.24}$$

The $\Gamma_{uv}^{(\alpha)}$ are components of a "generalized moment of inertia tensor" which depend on $Q_{\alpha s}$ $(s \geqslant 4)$. Since Λ_{us}, $\Gamma_{uv}^{(\alpha)}$, and $\tilde{\Gamma}_{uv}^{(\alpha)}$ are the elements of square matrices, which for nonlinear models are nonsingular, it follows from (2.19) and similar relations that

$$G_{st}^{(\alpha)} = \Sigma_{uv} \Lambda_{su}^{-1} \Gamma_{uv}^{(\alpha)-1} \Lambda_{tv}^{-1} \tag{3.25}$$

$$\tilde{G}_{st}^{(\alpha)} = \Sigma_{uv} \Lambda_{su}^{-1} \tilde{\Gamma}_{uv}^{(\alpha)-1} \Lambda_{tv}^{-1} \tag{3.26}$$

Note that $g_\alpha \equiv \det[g_{st}^{(\alpha)}] = \sin^2 \beta \det[\Gamma_{uv}^{(\alpha)}]$ and that $\tilde{g}_\alpha \equiv \det[\tilde{g}_{st}^{(\alpha)}] = \sin^2 \beta \det[\tilde{\Gamma}_{uv}^{(\alpha)}]$.

Next we give some special relations that will be needed for proving the symmetry of the stress tensor under certain assumptions. From (3.4) and (3.20) one readily finds that

$$\Sigma_\nu \zeta_{\alpha\nu}^{\frac{1}{2}} \tilde{\mathbf{b}}_{\nu t}^{(\alpha)} \tilde{\mathbf{R}}_{\alpha\nu} = \Sigma_{smnpq} \Lambda_{st} J_{smn}^{(\alpha)} S_{mp} S_{nq} \boldsymbol{\delta}_p \boldsymbol{\delta}_q \tag{3.27}$$

where

$$J_{smn}^{(\alpha)} = \Sigma_\nu \zeta_{\alpha\nu}^{\frac{1}{2}} \tilde{R}_{\alpha\nu n} \tilde{b}_{(\alpha)}^{\nu sm} \tag{3.28}$$

Next we multiply Eq. 3.27 by $G_{st}^{(\alpha)}$ and sum on t; when Eq. 3.26 is used, we then get

$$\Sigma_{\nu t} \zeta_{\alpha\nu}^{\frac{1}{2}} \tilde{G}_{st}^{(\alpha)} \tilde{\mathbf{b}}_{\nu t}^{(\alpha)} \tilde{\mathbf{R}}_{\alpha\nu} = \Sigma_{mnpquv} \Lambda_{su}^{-1} \tilde{\Gamma}_{uv}^{(\alpha)-1} J_{vmn}^{(\alpha)} S_{mp} S_{nq} \boldsymbol{\delta}_p \boldsymbol{\delta}_p \tag{3.29}$$

From (3.18) and (3.24) it follows immediately that

$$\tilde{\Gamma}_{uv}^{(\alpha)} = \Sigma_{\nu mn} \zeta_{\alpha\nu}^{\frac{1}{2}} \epsilon_{mnu} \tilde{R}_{\alpha\nu m} \tilde{b}_{(\alpha)}^{\nu vn} \qquad (u \leqslant 3) \tag{3.30}$$

and thus from (3.28)

$$J_{vmn}^{(\alpha)} - J_{vnm}^{(\alpha)} = \Sigma_u \epsilon_{nmu} \tilde{\Gamma}_{uv}^{(\alpha)} \tag{3.31}$$

A further relation involving the $J_{vmn}^{(\alpha)}$ is obtained from (3.17) and (3.28):

$$\Sigma_s \epsilon_{mns} J_{spq}^{(\alpha)} = \delta_{np} K_{qm}^{(\alpha)} - \delta_{mp} K_{qn}^{(\alpha)} \tag{3.32}$$

where

$$K_{nm}^{(\alpha)} = \Sigma_\nu \zeta_{\alpha\nu} \tilde{R}_{\alpha\nu n} \tilde{R}_{\alpha\nu m} \tag{3.33}$$

In the discussion in Section XIV, Eqs. 3.31 and 3.32 will play a central role. In addition, $K_{nm}^{(\alpha)}$ will appear in Sections XVI and XVII as a key structural parameter for computing viscosity, normal stresses, and the high-frequency limit of the dynamic viscosity.

We conclude this section giving several integrals that involve integrations over Euler angles.[7] The overbar means $\int\int\int \cdots \sin\beta \, d\alpha d\beta d\gamma$, and the $\boldsymbol{\lambda}_n$ are unit vectors in three-space that are functions of the Euler angles. Specifically, these may be the embedded unit vectors $\check{\boldsymbol{\delta}}_n$:

$$3\,\overline{(\boldsymbol{\lambda}_m\boldsymbol{\lambda}_n)}_{m'n'} = \overline{\boldsymbol{\lambda}_m\cdot\boldsymbol{\lambda}_n}\,\delta_{m'n'} \tag{3.34}$$

$$\begin{aligned}
30\,&\overline{(\boldsymbol{\lambda}_m\boldsymbol{\lambda}_n\boldsymbol{\lambda}_p\boldsymbol{\lambda}_q)}_{m'n'p'q'} \\
&= \left[4\,\overline{(\boldsymbol{\lambda}_m\cdot\boldsymbol{\lambda}_n)(\boldsymbol{\lambda}_p\cdot\boldsymbol{\lambda}_q)} - \overline{(\boldsymbol{\lambda}_m\cdot\boldsymbol{\lambda}_q)(\boldsymbol{\lambda}_n\cdot\boldsymbol{\lambda}_p)} - \overline{(\boldsymbol{\lambda}_m\cdot\boldsymbol{\lambda}_p)(\boldsymbol{\lambda}_n\cdot\boldsymbol{\lambda}_q)}\,\right]\delta_{m'n'}\delta_{p'q'} \\
&\quad + \left[4\,\overline{(\boldsymbol{\lambda}_m\cdot\boldsymbol{\lambda}_q)(\boldsymbol{\lambda}_n\cdot\boldsymbol{\lambda}_p)} - \overline{(\boldsymbol{\lambda}_m\cdot\boldsymbol{\lambda}_n)(\boldsymbol{\lambda}_p\cdot\boldsymbol{\lambda}_q)} - \overline{(\boldsymbol{\lambda}_m\cdot\boldsymbol{\lambda}_p)(\boldsymbol{\lambda}_n\cdot\boldsymbol{\lambda}_q)}\,\right]\delta_{m'q'}\delta_{n'p'} \\
&\quad + \left[4\,\overline{(\boldsymbol{\lambda}_m\cdot\boldsymbol{\lambda}_p)(\boldsymbol{\lambda}_n\cdot\boldsymbol{\lambda}_q)} - \overline{(\boldsymbol{\lambda}_m\cdot\boldsymbol{\lambda}_n)(\boldsymbol{\lambda}_p\cdot\boldsymbol{\lambda}_q)} - \overline{(\boldsymbol{\lambda}_m\cdot\boldsymbol{\lambda}_q)(\boldsymbol{\lambda}_n\cdot\boldsymbol{\lambda}_p)}\,\right]\delta_{m'p'}\delta_{n'q'}
\end{aligned} \tag{3.35}$$

Examples

At various points in the subsequent development, it is useful to illustrate intermediate results in terms of simple macromolecular models.

Rigid Dumbbell Model (For Rigid Linear Macromolecules)

This model consists of two beads joined by a rigid rod of length L; bead "1" has mass m_1, and bead "2" mass m_2. As internal coordinates we take $Q_1=\theta$, $Q_2=\phi$, where θ and ϕ are the spherical coordinates locating bead 2 with respect to the center of mass. We let $\check{\boldsymbol{\delta}}_R$ be a unit vector in the direction from the center of mass to bead 2; $\check{\boldsymbol{\delta}}_\theta$ and $\check{\boldsymbol{\delta}}_\phi$ are unit vectors perpendicular to $\check{\boldsymbol{\delta}}_R$ such that the vectors $\check{\boldsymbol{\delta}}_R$, $\check{\boldsymbol{\delta}}_\theta$, $\check{\boldsymbol{\delta}}_\phi$ form a right-handed coordinate system. We note the following properties of these vectors:

Unit vector	x,y,z-Components	Derivatives of unit vectors $\partial/\partial\theta$	$\partial/\partial\phi$
$\check{\boldsymbol{\delta}}_R$	$\sin\theta\cos\phi, \sin\theta\sin\phi, \cos\theta$	$\check{\boldsymbol{\delta}}_\theta$	$\check{\boldsymbol{\delta}}_\phi \sin\theta$
$\check{\boldsymbol{\delta}}_\theta$	$\cos\theta\cos\phi, \cos\theta\sin\phi, -\sin\theta$	$-\check{\boldsymbol{\delta}}_R$	$\check{\boldsymbol{\delta}}_\phi \cos\theta$
$\check{\boldsymbol{\delta}}_\phi$	$-\sin\phi, \cos\phi, 0$	0	$-\check{\boldsymbol{\delta}}_R\sin\theta - \check{\boldsymbol{\delta}}_\theta\cos\theta$

Then the locations of the beads are

$$\mathbf{R}_1 = -\left(\frac{\mu L}{m_1}\right)\breve{\boldsymbol{\delta}}_R; \qquad \mathbf{R}_2 = \left(\frac{\mu L}{m_2}\right)\breve{\boldsymbol{\delta}}_R \tag{3.36}$$

where $(1/\mu) = (1/m_1) + (1/m_2)$ is the reciprocal of the reduced mass μ. The quantities defined in (2.4), (2.13), and (2.19) are then

$$\mathbf{b}_{1\theta} = -\left(\frac{\mu L}{m_1^{1/2}}\right)\breve{\boldsymbol{\delta}}_\theta \tag{3.37}$$

$$\mathbf{b}_{2\theta} = \left(\frac{\mu L}{m_2^{1/2}}\right)\breve{\boldsymbol{\delta}}_\theta \tag{3.38}$$

$$\mathbf{b}_{1\phi} = -\left(\frac{\mu L}{m_1^{1/2}}\right)\breve{\boldsymbol{\delta}}_\phi \sin\theta \tag{3.39}$$

$$\mathbf{b}_{2\phi} = \left(\frac{\mu L}{m_2^{1/2}}\right)\breve{\boldsymbol{\delta}}_\phi \sin\theta \tag{3.40}$$

$$g_{\theta\theta} = \mu L^2 \tag{3.41}$$

$$g_{\phi\phi} = \mu L^2 \sin^2\theta \tag{3.42}$$

$$G_{\theta\theta} = \frac{1}{\mu L^2} \tag{3.43}$$

$$G_{\phi\phi} = \frac{1}{\mu L^2 \sin^2\theta} \tag{3.44}$$

The off-diagonal terms of g_{st} and G_{st} are zero, and

$$G = \det G_{st} = \frac{1}{\mu^2 L^4 \sin^2\theta} \tag{3.45}$$

Rouse Model[14] (For Flexible Macromolecules)

The Rouse model consists of N beads with mass m connected linearly by $N-1$ springs. The springs are here taken to be linear or nonlinear (in the original paper of Rouse, they are taken to be linear) and the tension in the νth connector, connecting beads ν and $\nu+1$, is called $\mathbf{F}_\nu^{(c)}$; this connector force is a function only of the separation between beads ν and $\nu+1$. This is an example of a model with no constraints.

As internal coordinates it is convenient to take the Cartesian components of the vectors that describe the connector orientations. We let $\mathbf{Q}_a$ be the vector from bead a to bead $a+1$, and its Cartesian components are Q_{an}, with $n=1,2,3$ and $a=1,2,\ldots,N-1$, so that

$$\mathbf{Q}_a=\Sigma_n\boldsymbol{\delta}_n Q_{an} \tag{3.46}$$

where $\boldsymbol{\delta}_n$ is the unit vector in the nth Cartesian coordinate direction. Then the position of the νth bead with respect to the center of mass is given in terms of the internal coordinates as

$$\mathbf{R}_\nu=\Sigma_a B_{\nu a}\mathbf{Q}_a=\Sigma_a\Sigma_n B_{\nu a}\boldsymbol{\delta}_n Q_{an} \tag{3.47}$$

in which $B_{\nu a}=a/N$ for $a<\nu$ and $B_{\nu a}=-[1-(a/N)]$ for $a\geqslant\nu$. Some properties of the $B_{\nu a}$ are[5]

$$\Sigma_\nu B_{\nu a}=0 \tag{3.48}$$

$$B_{\nu+1,a}-B_{\nu,a}=\delta_{\nu a} \tag{3.49}$$

$$\Sigma_\nu B_{\nu a}B_{\nu b}=C_{ab}=\begin{cases}\dfrac{a(N-b)}{N} & \text{if } a\leqslant b\\[2ex] \dfrac{b(N-a)}{N} & \text{if } b\leqslant a\end{cases} \tag{3.50}$$

It is further known that the eigenvalues of the $(N-1)\times(N-1)$ C-matrix, which is symmetric and nonsingular, are[26]

$$c_a=\frac{1}{4\sin^2(a\pi/2N)} \tag{3.51}$$

and that the C-matrix is the inverse of the "Rouse matrix" A:

$$\Sigma_c C_{ac}A_{cb}=\delta_{ab} \tag{3.52}$$

where

$$[A_{ab}]=\begin{bmatrix} 2 & -1 & 0 & \cdots\\ -1 & 2 & -1 & \cdots\\ 0 & -1 & 2 & \cdots\\ \vdots & \vdots & \vdots & \end{bmatrix} \tag{3.53}$$

Furthermore

$$\Sigma_a c_a = \Sigma_a C_{aa} = \frac{(N-1)(N+1)}{6} \tag{3.54}$$

and

$$\Sigma_a c_a^2 = \Sigma_{ab} C_{ab}^2 = \frac{(N-1)(N+1)(2N^2+7)}{180} \tag{3.55}$$

We can now write the various quantities needed for using the final kinetic theory results:

$$\mathbf{b}_{\nu,an} = m^{1/2} \frac{\partial}{\partial Q_{an}} \mathbf{R}_\nu = m^{1/2} B_{\nu a} \boldsymbol{\delta}_n \tag{3.56}$$

$$g_{an,bp} = \Sigma_\nu (\mathbf{b}_{\nu,an} \cdot \mathbf{b}_{\nu,bp}) = m C_{ab} \delta_{np} \tag{3.57}$$

$$\begin{aligned} g &= \det(g_{an,bp}) \\ &= \left[m^{N-1} \det(C_{ij}) \right]^3 \\ &= \frac{m^{3(N-1)}}{N^3} \end{aligned} \tag{3.58}$$

Elastic Rhombus Model (For Finitely Extendible Macromolecules)

This model, shown in Fig. 3, is a planar, four-bead model, with one opposing pair of beads connected by a Hookean spring whose equilibrium length is $L\sqrt{2}$; all beads have equal masses, and the rigid rods have length L. Let $\check{\boldsymbol{\delta}}_1$ be a unit vector in the direction from the center of mass to bead 1; $\check{\boldsymbol{\delta}}_2$ is the unit vector pointing toward bead 2. $\check{\boldsymbol{\delta}}_3$ is perpendicular to the

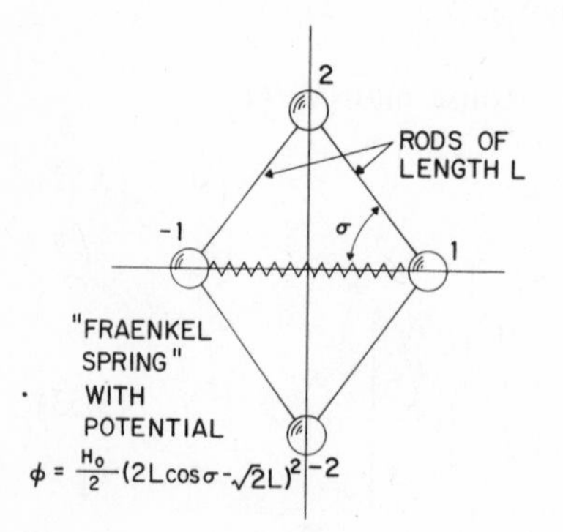

Fig. 3. The elastic rhombus model.

plane of the model such that $\check{\boldsymbol{\delta}}_1, \check{\boldsymbol{\delta}}_2, \check{\boldsymbol{\delta}}_3$ form a right-hand set. These embedded vectors are related to the unit vectors fixed in space by (3.1).

The locations of the beads are given by

$$\mathbf{R}_{\pm\nu} = \pm L(\delta_{\nu 1}\mathrm{K} + \delta_{\nu 2}\Sigma)\check{\boldsymbol{\delta}}_\nu \tag{3.59}$$

where $\mathrm{K} = \cos\sigma$ and $\Sigma = \sin\sigma$.

From (3.17) we get

$$b^{132} = -b^{123} = b^{-123} = -b^{-132} = b^{242} = -b^{-242} = m^{1/2}L\cos\sigma \tag{3.60}$$

$$b^{213} = -b^{231} = -b^{-213} = b^{-231} = -b^{141} = b^{-141} = m^{1/2}L\sin\sigma \tag{3.61}$$

Then from (3.23)

$$[\Gamma_{uv}] = 2mL^2 \begin{bmatrix} \Sigma^2 & 0 & 0 & 0 \\ 0 & \mathrm{K}^2 & 0 & 0 \\ 0 & 0 & 1 & 0 \\ 0 & 0 & 0 & 1 \end{bmatrix} \tag{3.62}$$

And from (3.12) and (3.21), we find the covariant metric tensor:

$$\frac{g_{st}}{2mL^2} = \begin{bmatrix} \mathrm{K}^2(C^2\text{¢}^2 + \$^2) + \Sigma^2(C^2\$^2 + \text{¢}^2) & (\mathrm{K}^2 - \Sigma^2)S\$\text{¢} & C & 0 \\ (\mathrm{K}^2 - \Sigma^2)S\$\text{¢} & \mathrm{K}^2\text{¢}^2 + \Sigma^2\$^2 & 0 & 0 \\ C & 0 & 1 & 0 \\ 0 & 0 & 0 & 1 \end{bmatrix} \tag{3.63}$$

with

$$\det g_{rs} = 4m^4L^8\sin^2 2\sigma\sin^2\beta. \tag{3.64}$$

The contravariant metric tensor is similarly obtained from (3.13) and (3.14):

$$\frac{G_{st}}{2mL^2S^2\Sigma^2\mathrm{K}^2} = \begin{bmatrix} \mathrm{K}^2\text{¢}^2 + \Sigma^2\$^2 & -(\mathrm{K}^2 - \Sigma^2)S\$\text{¢} & -(\mathrm{K}^2\text{¢}^2 + \Sigma^2\$^2)C & 0 \\ -(\mathrm{K}^2 - \Sigma^2)S\$\text{¢} & \mathrm{K}^2S^2\$^2 + \Sigma^2S^2\text{¢}^2 & (\mathrm{K}^2 - \Sigma^2)SC\$\text{¢} & 0 \\ -(\mathrm{K}^2\text{¢}^2 + \Sigma^2\$^2)C & (\mathrm{K}^2 - \Sigma^2)SC\$\text{¢} & \begin{matrix}\mathrm{K}^2C^2\text{¢}^2 + \Sigma^2C^2\$^2 \\ + \Sigma^2\mathrm{K}^2S^2\end{matrix} & 0 \\ 0 & 0 & 0 & \Sigma^2\mathrm{K}^2S^2 \end{bmatrix} \tag{3.65}$$

IV. THE TIME-EVOLUTION OF THE DYNAMICAL STATE OF THE SYSTEM

Next we turn to the velocities of the mass points $\mathbf{V}_{\alpha i\nu}$. Inasmuch as the mass-point position vectors $\mathbf{R}_{\alpha i\nu}$ can be expressed in terms of the generalized coordinates $Q_{\alpha is}$, the velocities can be written in terms of the time derivatives $\dot{Q}_{\alpha is}$:

$$\begin{aligned}
\mathbf{V}_{\alpha i\nu} &\equiv \dot{\mathbf{R}}_{\alpha i\nu} \\
&= \Sigma_s \dot{Q}_{\alpha is} \frac{\partial}{\partial Q_{\alpha is}} \mathbf{R}_{\alpha i\nu} \\
&= m_{\alpha\nu}^{-1/2} \Sigma_s \dot{Q}_{\alpha is} \mathbf{b}_{\nu s}^{(\alpha i)} \\
&= m_{\alpha\nu}^{-1/2} \Sigma_{st} G_{st}^{(\alpha i)} P_{\alpha it} \mathbf{b}_{\nu s}^{(\alpha i)}
\end{aligned} \tag{4.1}$$

In order to write the last line, we have used Eq. 4.5. From the definition of $\mathbf{V}_{\alpha i\nu}$, it is easily shown that

$$\Sigma_\nu m_{\alpha\nu} \mathbf{V}_{\alpha i\nu} = 0 \tag{4.2}$$

by using (2.9).

The kinetic energy associated with the internal motion of the molecule αi is

$$\begin{aligned}
T_{\alpha i} &= \frac{1}{2} \sum_{\nu=1}^{N_\alpha} m_{\alpha\nu} V_{\alpha i\nu}^2 \\
&= \frac{1}{2} \sum_{s=1}^{d_\alpha} \sum_{t=1}^{d_\alpha} g_{st}^{(\alpha i)} \dot{Q}_{\alpha is} \dot{Q}_{\alpha it}
\end{aligned} \tag{4.3}$$

where the third line of (4.1) has been used, as well as Eq. 2.13.

Since the momenta $P_{\alpha is}$ conjugate to $Q_{\alpha is}$ are given by $\partial T_{\alpha i}/\partial \dot{Q}_{\alpha is}$, we have,

$$P_{\alpha is} = \Sigma_t g_{st}^{(\alpha i)} \dot{Q}_{\alpha it} \tag{4.4}$$

$$\dot{Q}_{\alpha is} = \Sigma_t G_{st}^{(\alpha i)} P_{\alpha it} \tag{4.5}$$

Equation 4.5 can now be used to rewrite $T_{\alpha i}$ in terms of the momenta.

The Hamiltonian for the system consisting of a mixture of several types of molecules is

$$H = \sum_{\alpha i} \left(\frac{p_{\alpha i}^2}{2m_\alpha} + T_{\alpha i} \right) + \Phi \tag{4.6}$$

in which $T_{\alpha i}=\frac{1}{2}\Sigma_{st} G_{st}^{(\alpha i)} P_{\alpha is} P_{\alpha it}$ and the potential energy Φ is a function of the configurational coordinates.

The present discussion is restricted to the special case in which Φ is a sum of pair potentials between all mass points in the same and different molecules:

$$\Phi=\frac{1}{2} \sum_{\substack{\alpha i \nu \\ \beta j \mu}} \phi_{\alpha i \nu, \beta j \mu} \tag{4.7}$$

It is understood that $\phi_{\alpha i \nu, \alpha i \nu}=0$. We note that $\phi_{\alpha i \nu, \beta j \mu}$ is a function of the distance $r_{\alpha i \nu, \beta j \mu}$, which is the magnitude of the vector

$$\mathbf{r}_{\alpha i \nu, \beta j \mu}=(\mathbf{r}_{\beta j}+\mathbf{R}_{\beta j \mu})-(\mathbf{r}_{\alpha i}+\mathbf{R}_{\alpha i \nu}) \tag{4.8}$$

The restriction to the sum of pair potentials limits rather severely the dependence of the intramolecular potential of bead–rod–spring models on the bond angles. For example, in a three–bead–two–spring model, this restriction does not allow us in general to consider the potential as the sum of two spring potentials and a potential dependent on the included angle.

We now define $\mathbf{F}_{\nu\mu}^{(\alpha i, \beta j)}$ as the force on mass point ν of molecule αi exerted by mass point μ of molecule βj; this force, which depends on $\mathbf{r}_{\beta j}-\mathbf{r}_{\alpha i}$ and on some of the $Q_{\alpha is}$ and $Q_{\beta js}$, is related to $\phi_{\alpha i \nu, \beta j \mu}$ by

$$\begin{aligned}\mathbf{F}_{\nu\mu}^{(\alpha i, \beta j)} &= \frac{\partial}{\partial \mathbf{r}_{\alpha i \nu, \beta j \mu}} \phi_{\alpha i \nu, \beta j \mu} \\ &= \frac{\mathbf{r}_{\alpha i \nu, \beta j \mu}}{r_{\alpha i \nu, \beta j \mu}} \phi'_{\alpha i \nu, \beta j \mu}\end{aligned} \tag{4.9}$$

where the prime denotes differentiation with respect to $r_{\alpha i \nu, \beta j \mu}$. Note that with these definitions, the force between two mass points constrained to a fixed separation is not defined.

For subsequent discussions we introduce the following symbol:

$$\mathbf{F}_{\nu}^{(\alpha i)}=\sum_{\beta j \mu} \mathbf{F}_{\nu\mu}^{(\alpha i, \beta j)} \tag{4.10}$$

which is the net force on mass point ν of molecule αi due to the other mass points of the same molecule and those of all the other molecules in the system. We also introduce the symbol:

$$\mathbf{F}^{(\alpha i, \beta j)}=\Sigma_{\nu\mu} \mathbf{F}_{\nu\mu}^{(\alpha i, \beta j)} \tag{4.11}$$

which is the force exerted by molecule βj on molecule αi. These various forces have the following properties:

$$\mathbf{F}_{\mu\nu}^{(\beta j,\alpha i)}(\mathbf{r}_{\alpha i}-\mathbf{r}_{\beta j}, Q_{\beta j}, Q_{\alpha i}) = -\mathbf{F}_{\nu\mu}^{(\alpha i,\beta j)}(\mathbf{r}_{\beta j}-\mathbf{r}_{\alpha i}, Q_{\alpha i}, Q_{\beta j}) \tag{4.12}$$

$$\mathbf{F}^{(\beta j,\alpha i)}(\mathbf{r}_{\alpha i}-\mathbf{r}_{\beta j}, Q_{\beta j}, Q_{\alpha i}) = -\mathbf{F}^{(\alpha i,\beta j)}(\mathbf{r}_{\beta j}-\mathbf{r}_{\alpha i}, Q_{\alpha i}, Q_{\beta j}) \tag{4.13}$$

$$\Sigma_{\mu\nu}\mathbf{F}_{\nu\mu}^{(\alpha i,\alpha i)} = \mathbf{F}^{(\alpha i,\alpha i)} = 0 \tag{4.14}$$

The latter states the fact that all the internal forces within one molecule must sum to zero.

The dynamical state of the system is described by the set of position and momentum coordinates: $\mathbf{r}_{\alpha i}, Q_{\alpha is}, \mathbf{p}_{\alpha i}, P_{\alpha is}$. The time evolution of the dynamical state is described by Hamilton's equations of motion:

$$\dot{\mathbf{r}}_{\alpha i} = \frac{\partial H}{\partial \mathbf{p}_{\alpha i}} = \frac{1}{m_\alpha}\mathbf{p}_{\alpha i} \tag{4.15}$$

$$\dot{Q}_{\alpha is} = \frac{\partial H}{\partial P_{\alpha is}} = \frac{\partial T_{\alpha i}}{\partial P_{\alpha is}} \tag{4.16}$$

$$\dot{\mathbf{p}}_{\alpha i} = -\frac{\partial H}{\partial \mathbf{r}_{\alpha i}} = \Sigma_\nu \mathbf{F}_\nu^{(\alpha i)} \tag{4.17}$$

$$\dot{P}_{\alpha is} = -\frac{\partial H}{\partial Q_{\alpha is}} = -\frac{\partial T_{\alpha i}}{\partial Q_{\alpha is}} + \Sigma_\nu m_{\alpha\nu}^{-1/2}(\mathbf{b}_{\nu s}^{(\alpha i)}\cdot\mathbf{F}_\nu^{(\alpha i)}) \tag{4.18}$$

where $T_{\alpha i}$ is understood to be expressed in terms of the momenta using Eqs. 4.3 and 4.5. This completes the discussion of the time evolution of the dynamical state of the system.

Before concluding this section we now give an expression for the time rate of change of the velocities $\mathbf{V}_{\alpha i\nu}$ that will be needed later:

$$\begin{aligned}\dot{\mathbf{V}}_{\alpha i\nu} &= \sum_{st\mu}(m_{\alpha\nu}m_{\alpha\mu})^{-1/2}G_{st}^{(\alpha i)}\mathbf{b}_{\nu s}^{(\alpha i)}\mathbf{b}_{\mu t}^{(\alpha i)}\cdot\mathbf{F}_\mu^{(\alpha i)} \\ &+ (m_{\alpha\nu})^{-1/2}\sum_{stuv}\left[G_{tv}^{(\alpha i)}\frac{\partial}{\partial Q_{\alpha it}}(G_{su}^{(\alpha i)}\mathbf{b}_{\nu s}^{(\alpha i)})\right. \\ &\left.-\frac{1}{2}G_{st}^{(\alpha i)}\mathbf{b}_{\nu s}^{(\alpha i)}\frac{\partial}{\partial Q_{\alpha it}}G_{uv}^{(\alpha i)}\right]P_{\alpha iu}P_{\alpha iv}\end{aligned} \tag{4.19}$$

To get this result one uses the last line of (4.1) and the expressions in (4.5) and (4.18) for $\dot{Q}_{\alpha is}$ and $\dot{P}_{\alpha is}$. It is convenient to rewrite this expression in the form:

$$m_{\alpha\nu}\dot{\mathbf{V}}_{\alpha i\nu} = \mathfrak{F}^{(I)}_{\alpha i\nu} + \sum_{\beta j} \tilde{\mathfrak{F}}^{(II)}_{\alpha i\beta j\nu} \tag{4.20}$$

where $\mathfrak{F}^{(\mathrm{I})}_{\alpha i\nu}$ describes the intramolecular effects and $\tilde{\mathfrak{F}}^{(\mathrm{II})}_{\alpha i\beta j\nu}$ describes the intermolecular effects:

$$\begin{aligned}\mathfrak{F}^{(\mathrm{I})}_{\alpha i\nu} = &\sum_{st\mu\eta} \left(\frac{m_{\alpha\nu}}{m_{\alpha\mu}}\right)^{1/2} G^{(\alpha i)}_{st}\mathbf{b}^{(\alpha i)}_{\nu s}\mathbf{b}^{(\alpha i)}_{\mu t}\cdot\mathbf{F}^{(\alpha i,\alpha i)}_{\mu\eta}\\ &+ m_{\alpha\nu}^{1/2}\sum_{stuv}\left[G^{(\alpha i)}_{tv}\frac{\partial}{\partial Q_{\alpha it}}\left(G^{(\alpha i)}_{su}\mathbf{b}^{(\alpha i)}_{\nu s}\right)\right.\\ &\left.-\frac{1}{2}G^{(\alpha i)}_{st}\mathbf{b}^{(\alpha i)}_{\nu s}\frac{\partial}{\partial Q_{\alpha it}}G^{(\alpha i)}_{uv}\right]P_{\alpha iu}P_{\alpha iv}\end{aligned} \tag{4.21}$$

$$\tilde{\mathfrak{F}}^{(II)}_{\alpha i\beta j\nu} = \begin{cases} 0 & \text{if } \beta=\alpha \text{ and } j=i \\ \displaystyle\sum_{st\mu\eta}\left(\frac{m_{\alpha\nu}}{m_{\alpha\mu}}\right)^{1/2} G^{(\alpha i)}_{st}\mathbf{b}^{(\alpha i)}_{\nu s}\mathbf{b}^{(\alpha i)}_{\mu t}\cdot\mathbf{F}^{(\alpha i,\beta j)}_{\mu\eta} & \text{otherwise}\end{cases} \tag{4.22}$$

These expressions for $\mathfrak{F}^{(\mathrm{I})}_{\alpha i\nu}$ and $\tilde{\mathfrak{F}}^{(\mathrm{II})}_{\alpha i\beta j\nu}$ arise in the next section in the expression for the stress tensor.

V. TIME-EVOLUTION OF ENSEMBLE AVERAGES AND THE EQUATIONS OF CHANGE

Let χ represent the entire set of position, configuration, and momentum coordinates of a single system, that is, a mixture of several chemical species $\alpha = 0, 1, 2, \ldots, N_\alpha$. Then the state of an ensemble of systems is described by a distribution function $f(\chi, t)$ in the phase space. This distribution function is normalized so that

$$\int f(\chi, t)\,d\chi = 1 \tag{5.1}$$

As the states of the various systems in the ensemble change with time, the state of the ensemble changes according to the Liouville equation of classical statistical mechanics:

$$\frac{\partial f}{\partial t} + \mathfrak{L} f = 0 \tag{5.2}$$

where $\mathfrak{L}$ is the Liouville operator:

$$\mathfrak{L} = \sum_{\alpha i} \left[\left(\frac{\partial}{\partial \mathbf{p}_{\alpha i}} H \cdot \frac{\partial}{\partial \mathbf{r}_{\alpha i}} \right) - \left(\frac{\partial}{\partial \mathbf{r}_{\alpha i}} H \cdot \frac{\partial}{\partial \mathbf{p}_{\alpha i}} \right) \right]$$
$$+ \sum_{\alpha i s} \left[\frac{\partial H}{\partial P_{\alpha i s}} \frac{\partial}{\partial Q_{\alpha i s}} - \frac{\partial H}{\partial Q_{\alpha i s}} \frac{\partial}{\partial P_{\alpha i s}} \right]$$
$$= \sum_{\alpha i} \left[\left(\dot{\mathbf{r}}_{\alpha i} \cdot \frac{\partial}{\partial \mathbf{r}_{\alpha i}} \right) + \left(\dot{\mathbf{p}}_{\alpha i} \cdot \frac{\partial}{\partial \mathbf{p}_{\alpha i}} \right) \right]$$
$$+ \sum_{\alpha i s} \left[\dot{Q}_{\alpha i s} \frac{\partial}{\partial Q_{\alpha i s}} + \dot{P}_{\alpha i s} \frac{\partial}{\partial P_{\alpha i s}} \right] \tag{5.3}$$

Let $A(\chi)$ be an arbitrary time-independent dynamical variable of a system, that is, an arbitrary function defined in the phase space. The average over the systems of the ensemble of the value of $A(\chi)$ for a single system is then

$$\langle A \rangle = \int A(\chi) f(\chi, t) \, d\chi \tag{5.4}$$

This average value changes with time due to the time evolution of the distribution function. It follows directly from the Liouville equation that

$$\frac{\partial}{\partial t} \langle A \rangle = \int A(\chi) \left(\frac{\partial}{\partial t} f(\chi, t) \right) d\chi$$
$$= \int f(\chi, t) \mathfrak{L} A(\chi) \, d\chi$$
$$= \langle \mathfrak{L} A \rangle \tag{5.5}$$

which is the *general equation of change*. For special choices of A, we obtain the equations of hydrodynamics; we are guided in this by the earlier work of Irving and Kirkwood.[24] In obtaining the final result in (5.5), integrations by parts are performed, and it has been assumed that all surface integrals will vanish or cancel.

First, we consider the *equation of continuity* for species α. This is an equation describing the time evolution of the mass concentration ρ_α. This quantity depends not only on the distribution of molecules throughout the space, but also on the mass distribution within the molecules. Thus, let us consider the general equation of change (Eq. 5.5), with

$$A(\chi)=\sum_{i\nu} m_{\alpha\nu}\delta(\mathbf{r}_{\alpha i}+\mathbf{R}_{\alpha i\nu}-\mathbf{r}) \tag{5.6}$$

a dynamical variable that depends parametrically on $\mathbf{r}$ and the index α. Each term in this sum is zero unless the corresponding mass point is at $\mathbf{r}$. Thus, the average over the ensemble of this dynamical variable

$$\langle A\rangle=\rho_\alpha(\mathbf{r},t) \tag{5.7}$$

defines the mass density at $\mathbf{r}$ associated with molecules of species α.

From the form of the Liouville operator (Eq. 5.3), it follows that

$$\begin{aligned}
\mathfrak{L}A(\chi)&=\sum_{i\nu} m_{\alpha\nu}\left[\dot{\mathbf{r}}_{\alpha i}\cdot\frac{\partial}{\partial\mathbf{r}_{\alpha i}}+\sum_s \dot{Q}_{\alpha is}\frac{\partial}{\partial Q_{\alpha is}}\right]\delta(\mathbf{r}_{\alpha i}+\mathbf{R}_{\alpha i\nu}-\mathbf{r})\\
&=\sum_{i\nu} m_{\alpha\nu}\left[\frac{1}{m_\alpha}\mathbf{p}_{\alpha i}\cdot\frac{\partial}{\partial\mathbf{r}_{\alpha i}}+\sum_s \dot{Q}_{\alpha is}\left(\frac{\partial}{\partial Q_{\alpha is}}\mathbf{R}_{\alpha i\nu}\right)\cdot\frac{\partial}{\partial\mathbf{R}_{\alpha i\nu}}\right]\\
&\quad\times\delta(\mathbf{r}_{\alpha i}+\mathbf{R}_{\alpha i\nu}-\mathbf{r})\\
&=-\left(\frac{\partial}{\partial\mathbf{r}}\cdot\sum_{i\nu} m_{\alpha\nu}\left[\frac{1}{m_\alpha}\mathbf{p}_{\alpha i}+\mathbf{V}_{\alpha i\nu}\right]\delta(\mathbf{r}_{\alpha i}+\mathbf{R}_{\alpha i\nu}-\mathbf{r})\right)
\end{aligned} \tag{5.8}$$

where the $\mathbf{V}_{\alpha i\nu}$ are the velocities defined in (4.1). Note that in going from the next-to-last to the last line of (5.8), use has been made of the fact that $\partial\delta(x-x_0)/\partial x=-\partial\delta(x-x_0)/\partial x_0$. Note further that the Dirac delta function shown with a vector argument is in fact a product of three delta functions each one of which has a component of the vector as its argument (i.e., $\delta(\mathbf{A})\equiv\delta(A_x)\delta(A_y)\delta(A_z)$).

Next we take the ensemble average of $\mathfrak{L}A(\chi)$, which gives

$$\langle\mathfrak{L}A\rangle=-\frac{\partial}{\partial\mathbf{r}}\cdot\rho_\alpha\mathbf{u}_\alpha \tag{5.8a}$$

where

$$\rho_\alpha\mathbf{u}_\alpha=\sum_{i\nu} m_{\alpha\nu}\left\langle\left(\frac{1}{m_\alpha}\mathbf{p}_{\alpha i}+\mathbf{V}_{\alpha i\nu}\right)\delta(\mathbf{r}_{\alpha i}+\mathbf{R}_{\alpha i\nu}-\mathbf{r})\right\rangle \tag{5.9}$$

defines the density of linear momentum associated with molecules of species α. It then follows from the general equation of change (Eq. (5.5)) that

$$\frac{\partial}{\partial t}\rho_\alpha = -\frac{\partial}{\partial \mathbf{r}}\cdot\rho_\alpha\mathbf{u}_\alpha \tag{5.10}$$

which is the equation of continuity for molecules of species α.

By comparison of (5.9) and (5.10) with (1.1), it may be seen that the *mass flux of species* α with respect to the mass-average velocity is

$$\mathbf{j}_\alpha = \sum_{i\nu} m_{\alpha\nu}\left\langle\left(\frac{1}{m_\alpha}\mathbf{p}_{\alpha i} + \mathbf{V}_{\alpha i\nu} - \mathbf{u}\right)\delta(\mathbf{r}_{\alpha i} + \mathbf{R}_{\alpha i\nu} - \mathbf{r})\right\rangle \tag{5.11}$$

where use has been made of the fact that $\rho_\alpha\mathbf{u}_\alpha = \rho_\alpha\mathbf{u} + \mathbf{j}_\alpha$.

Next, we proceed to the *equation of motion*, which is an equation for the time evolution of the total momentum density $\rho\mathbf{u} = \Sigma_\alpha\rho_\alpha\mathbf{u}_\alpha$. This equation may be obtained as a special case of the general equation of change in a manner similar to that used previously to develop the equation of continuity. To derive this equation let us consider the dynamical variable

$$\mathbf{A}(\chi) = \sum_{\alpha i\nu} m_{\alpha\nu}\left(\frac{1}{m_\alpha}\mathbf{p}_{\alpha i} + \mathbf{V}_{\alpha i\nu}\right)\delta(\mathbf{r}_{\alpha i} + \mathbf{R}_{\alpha i\nu} - \mathbf{r}) \tag{5.12}$$

so that from (5.9) and the definition of $\mathbf{u}$

$$\langle A\rangle = \rho\mathbf{u} \tag{5.13}$$

With this dynamical variable it follows from the expression for the Liouville operator (Eq. 5.3) that

$$\mathfrak{L}\mathbf{A} = -\left(\frac{\partial}{\partial \mathbf{r}}\cdot\sum_{\alpha i\nu} m_{\alpha\nu}\left(\frac{1}{m_\alpha}\mathbf{p}_{\alpha i} + \mathbf{V}_{\alpha i\nu}\right)\left(\frac{1}{m_\alpha}\mathbf{p}_{\alpha i} + \mathbf{V}_{\alpha i\nu}\right)\delta(\mathbf{r}_{\alpha i} + \mathbf{R}_{\alpha i\nu} - \mathbf{r})\right)$$
$$+ \sum_{\alpha i\nu}\mathfrak{F}_\nu^{(\alpha i)}\delta(\mathbf{r}_{\alpha i} + \mathbf{R}_{\alpha i\nu} - \mathbf{r}) \tag{5.14}$$

where $\mathfrak{F}_\nu^{(\alpha i)}$ is an "effective force" on mass point ν of molecule αi, given by

$$\mathfrak{F}_\nu^{(\alpha i)} = m_{\alpha\nu}\left(\frac{1}{m_\alpha}\dot{\mathbf{p}}_{\alpha i} + \dot{\mathbf{V}}_{\alpha i\nu}\right)$$
$$= \mathfrak{F}_{\alpha i\nu}^{(\mathrm{I})} + \sum_{\beta j}\mathfrak{F}_{\alpha i\beta j\nu}^{(\mathrm{II})} \tag{5.15}$$

where

$$\mathfrak{F}^{(\mathrm{II})}_{\alpha i\beta j\nu} = \tilde{\mathfrak{F}}^{(\mathrm{II})}_{\alpha i\beta j\nu} + \frac{m_{\alpha\nu}}{m_\alpha}\sum_{\mu\eta}\mathbf{F}^{(\alpha i,\beta j)}_{\mu\eta} \tag{5.16}$$

If there are no constraints within molecule αi, then $\mathfrak{F}^{(\alpha i)}_\nu$ is the same as $\mathbf{F}^{(\alpha i)}_\nu$. In Table II we give the various contributions to the effective force for rigid and elastic dumbbells.

TABLE II. Terms in Effective Force Expression for $\mathfrak{F}^{(\alpha i)}_1$ for Dumbbells (5.16)

	Rigid dumbbells	Elastic dumbbells
Intramolecular		
First term of $\mathfrak{F}^{(\mathrm{I})}_{\alpha_i 1}$, (4.21)	0	$\mathbf{F}^{(\alpha i)}_{12}$
Second term of $\mathfrak{F}^{(\mathrm{I})}_{\alpha_i 1}$ (4.21)	$\mu L(\dot{\theta}^2_{\alpha i} + \sin^2\theta_{\alpha i}\dot{\phi}^2_{\alpha i})\boldsymbol{\delta}_R$	0
Intermolecular		
First term of $\sum_{\beta j}\mathfrak{F}^{(\mathrm{II})}_{\alpha i\beta j1}$, (5.15) or (4.22)	$(\check{\boldsymbol{\delta}}_\theta\check{\boldsymbol{\delta}}_\theta + \check{\boldsymbol{\delta}}_\phi\check{\boldsymbol{\delta}}_\phi)\cdot\left(\frac{m_1}{m}\mathbf{F}^{(\alpha i)}_1 - \frac{m_2}{m}\mathbf{F}^{(\alpha i)}_2\right)$	$\frac{m_2}{m}\mathbf{F}^{(\alpha i)}_1 - \frac{m_1}{m}\mathbf{F}^{(\alpha i)}_2$
Second term of $\sum_{\beta j}\mathfrak{F}^{(\mathrm{II})}_{\alpha i\beta j1}$, (5.15)	$\frac{m_1}{m}(\mathbf{F}^{(\alpha i)}_1 + \mathbf{F}^{(\alpha i)}_2)$	$\frac{m_1}{m}(\mathbf{F}^{(\alpha i)}_1 + \mathbf{F}^{(\alpha i)}_2)$

It now follows from the general equation of change (Eq. 5.5) and the definitions of the mass density ρ and the mass-average velocity $\mathbf{u}$ that

$$\frac{\partial}{\partial t}\rho\mathbf{u} = -\frac{\partial}{\partial \mathbf{r}}\cdot\rho\mathbf{u}\mathbf{u} - \frac{\partial}{\partial \mathbf{r}}\cdot\mathbf{p}^{(k)} + \mathfrak{F} \tag{5.17}$$

where

$$\mathbf{p}^{(k)} = \sum_{\alpha i\nu} m_{\alpha\nu}\left\langle\left(\frac{1}{m_\alpha}\mathbf{p}_{\alpha i} + \mathbf{V}_{\alpha i\nu} - \mathbf{u}\right)\left(\frac{1}{m_\alpha}\mathbf{p}_{\alpha i} + \mathbf{V}_{\alpha i\nu} - \mathbf{u}\right)\right.$$
$$\left.\times\,\delta(\mathbf{r}_{\alpha i} + \mathbf{R}_{\alpha i\nu} - \mathbf{r})\right\rangle \tag{5.18}$$

is the *kinetic contribution to the momentum flux* and

$$\mathfrak{F} = \sum_{\alpha i \nu} \langle \mathfrak{F}_\nu^{(\alpha i)} \delta(\mathbf{r}_{\alpha i} + \mathbf{R}_{\alpha i \nu} - \mathbf{r}) \rangle \tag{5.19}$$

is an *average effective force density*; $\mathfrak{F}_\nu^{(\alpha i)}$ is given by (5.15). Equation 5.17 is not quite in the form of the equation of motion given in (1.2). It is shown subsequently that the effective force density term can be written as the divergence of $\mathsf{p}^{(\mathrm{I})} + \mathsf{p}^{(\mathrm{II})}$, which quantities then describe two other contributions to the momentum-flux tensor. Then we will have Eq. (1.2) with

$$\mathsf{p} = \mathsf{p}^{(k)} + \mathsf{p}^{(\mathrm{I})} + \mathsf{p}^{(\mathrm{II})} \tag{5.20}$$

VI. CONTRACTED DISTRIBUTION FUNCTIONS AND AVERAGE VALUES

In the foregoing section a distribution function $f(\mathbf{r}_{\alpha i}, Q_{\alpha is}, \mathbf{p}_{\alpha i}, P_{\alpha is}, t)$ was defined. In this section we define several contracted distribution functions needed subsequently. These will be:

f_α	Singlet distribution function in phase space
$f_{\alpha\beta}, \tilde{f}_{\alpha\beta}$	Pair distribution functions in phase space
ψ_α	Singlet distribution function in configuration space
$\psi_{\alpha\beta}, \tilde{\psi}_{\alpha\beta}$	Pair distribution functions in configuration space

We now discuss these *seriatim*.

First, we define the singlet distribution function for molecules of species α

$$f_\alpha(\mathbf{r}, Q_\alpha, \mathbf{p}_\alpha, P_\alpha, t) = \sum_i \langle \delta(\mathbf{r}_{\alpha i} - \mathbf{r}) \delta(Q_{\alpha i} - Q_\alpha) \delta(\mathbf{p}_{\alpha i} - \mathbf{p}_\alpha) \delta(P_{\alpha i} - P_\alpha) \rangle \tag{6.1}$$

The function f_α is the probability density for molecules of species α in the one-molecule phase space, and it is normalized in such a way that at equilibrium

$$\int f_\alpha dQ_\alpha d\mathbf{p}_\alpha dP_\alpha = n_\alpha(\mathbf{r}, t) \tag{6.2}$$

in which n_α is the number density of species α. The argument Q_α stands for the set of generalized coordinates $Q_{\alpha 1}, Q_{\alpha 2}, \ldots, Q_{\alpha d_\alpha}$ and $\delta(Q_{\alpha i} - Q_\alpha)$ is shorthand for $\Pi_{s=1}^{d_\alpha} \delta(Q_{\alpha is} - Q_{\alpha s})$. Similar comments apply to the P_α. The quantity $f_\alpha d\mathbf{r}\, dQ_\alpha\, d\mathbf{p}_\alpha\, dP_\alpha$ is the probability that there is a molecule of species α with center of mass in the region $d\mathbf{r}$ about $\mathbf{r}$, with internal coordinates in the range $dQ_{\alpha s}$ about $Q_{\alpha s}$ $(s=1,2,\ldots,d_\alpha)$ and so on.

The second contracted distribution function is defined in an analogous manner:

$$f_{\alpha\beta}(\mathbf{r}_\alpha,\mathbf{r}_\beta,Q_\alpha,Q_\beta,\mathbf{p}_\alpha,\mathbf{p}_\beta,P_\alpha,P_\beta,t)=\sum_{ij}\langle\delta(\mathbf{r}_{\alpha i}-\mathbf{r}_\alpha)\delta(\mathbf{r}_{\beta j}-\mathbf{r}_\beta)\delta(Q_{\alpha i}-Q_\alpha)$$

$$\times\delta(Q_{\beta j}-Q_\beta)\delta(\mathbf{p}_{\alpha i}-\mathbf{p}_\alpha)\delta(\mathbf{p}_{\beta j}-\mathbf{p}_\beta)\delta(P_{\alpha i}-P_\alpha)\delta(P_{\beta j}-P_\beta)\rangle \quad (6.3)$$

and is a function in the phase space of two molecules. This is the joint probability density of finding two molecules at the particular points in the phase spaces. The last expression applies both when $\alpha=\beta$ and when $\alpha\neq\beta$, except that when $\alpha=\beta$ the term in which $j=i$ is to be omitted from the sum.

The pair distribution, as defined by (6.3), is a strongly varying function of the distance $\mathbf{R}_{\alpha\beta}$ between the centers of mass of the two molecules, but is a slowly varying function of the position $\mathbf{r}_{\alpha\beta}$ of the center of mass of the two-molecule system. Thus, we introduce a change of variables and define

$$\tilde{f}_{\alpha\beta}(\mathbf{r}_{\alpha\beta},\mathbf{R}_{\alpha\beta},Q_\alpha,Q_\beta,\mathbf{p}_\alpha,\mathbf{p}_\beta,P_\alpha,P_\beta,t)=f_{\alpha\beta}(\mathbf{r}_\alpha,\mathbf{r}_\beta,Q_\alpha,Q_\beta,\mathbf{p}_\alpha,\mathbf{p}_\beta,P_\alpha,P_\beta,t) \quad (6.4)$$

where

$$\mathbf{r}_{\alpha\beta}=\frac{m_\alpha\mathbf{r}_\alpha+m_\beta\mathbf{r}_\beta}{m_\alpha+m_\beta} \quad (6.5)$$

$$\mathbf{R}_{\alpha\beta}=\mathbf{r}_\beta-\mathbf{r}_\alpha \quad (6.6)$$

The function $\tilde{f}_{\alpha\beta}$ will be more useful in the kinetic theory presentation. Note that $\tilde{f}_{\alpha\beta}(\mathbf{r}_{\alpha\beta},\mathbf{R}_{\alpha\beta},\ldots)=\tilde{f}_{\beta\alpha}(\mathbf{r}_{\beta\alpha},\mathbf{R}_{\beta\alpha},\ldots)$; also note that the Jacobian of the transformation in (6.5) and (6.6) is unity.

Next, we define a distribution function ψ_α in the configuration space of a single molecule:

$$\psi_\alpha(\mathbf{r},Q_\alpha,t)=\int f_\alpha(\mathbf{r},Q_\alpha,\mathbf{p}_\alpha,P_\alpha,t)\,d\mathbf{p}_\alpha\,dP_\alpha \quad (6.7)$$

According to (6.2), ψ_α is normalized so that at equilibrium

$$\int\psi_\alpha\,dQ_\alpha=n_\alpha(\mathbf{r},t) \quad (6.8)$$

Similarly

$$\psi_{\alpha\beta}(\mathbf{r}_\alpha, \mathbf{r}_\beta, Q_\alpha, Q_\beta, t) = \int f_{\alpha\beta}\, d\mathbf{p}_\alpha\, d\mathbf{p}_\beta\, dP_\alpha\, dP_\beta \tag{6.9}$$

$$\tilde{\psi}_{\alpha\beta}(\mathbf{r}_{\alpha\beta}, \mathbf{R}_{\alpha\beta}, Q_\alpha, Q_\beta, t) = \int \tilde{f}_{\alpha\beta}\, d\mathbf{p}_\alpha\, d\mathbf{p}_\beta\, dP_\alpha\, dP_\beta \tag{6.10}$$

are also useful configuration-space distribution functions.

Before leaving the subject of distribution functions, we discuss the notation to be used for several types of averages. In (5.4) the notation $\langle\ \rangle$ was introduced as an average value in the complete phase space of a system composed of many molecules of several different chemical species. Two additional averages $\langle\ \rangle^{(\alpha)}$ and $[\ ]^{(\alpha)}$ are now introduced:

$$\langle A\rangle^{(\alpha)} f_\alpha(\mathbf{r}, Q_\alpha, \mathbf{p}_\alpha, P_\alpha, t) = \sum_i \langle A(\chi)\delta(\mathbf{r}_{\alpha i} - \mathbf{r})\delta(Q_{\alpha i} - Q_\alpha) \times \delta(\mathbf{p}_{\alpha i} - \mathbf{p}_\alpha)\delta(P_{\alpha i} - P_\alpha)\rangle \tag{6.11}$$

$$[a]^{(\alpha)}\psi_\alpha(\mathbf{r}, Q_\alpha, t) = \int a(\mathbf{r}, Q_\alpha, \mathbf{p}_\alpha, P_\alpha) f_\alpha(\mathbf{r}, Q_\alpha, \mathbf{p}_\alpha, P_\alpha, t)\, d\mathbf{p}_\alpha\, dP_\alpha \tag{6.12}$$

$\langle A\rangle^{(\alpha)}$ depends on the coordinates and momenta of the single-molecule phase space; $[a]^{(\alpha)}$ depends only on the coordinates of the single-molecule configuration space. We adopt the convention that $\langle A\rangle^{(\alpha)}$ and $[a]^{(\alpha)}$ have the same arguments as the distribution function immediately following. In addition, ψ_α and $\mathbf{u}_\alpha$ with no argument implies $\psi_\alpha(\mathbf{r}, Q_\alpha, t)$ and $\mathbf{u}_\alpha(\mathbf{r}, t)$.

VII. GENERAL EXPRESSIONS FOR THE FLUXES IN TERMS OF THE DISTRIBUTION FUNCTIONS f_α AND $\tilde{f}_{\alpha\beta}$

In (5.11) we gave the expression for the mass flux $\mathbf{j}_\alpha$ in terms of an integral involving the distribution function f. In (5.18) and (5.19) we gave expressions for $\mathsf{p}^{(k)}$ and $\mathfrak{F}$ in terms of f. In this section we want to get $\mathbf{j}_\alpha$ and $\mathsf{p}^{(k)}$ in terms of the singlet distribution function and also to rewrite $\mathfrak{F}$ as $(\partial/\partial\mathbf{r})\cdot(\mathsf{p}^{(\mathrm{I})} + \mathsf{p}^{(\mathrm{II})})$ with $\mathsf{p}^{(\mathrm{I})}$ and $\mathsf{p}^{(\mathrm{II})}$ in terms of integrals involving f.

It follows directly from the definitions (Eqs. 5.6 and 5.7) that the mass concentration and momentum density for species α are

$$\rho_\alpha(\mathbf{r}, t) = \sum_\nu m_{\alpha\nu} \int f_\alpha(\mathbf{r} - \mathbf{R}_{\alpha\nu}, Q_\alpha, \mathbf{p}_\alpha, P_\alpha, t)\, dQ_\alpha\, d\mathbf{p}_\alpha\, dP_\alpha \tag{7.1}$$

$$\rho_\alpha \mathbf{u}_\alpha = \sum_\nu m_{\alpha\nu} \int \left(\frac{1}{m_\alpha}\mathbf{p}_\alpha + \mathbf{V}_{\alpha\nu}\right) f_\alpha(\mathbf{r} - \mathbf{R}_{\alpha\nu}, Q_\alpha, \mathbf{p}_\alpha, P_\alpha, t)\, dQ_\alpha\, d\mathbf{p}_\alpha\, dP_\alpha. \tag{7.2}$$

These are easily verified by substituting f_α from (6.1) into these equations and performing the integrations over $Q_\alpha, \mathbf{p}_\alpha, P_\alpha$, remembering that the $\mathbf{R}_{\alpha i\nu}$ are functions of the $Q_{\alpha is}$.

In a similar fashion one can verify that the *mass flux of species* α, and the *kinetic contribution to the momentum flux* may be written in the form:

$$\mathbf{j}_\alpha = \sum_\nu m_{\alpha\nu} \int \left(\frac{1}{m_\alpha}\mathbf{p}_\alpha + \mathbf{V}_{\alpha\nu} - \mathbf{u} \right) f_\alpha(\mathbf{r} - \mathbf{R}_{\alpha\nu}, Q_\alpha, \mathbf{p}_\alpha, P_\alpha, t)\, dQ_\alpha d\mathbf{p}_\alpha dP_\alpha \quad (7.3)$$

$$\mathsf{p}^{(k)} = \sum_{\alpha\nu} m_{\alpha\nu} \int \left(\frac{1}{m_\alpha}\mathbf{p}_\alpha + \mathbf{V}_{\alpha\nu} - \mathbf{u} \right)\left(\frac{1}{m_\alpha}\mathbf{p}_\alpha + \mathbf{V}_{\alpha\nu} - \mathbf{u} \right)$$

$$\times f_\alpha(\mathbf{r} - \mathbf{R}_{\alpha\nu}, Q_\alpha, \mathbf{p}_\alpha, P_\alpha, t)\, dQ_\alpha d\mathbf{p}_\alpha dP_\alpha \quad (7.4)$$

Furthermore the average effective force density $\mathfrak{F}$ in (5.19) may be re-written using the second expression in (5.16). This gives

$$\mathfrak{F} = \mathfrak{F}^{(\mathrm{I})} + \mathfrak{F}^{(\mathrm{II})} \quad (7.5)$$

in which

$$\mathfrak{F}^{(\mathrm{I})} = \sum_{\alpha\nu} \int \mathfrak{F}^{(\mathrm{I})}_{\alpha\nu} f_\alpha(\mathbf{r} - \mathbf{R}_{\alpha\nu}, Q_\alpha, \mathbf{p}_\alpha, P_\alpha, t)\, dQ_\alpha d\mathbf{p}_\alpha dP_\alpha \quad (7.6)$$

$$\mathfrak{F}^{(\mathrm{II})} = \sum_{\alpha\beta\nu} \int \mathfrak{F}^{(\mathrm{II})}_{\alpha\beta\nu} f_{\alpha\beta}(\mathbf{r} - \mathbf{R}_{\alpha\nu}, \mathbf{r}_\beta, Q_\alpha, Q_\beta, \mathbf{p}_\alpha, \mathbf{p}_\beta, P_\alpha, P_\beta, t)$$

$$\times d\mathbf{r}_\beta dQ_\alpha dQ_\beta d\mathbf{p}_\alpha d\mathbf{p}_\beta dP_\alpha dP_\beta$$

$$= \sum_{\alpha\beta\nu} \int \mathfrak{F}^{(\mathrm{II})}_{\alpha\beta\nu} \tilde{f}_{\alpha\beta}\left(\mathbf{r} - \mathbf{R}_{\alpha\nu} + \frac{m_\beta}{m_\alpha + m_\beta}\mathbf{R}_{\alpha\beta}, \mathbf{R}_{\alpha\beta}, Q_\alpha, Q_\beta, \mathbf{p}_\alpha, \mathbf{p}_\beta, P_\alpha, P_\beta, t \right)$$

$$\times d\mathbf{R}_{\alpha\beta} dQ_\alpha dQ_\beta d\mathbf{p}_\alpha d\mathbf{p}_\beta dP_\alpha dP_\beta \quad (7.7)$$

The $\mathfrak{F}^{(\mathrm{I})}_{\alpha\nu}$ and $\mathfrak{F}^{(\mathrm{II})}_{\alpha\beta\nu}$ in these equations are given in (4.21), (4.22), and (5.15).

The next task is to write $\mathfrak{F}^{(\mathrm{I})}$ and $\mathfrak{F}^{(\mathrm{II})}$ as divergences of tensors:

$$\mathfrak{F}^{(\mathrm{I})} = -\frac{\partial}{\partial \mathbf{r}} \cdot \mathsf{p}^{(\mathrm{I})} \quad (7.8)$$

$$\mathfrak{F}^{(\mathrm{II})} = -\frac{\partial}{\partial \mathbf{r}} \cdot \mathsf{p}^{(\mathrm{II})} \quad (7.9)$$

This will be accomplished by the use of a "modified Taylor theorem": If $L(\mathbf{r})$ is any sufficiently well behaved function, then

$$L(\mathbf{r}+\mathbf{R})=L(\mathbf{r})+\mathbf{R}\cdot\frac{\partial}{\partial\mathbf{r}}\int_0^1 L(\mathbf{r}+\xi\mathbf{R})\,d\xi \tag{7.10}$$

Application of this theorem to (7.6) gives

$$\mathfrak{F}^{(\mathrm{I})}=\sum_{\alpha\nu}\left[\int \mathfrak{F}^{(\mathrm{I})}_{\alpha\nu}f_\alpha(\mathbf{r},Q_\alpha,\mathbf{p}_\alpha,P_\alpha,t)\,dQ_\alpha d\mathbf{p}_\alpha dP_\alpha\right.$$
$$\left.-\frac{\partial}{\partial\mathbf{r}}\cdot\int\int_0^1\mathbf{R}_{\alpha\nu}\mathfrak{F}^{(\mathrm{I})}_{\alpha\nu}f_\alpha(\mathbf{r}-\xi\mathbf{R}_{\alpha\nu},Q_\alpha,\mathbf{p}_\alpha,P_\alpha,t)\,d\xi dQ_\alpha d\mathbf{p}_\alpha dP_\alpha\right] \tag{7.11}$$

It follows from (2.11) and (4.21) that the first integral is zero. When the second term is compared with (7.8), one finds that

$$\mathsf{p}^{(\mathrm{I})}=\sum_{\alpha\nu}\int\int_0^1\mathbf{R}_{\alpha\nu}\mathfrak{F}^{(\mathrm{I})}_{\alpha\nu}f_\alpha(\mathbf{r}-\xi\mathbf{R}_{\alpha\nu},Q_\alpha,\mathbf{p}_\alpha,P_\alpha,t)\,d\xi dQ_\alpha d\mathbf{p}_\alpha dP_\alpha \tag{7.12}$$

This is the *intramolecular force contribution to the momentum flux*. (Equation 7.10 appears to have been used first in kinetic theory by Imam-Rahajoe[27] in his work on dense gases.) The modified Taylor theorem may be applied in the same way to (7.7), and once again the first term vanishes: The contribution of the $\mathfrak{F}^{(\mathrm{II})}_{\alpha\beta\nu}$ term in (5.16) vanishes because of (2.11); the contribution associated with the second term in (5.16) vanishes because the distribution function is symmetric in the indices α and β, and $\mathbf{F}^{(\alpha,\beta)}$ is antisymmetric. This leads to

$$\mathsf{p}^{(\mathrm{II})}=\sum_{\alpha\beta\nu}\int\int_0^1\left(\mathbf{R}_{\alpha\nu}-\frac{m_\beta}{m_\alpha+m_\beta}\mathbf{R}_{\alpha\beta}\right)\mathfrak{F}^{(\mathrm{II})}_{\alpha\beta\nu}$$
$$\times\tilde{f}_{\alpha\beta}\left(\mathbf{r}-\xi\mathbf{R}_{\alpha\nu}+\frac{\xi m_\beta}{m_\alpha+m_\beta}\mathbf{R}_{\alpha\beta},\mathbf{R}_{\alpha\beta},Q_\alpha,Q_\beta,\mathbf{p}_\alpha,\mathbf{p}_\beta,P_\alpha,P_\beta,t\right)$$
$$\times d\xi d\mathbf{R}_{\alpha\beta}dQ_\alpha dQ_\beta d\mathbf{p}_\alpha d\mathbf{p}_\beta dP_\alpha dP_\beta \tag{7.13}$$

This is the *intermolecular force contribution to the momentum flux*.

The various contributions to the momentum flux are summarized in Table III. Up to this point the only assumption that has been made is the pairwise additivity of the interparticle potentials.

TABLE III. Contributions to the Momentum Flux in a System Composed of a Structureless Solvent and Several Macromolecular Species

$$\frac{\partial}{\partial t}\rho\mathbf{u} = -\frac{\partial}{\partial\mathbf{r}}\cdot\rho\mathbf{u}\mathbf{u} - \frac{\partial}{\partial\mathbf{r}}\cdot\mathsf{p}; \qquad \mathsf{p} = \mathsf{p}^{(k)} + \mathsf{p}^{(\mathrm{I})} + \mathsf{p}^{(\mathrm{II})}$$

Contribution	Final expression in equation	Pictorial significance, showing mechanism for momentum transport across an arbitrary plane	
			Structureless solvent molecules; Polymer molecules
$\mathsf{p}^{(k)}$	(7.4) (uses f_α)	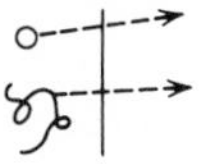	Solvent and polymer molecules move across plane
$\mathsf{p}^{(\mathrm{I})}$	(7.12) (uses f_α)	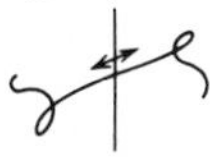	Intramolecular forces give rise to tension in polymer molecule straddling plane
			Solvent–solvent interaction
		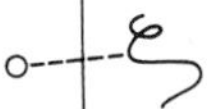	Solvent–polymer interaction
$\mathsf{p}^{(\mathrm{II})}$	(7.13) (uses $\tilde{f}_{\alpha\beta}$)		Polymer–polymer interaction (both polymers same)
		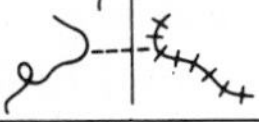	Polymer–polymer interaction (two different species)

We have shown that $(\partial/\partial\mathbf{r})\cdot\mathsf{p}$, where p is given by (5.20), (7.4), (7.12), and (7.13), gives the correct equation of motion. There remains however the question as to whether $\mathbf{n}\cdot\mathsf{p}$ gives the force on a wall of orientation $\mathbf{n}$, since this quantity may appear in the boundary conditions needed to solve the fluid dynamic equations. In order to discuss this question, let us consider the wall as part of the complete system described by the ensemble distribution function of (5.1). The molecules of the wall make up a continuum that exerts a force on the remaining molecules of the system derivable from a potential ϕ_w.

It follows from (5.17) that for the system including the wall

$$-\frac{\partial}{\partial\mathbf{r}}\cdot\mathsf{p}^{(k)} + \mathcal{F} = -\frac{\partial}{\partial\mathbf{r}}\cdot\mathsf{p} \tag{7.14}$$

may be interpreted as the statistical average body force per unit volume anywhere within the system.

Now consider a region of space in the complete system that has the shape of a thin flat "pill box" (see Fig. 4). One surface of the pill box is just

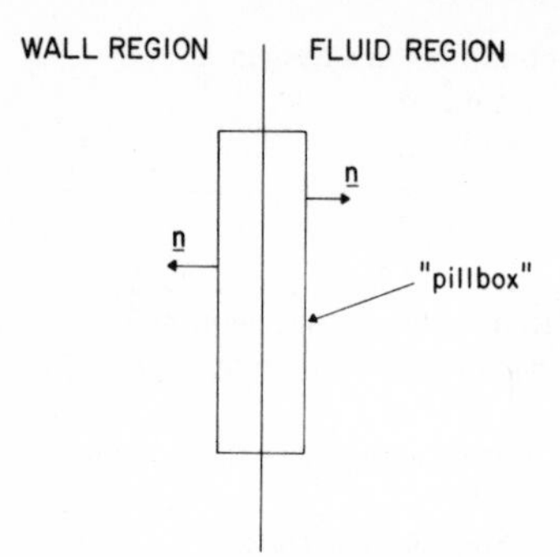

Fig. 4. The "complete system" including the "wall region" and the "fluid region."

inside the wall region, and one surface is just inside the fluid region. From the foregoing expression the total body force on this region is

$$-\int_{\substack{\text{pill}\\\text{box}}} \left(\frac{\partial}{\partial \mathbf{r}}\cdot \mathsf{p}\right) d\mathbf{r} = -\int_{\substack{\text{pill}\\\text{box}}} (\mathbf{n}\cdot \mathsf{p})\, dS \tag{7.15}$$

since p is well behaved.

We next consider the limit in which the potential due to the wall becomes large sufficiently rapidly that there is negligible probability that a molecule of the fluid is in the wall region; call this the "rigid barrier limit" (RBL). In the RBL the distribution function $f(q,p,t)$ becomes zero if the coordinates of any molecule are within the wall region. Thus the expression for p, which is given by (5.20), (7.4), (7.12), and (7.13), is zero in this wall region.

In the limit that the pill box is sufficiently thin. The edge does not contribute significantly to the surface integral in (7.15). Since in the RBL, p is zero inside the wall region, this portion of the surface does not contribute, and hence the total body force is simply the integral over the surface of the pill box in the fluid region of $-(\mathbf{n}\cdot\mathsf{p})$. Thus the force per unit area on the "complete system" is $-(\mathbf{n}\cdot\mathsf{p})$. Therefore, $-(\mathbf{n}\cdot\mathsf{p})$ is the force per unit area on the wall.

VIII. TIME-EVOLUTION OF THE SINGLET DISTRIBUTION FUNCTION f_α

An equation for the time evolution of the singlet distribution function may be obtained from the general equation of change (Eq. 5.5) by taking as the dynamical variable

$$A(\chi) = \sum_i A_i = \sum_i \delta(\mathbf{r}_{\alpha i} - \mathbf{r})\delta(Q_{\alpha i} - Q_\alpha)\delta(\mathbf{p}_{\alpha i} - \mathbf{p}_\alpha)\delta(P_{\alpha i} - P_\alpha) \tag{8.1}$$

It follows directly from the definition of the singlet distribution function

(Eq. 6.1) that with this dynamical variable

$$\langle A\rangle = f_\alpha(\mathbf{r}, Q_\alpha, \mathbf{p}_\alpha, P_\alpha, t) \tag{8.2}$$

and

$$\mathfrak{L}A = -\sum_i \left\{ \frac{1}{m_\alpha}\frac{\partial}{\partial \mathbf{r}}\cdot\mathbf{p}_{\alpha i}A_i + \frac{\partial}{\partial \mathbf{p}_\alpha}\cdot\dot{\mathbf{p}}_{\alpha i}A_i + \sum_s \left[\frac{\partial}{\partial Q_{\alpha s}}\dot{Q}_{\alpha is}A_i + \frac{\partial}{\partial P_{\alpha s}}\dot{P}_{\alpha is}A_i \right]\right\} \tag{8.3}$$

Thus, with this dynamical variable, the general equation of change (Eq. 5.5) becomes

$$\frac{\partial}{\partial t}f_\alpha + \frac{1}{m_\alpha}\frac{\partial}{\partial \mathbf{r}}\cdot\mathbf{p}_\alpha f_\alpha + \frac{\partial}{\partial \mathbf{p}_\alpha}\cdot\langle\dot{\mathbf{p}}_\alpha\rangle^{(\alpha)}f_\alpha + \sum_s \left[\frac{\partial}{\partial Q_{\alpha s}}\dot{Q}_{\alpha s}f_\alpha + \frac{\partial}{\partial P_{\alpha s}}\langle\dot{P}_{\alpha s}\rangle^{(\alpha)}f_\alpha\right] = 0 \tag{8.4}$$

in which the $\langle\ \rangle^{(\alpha)}$ notation of (6.11) has been introduced. These quantities can be written as

$$\langle\dot{\mathbf{p}}_\alpha\rangle^{(\alpha)} = \sum_\beta \langle\mathbf{F}^{(\alpha,\beta)}\rangle^{(\alpha)} \tag{8.5}$$

$$\langle\dot{P}_{\alpha s}\rangle^{(\alpha)} = -\frac{1}{2}\sum_{uv}\left(\frac{\partial}{\partial Q_{\alpha s}}G_{uv}^{(\alpha)}\right)P_{\alpha u}P_{\alpha v} + \sum_\nu m_{\alpha\nu}^{-\frac{1}{2}}\mathbf{b}_{\nu s}^{(\alpha)}\cdot\langle\mathbf{F}_\nu^{(\alpha)}\rangle^{(\alpha)} \tag{8.6}$$

The first of these relations was obtained using Eqs. 4.10, 4.11, and 4.17. The second equation requires use of (4.3), reexpressed in terms of the $P_{\alpha u}$, and Eq. 4.18.

Equations 8.5 and 8.6 may also be written in terms of the pair distribution function:

$$\langle\dot{\mathbf{p}}_\alpha\rangle^{(\alpha)}f_\alpha = \sum_\beta \int \mathbf{F}^{(\alpha,\beta)}f_{\alpha\beta}(\mathbf{r},\mathbf{r}_\beta,Q_\alpha,Q_\beta,\mathbf{p}_\alpha,\mathbf{p}_\beta,P_\alpha,P_\beta,t)\,d\mathbf{r}_\beta\,dQ_\beta\,d\mathbf{p}_\beta\,dP_\beta \tag{8.7}$$

$$\begin{aligned}\langle\dot{P}_{\alpha s}\rangle^{(\alpha)}f_\alpha = &-\frac{1}{2}f_\alpha\sum_{uv}\left(\frac{\partial}{\partial Q_{\alpha s}}G_{uv}^{(\alpha)}\right)P_{\alpha u}P_{\alpha v} + f_\alpha\sum_{\nu\mu}m_{\alpha\nu}^{-\frac{1}{2}}\mathbf{b}_{\nu s}^{(\alpha)}\cdot\mathbf{F}_{\nu\mu}^{(\alpha)}\\ &+\sum_{\beta\mu\nu}m_{\alpha\nu}^{-\frac{1}{2}}\mathbf{b}_{\nu s}^{(\alpha)}\cdot\int\mathbf{F}_{\nu\mu}^{(\alpha,\beta)}f_{\alpha\beta}(\mathbf{r},\mathbf{r}_\beta,Q_\alpha,Q_\beta,\mathbf{p}_\alpha,\mathbf{p}_\beta,P_\alpha,P_\beta,t)\\ &\times d\mathbf{r}_\beta\,dQ_\beta\,d\mathbf{p}_\beta\,dP_\beta\end{aligned} \tag{8.8}$$

In the next section we use these results to get two important kinetic equations in configuration space.

IX. KINETIC EQUATIONS IN CONFIGURATION SPACE

Most previous macromolecular kinetic theories have been formulated in configuration space using a mixture of dynamical and statistical quantities; in these theories Brownian motion is introduced on an *ad hoc* basis. The theories involve writing down an *equation of continuity for the configurational space distribution function* and an *equation of motion for each internal degree of freedom*. These two "kinetic equations" are then combined, after omitting the acceleration terms in the equation of motion, to give the "*diffusion equation*" for the distribution function.

Here the kinetic equations are derived from the equation for the time evolution of the distribution function f_α. An equation of continuity for ψ_α is obtained directly by integrating Eq. 8.4 over the momenta. Equations of motion for the center of mass and for each internal degree of freedom are obtained by performing a similar integration after multiplying Eq. 8.4 by $\mathbf{p}_\alpha$ and $P_{\alpha s}$, respectively. These equations involve only statistical quantities; however, it is not difficult to relate the terms in them to the corresponding terms in older theories.

A. Equation of Continuity in Configuration Space

We first integrate Eq 8.4 over the momenta, using the definition of $[\ ]^{(\alpha)}$ in (6.12), to obtain

$$\frac{\partial}{\partial t}\psi_\alpha + \frac{1}{m_\alpha}\frac{\partial}{\partial \mathbf{r}}\cdot\left([\mathbf{p}_\alpha]^{(\alpha)}\psi_\alpha\right) + \sum_s \frac{\partial}{\partial Q_{\alpha s}}\left([\dot{Q}_{\alpha s}]^{(\alpha)}\psi_\alpha\right) = 0. \tag{9.1}$$

Using Eq. 4.5 one may rewrite this equation in the form

$$\underset{(1)}{\frac{\partial}{\partial t}\psi_\alpha} + \underset{(2)}{\frac{1}{m_\alpha}\frac{\partial}{\partial \mathbf{r}}\cdot\left([\mathbf{p}_\alpha]^{(\alpha)}\psi_\alpha\right)} + \underset{(3)}{\sum_{st} \frac{\partial}{\partial Q_{\alpha s}}\left(G_{st}^{(\alpha)}[P_{\alpha t}]^{(\alpha)}\psi_\alpha\right)} = 0 \tag{9.2}$$

This is an *equation of continuity in the configuration space*, describing the change of the probability density with position and time.

Equation 9.2 can be compared with Eq. 18 of Ref. 28, by noting that our ψ_α is equivalent to Riseman and Kirkwood's $g^{1/2}f$. They do not have a term corresponding to our term (2); this term vanishes if we assume that ψ_α is independent of position, that there is equilibration in momentum space, and that the macromolecular fluid is incompressible.

B. Equation of Motion for Center of Mass

We multiply Eq. 8.4 by $\mathbf{p}_\alpha$ and integrate over all momenta to obtain

$$\frac{\partial}{\partial t}\left[\mathbf{p}_\alpha\right]^{(\alpha)}\psi_\alpha+\frac{1}{m_\alpha}\frac{\partial}{\partial \mathbf{r}}\cdot\left[\mathbf{p}_\alpha\mathbf{p}_\alpha\right]^{(\alpha)}\psi_\alpha+\sum_s\frac{\partial}{\partial Q_{\alpha s}}\left[\mathbf{p}_\alpha\dot{Q}_{\alpha s}\right]^{(\alpha)}\psi_\alpha$$
$$=\left[\langle\dot{\mathbf{p}}_\alpha\rangle^{(\alpha)}\right]^{(\alpha)}\psi_\alpha \tag{9.3}$$

The term in $\dot{Q}_{\alpha s}$ can be transformed using Eq. 4.5. The term with $\langle\dot{\mathbf{p}}_\alpha\rangle^{(\alpha)}$ can be rewritten using Eqs. 4.11, 6.9, 6.12, and 8.7:

$$\frac{\partial}{\partial t}\left[\mathbf{p}_\alpha\right]^{(\alpha)}\psi_\alpha+\frac{1}{m_\alpha}\frac{\partial}{\partial \mathbf{r}}\cdot\left[\mathbf{p}_\alpha\mathbf{p}_\alpha\right]^{(\alpha)}\psi_\alpha+\sum_{st}\frac{\partial}{\partial Q_{\alpha s}}G_{st}^{(\alpha)}\left[\mathbf{p}_\alpha P_{\alpha t}\right]^{(\alpha)}\psi_\alpha$$
$$=\sum_{\beta\nu\mu}v_{\alpha\beta}\psi_\alpha(\mathbf{r}+\mathbf{R}_{\alpha\nu},Q_\alpha)\int\overline{\mathbf{F}}_{\nu\mu}^{(\alpha,\beta)}(\mathbf{r}+\mathbf{R}_{\alpha\nu},Q_\alpha,Q_\beta)\psi_\beta(\mathbf{r}+\mathbf{R}_{\alpha\nu},Q_\beta)\,dQ_\beta \tag{9.4}$$

Here $\overline{\mathbf{F}}_{\nu\mu}^{(\alpha,\beta)}(\mathbf{r},Q_\alpha,Q_\beta)$ is an average force on mass points ν of molecules of species α, defined by

$$\overline{\mathbf{F}}_{\nu\mu}^{(\alpha,\beta)}(\mathbf{r},Q_\alpha,Q_\beta)\psi_\alpha(\mathbf{r},Q_\alpha)\psi_\beta(\mathbf{r},Q_\beta)$$
$$=\frac{1}{v_{\alpha\beta}}\int\mathbf{F}_{\nu\mu}^{(\alpha,\beta)}(\mathbf{r}-\mathbf{R}_{\alpha\nu}-\mathbf{r}_\beta,Q_\alpha,Q_\beta)\psi_{\alpha\beta}(\mathbf{r}-\mathbf{R}_{\alpha\nu},\mathbf{r}_\beta,Q_\alpha,Q_\beta)\,d\mathbf{r}_\beta \tag{9.5}$$

The quantity $v_{\alpha\beta}$ is a volume included for dimensional reasons only; if we define $v_{\alpha\beta}=v_{\beta\alpha}$, then $\overline{\mathbf{F}}_{\nu\mu}^{(\alpha,\beta)}=-\overline{\mathbf{F}}_{\mu\nu}^{(\beta,\alpha)}$, which has the same form as Eq. 4.12. In (9.4) the sum on β includes the term in which β equals α. The force $\overline{\mathbf{F}}_{\nu\mu}^{(\alpha,\beta)}$ may be interpreted as the average force due to mass points $\beta\mu$ on mass points $\alpha\nu$ located at $\mathbf{r}$, the center of mass of molecule α being at $\mathbf{r}-\mathbf{R}_{\alpha\nu}$ and the configuration being specified by Q_α.

C. Equations of Motion for Each Degree of Freedom

Next we multiply Eq. 8.4 by $P_{\alpha s}$ and then integrate over all momenta, to get an *equation of motion* for the sth degree of freedom:

$$\frac{\partial}{\partial t}\left(\left[P_{\alpha s}\right]^{(\alpha)}\psi_\alpha\right)+\frac{1}{m_\alpha}\left(\frac{\partial}{\partial \mathbf{r}}\cdot\left[\mathbf{p}_\alpha P_{\alpha s}\right]^{(\alpha)}\psi_\alpha\right)+\sum_t\frac{\partial}{\partial Q_{\alpha t}}\left(\left[\dot{Q}_{\alpha t}P_{\alpha s}\right]^{(\alpha)}\psi_\alpha\right)$$
$$=\left[\langle\dot{P}_{\alpha s}\rangle^{(\alpha)}\right]^{(\alpha)}\psi_\alpha \tag{9.6}$$

or using Eqs. 4.5 and 7.12:

$$\overset{(1)}{\frac{\partial}{\partial t}\left([P_{\alpha s}]^{(\alpha)}\psi_\alpha\right)} + \overset{(2)}{\frac{1}{m_\alpha}\left(\frac{\partial}{\partial \mathbf{r}}\cdot[\mathbf{p}_\alpha P_{\alpha s}]^{(\alpha)}\psi_\alpha\right)}$$

$$+\overset{(3)}{\sum_{tu}\frac{\partial}{\partial Q_{\alpha t}}\left(G_{tu}^{(\alpha)}[P_{\alpha u}P_{\alpha s}]^{(\alpha)}\psi_\alpha\right)} + \overset{(4)}{\tfrac{1}{2}\psi_\alpha\sum_{uv}\left(\frac{\partial}{\partial Q_{\alpha s}}G_{uv}^{(\alpha)}\right)[P_{\alpha u}P_{\alpha v}]^{(\alpha)}}$$

$$=\overset{(5)}{\psi_\alpha\sum_{\nu\mu}m_{\alpha\nu}^{-\frac{1}{2}}\mathbf{b}_{\nu s}^{(\alpha)}\cdot\mathbf{F}_{\nu\mu}^{(\alpha)}} + \overset{(6)}{\sum_{\beta\nu\mu}v_{\alpha\beta}m_{\alpha\nu}^{-1/2}\psi_\alpha(\mathbf{r}+\mathbf{R}_{\alpha\nu},Q_\alpha)\mathbf{b}_{\nu s}^{(\alpha)}}$$

$$\cdot\int\overline{\mathbf{F}}_{\nu\mu}^{(\alpha,\beta)}(\mathbf{r}+\mathbf{R}_{\alpha\nu},Q_\alpha,Q_\beta)\psi_\beta(\mathbf{r}+\mathbf{R}_{\alpha\nu},Q_\beta)\,dQ_\beta \qquad (9.7)$$

where $\overline{\mathbf{F}}_{\nu\mu}^{(\alpha,\beta)}(\mathbf{r},Q_\alpha,Q_\beta)$ is the quantity defined in (9.5).

Let us now comment on the physical interpretation of (9.7):

Terms (1) and (2) are the *acceleration terms*; they are generally believed to be small and have usually been neglected in earlier configuration space theories. Specifically, they do not appear in Eq. 16 of Ref. 28.

Terms (3) and (4) can be identified as *Brownian motion* terms, by comparison with the older theories. Note that in the present theory Brownian motion terms are not added to the theory on an *ad hoc* basis, but arise naturally in the integrations over momenta. In the Kirkwood–Riseman theory the Brownian motion term appears as $kT(\partial f/\partial q^\alpha)$ in Eq. 16 of Ref. 28.

Term (5) is associated with the *connector forces in nonrigid connectors*; in other words, these terms arise because of the intramolecular forces acting between the mass points in one molecule. This term corresponds to the term $(\partial V^0/\partial q^\alpha)f$ in Eq. 16 of Ref. 28.

Term (6) contains: (a) the *solvent-macromolecule forces*, which are generally approximated (as we also do later) in terms of a "friction coefficient" that describes the hydrodynamic drag on some part of the macromolecule as it moves through the solvent; and (b) the *entanglement forces* describing the polymer–polymer interaction, which is neglected in most theories (notable exceptions being the works of Fixman[20,21] and Williams[13]). The solvent macromolecule force appears in the Kirkwood–Riseman theory as $F_\alpha f$ in Eq. 16 of Ref. 28.

We conclude by emphasizing that the equation of continuity (Eq. 9.2) and the equations of motion (Eqs. 9.4 and 9.7) contain no approximations other than those associated with pairwise additivity of the forces between mass points. We also emphasize that these equations are different from those in the older theories in that they do not contain a mixture of dynamical and statistical quantities. To parallel the older configuration-space theories, we have to combine Eq. 9.7 for $[P_\alpha T]^{(\alpha)}$ with Eq. (9.2) to get the "diffusion equation" for ψ_α. Before doing this we discuss several important assumptions that are introduced at various points in the discussion.

X. SEVERAL ASSUMPTIONS

In this section we introduce several assumptions in order to show how the theory given here is related to earlier kinetic theories and in order to develop some new results. The first assumption we discuss is that of equilibration in momentum space.

At equilibrium the phase-space distribution function is of the form:

$$f_\alpha(\mathbf{r}, Q_\alpha, \mathbf{p}_\alpha, P_\alpha, t) = \text{const} \times \exp\left[-\frac{H_\alpha}{kT} - \frac{(\mathbf{p}_\alpha - m_\alpha \mathbf{u})^2}{2m_\alpha kT}\right] \tag{10.1}$$

where $\mathbf{u}$ and T are the values of solution velocity and temperature, and the constant is determined by the normalization. Here H_α is the portion of the Hamiltonian H that is associated with species α only, and that has the form:

$$H_\alpha = \frac{1}{2}\sum_{st} G_{st}^{(\alpha)} P_{\alpha s} P_{\alpha t} + \phi_\alpha \tag{10.2}$$

where $\phi_\alpha = (\frac{1}{2})\Sigma_{\nu\mu}\phi_{\alpha\nu,\alpha\mu}$ is the intramolecular potential energy for a molecule of species α; ϕ_α depends only on the $Q_{\alpha s}$.

Now for a system not at equilibrium, f_α would have to be determined by solving (8.4). For polymer systems this has never been done, although the analogous problem for gases has been attacked by several methods, the oldest being that of Chapman and Enskog. In the simplification of the Brownian motion terms, it is necessary to introduce the following assumption in order to avoid having to evaluate the time-dependent f_α.

Assumption E. It is assumed that in a system away from equilibrium, the momentum distribution is identical to that contained in (10.1) with $\mathbf{u}(\mathbf{r}, t)$ and $T(\mathbf{r}, t)$ being the local, instantaneous values. This is the assumption of "equilibration in momentum space."

All subsequent equations that contain this assumption will show the letter "E" before the equation number. Assumption E thus allows us to write f_α approximately as

$$f_\alpha(\mathbf{r}, Q_\alpha, \mathbf{p}_\alpha, P_\alpha, t) = \frac{1}{Z_\alpha(Q_\alpha)}\psi_\alpha(\mathbf{r}, Q_\alpha, t)\exp\left[-\frac{H_\alpha}{kT} - \frac{(\mathbf{p}_\alpha - m_\alpha \mathbf{u})^2}{2m_\alpha kT}\right] \qquad \text{E:(10.3)}$$

where $Z_\alpha(Q_\alpha)$ is a normalization factor:

$$Z_\alpha = \int\int \exp\left[-\frac{H_\alpha}{kT} - \frac{(\mathbf{p}_\alpha - m_\alpha \mathbf{u})^2}{2m_\alpha kT}\right] d\mathbf{p}_\alpha dP_\alpha$$

$$= (2\pi m_\alpha kT)^{\frac{3}{2}} (2\pi kT)^{d_\alpha/2} G_\alpha^{-\frac{1}{2}} \exp\frac{-\phi_\alpha}{kT} \qquad \text{E:(10.4)}$$

and G_α is the determinant of the $G_{st}^{(\alpha)}$.

When the assumption of equilibration in momentum space is used, the various $[\ ]^{(\alpha)}$ quantities, defined in (6.12), that arise in (8.1) and (8.3) and elsewhere may be given explicitly:

$$\begin{aligned} [\mathbf{p}_\alpha]^{(\alpha)} &= m_\alpha \mathbf{u} \\ [\mathbf{p}_\alpha \mathbf{p}_\alpha]^{(\alpha)} &= m_\alpha kT \mathsf{U} + m_\alpha^2 \mathbf{u}\mathbf{u} \\ [P_{\alpha s}]^{(\alpha)} &= 0 \\ [\mathbf{p}_\alpha P_{\alpha s}]^{(\alpha)} &= 0 \\ [P_{\alpha s} P_{\alpha t}]^{(\alpha)} &= kT g_{st}^{(\alpha)} \end{aligned} \qquad \text{E:(10.5)}$$

We shall have occasion to refer to these results later.

Throughout most of the subsequent development, we shall restrict the macromolecular system to one with a single solvent that is structurally simple. The solvent consists of a single chemical species, labeled $\alpha = 0$, that has no internal structure; that is, species "0" will consist of a single mass point that will be designated by $\nu = 0$. For this system we introduce the following approximation for the polymer–solvent interaction.

Assumption F. The mean force on particle $\alpha\nu$ as it moves through a single structureless solvent "0" is

$$\begin{aligned} \overline{\mathbf{F}}_{\nu 0}^{(\alpha,0)} \psi_\alpha(\mathbf{r}, Q_\alpha) &= -\overline{\mathbf{F}}_{0\nu}^{(0,\alpha)} \psi_\alpha(\mathbf{r}, Q_\alpha) \\ &= -\frac{\zeta_{\alpha\nu}}{v_{\alpha 0} n_0}\left[\mathbf{u}_{\alpha\nu}(\mathbf{r}, Q_\alpha)\psi_\alpha(\mathbf{r}, Q_\alpha) - \mathbf{u}_0(\mathbf{r})\psi_\alpha(\mathbf{r} - \mathbf{R}_{\alpha\nu}, Q_\alpha)\right] \end{aligned} \qquad \text{F:(10.6)}$$

in which the "friction coefficients" $\zeta_{\alpha\nu}$ are constants with dimensions (mass)/(time). It is further assumed that the solvent concentration n_0 is constant throughout the solution.

The volume $v_{\alpha 0}$ was defined in (9.5). The quantity $\mathbf{u}_{\alpha\nu}(\mathbf{r}, Q_\alpha)$ is the average velocity, at position $\mathbf{r}$, of mass points ν of molecules of species α, defined by

$$\mathbf{u}_{\alpha\nu}\psi_\alpha(\mathbf{r}, Q_\alpha) = \int \left(\frac{1}{m_\alpha}\mathbf{p}_\alpha + \mathbf{V}_{\alpha\nu} \right) f_\alpha(\mathbf{r} - \mathbf{R}_{\alpha\nu}, Q_\alpha, \mathbf{p}_\alpha, P_\alpha, t)\, d\mathbf{p}_\alpha dP_\alpha$$

$$= \left[\frac{1}{m_\alpha}\mathbf{p}_\alpha + \mathbf{V}_{\alpha\nu} \right]^{(\alpha)} \psi_\alpha(\mathbf{r} - \mathbf{R}_{\alpha\nu}, Q_\alpha) \qquad (10.7)$$

This velocity is defined such that if $\mathbf{u}_{\alpha\nu}$ is multiplied by $m_{\alpha\nu}\psi_\alpha(\mathbf{r}, Q_\alpha)$, summed on ν, and integrated with respect to Q_α, Eq. 7.2 for $\rho_\alpha \mathbf{u}_\alpha$ is obtained. The quantities $\overline{\mathbf{F}}_{\nu 0}^{(\alpha,0)}$ and $\mathbf{u}_{\alpha\nu}$ are understood to have the same position arguments as those of the ψ_α that follow. When no arguments are shown for ψ_α, it is intended that $\psi_\alpha(\mathbf{r}, Q_\alpha, t)$ be understood. The idea in (10.6) that the force is proportional to the relative velocity of the two species is reminiscent of Stokes' drag law for a sphere moving unidirectionally through a viscous liquid with constant speed. Such a concept has also been developed by Kirkwood[29] in his study of the theory of transport processes in monatomic liquids. Indeed if we put $\zeta_{\alpha\nu} = 6\pi\eta_0 r_{\alpha\nu}^{(0)}$, where η_0 is the solvent viscosity and $r_{\alpha\nu}^{(0)}$ is the radius of bead ν of the macromolecule, then Eq. 10.6 looks very much like Stokes' law. However, Eq. 10.6 is an empirical relation among *statistical* quantities, whereas in the older configuration-space kinetic theories the analogous relations among *dynamical* quantities were introduced.

In order to simplify some results an additional assumption regarding the hydrodynamic drag can be introduced.

Assumption P. It is assumed that the friction coefficient for particle ν of species α as it moves through a single structureless solvent, is proportional to the mass of the particle:

$$\zeta_{\alpha\nu} = C_\alpha m_{\alpha\nu} \qquad \text{P}:(10.8)$$

in which the C_α are constants.

This assumption is somewhat weaker than than assuming that all the particles have the same mass and the same friction coefficient. Most of the development does not make use of Assumption P, but the final results are considerably simplified by its introduction.

In connection with integrals of the form given in (7.13), it is possible to effect important simplifications by introducing the following crucial assumption.

Assumption S. If the forces between the mass points are of sufficiently short range, the integrand of

$$\int \mathbf{R}_{\alpha\beta}\mathbf{F}_{\mu\eta}^{(\alpha,\beta)}\tilde{\psi}_{\alpha\beta}\left(\mathbf{r}-\xi\mathbf{R}_{\alpha\nu}+\frac{\xi m_\beta}{m_\alpha+m_\beta}\mathbf{R}_{\alpha\beta},\mathbf{R}_{\alpha\beta},Q_\alpha,Q_\beta,t\right)d\mathbf{R}_{\alpha\beta} \qquad \text{S}:(10.9)$$

is very nearly zero unless the two particles $\alpha\mu$ and $\beta\eta$ are close together, that is, unless $\mathbf{R}_{\alpha\beta}$ is approximately equal to $\mathbf{R}_{\alpha\mu}-\mathbf{R}_{\beta\eta}$. Since $\tilde{\psi}_{\alpha\beta}$ depends only slightly on the first argument (although strongly on the second argument), we can therefore approximate $\mathbf{R}_{\alpha\beta}$ by $\mathbf{R}_{\alpha\mu}-\mathbf{R}_{\beta\eta}$ in integrals of this form everywhere except in the second argument of $\tilde{\psi}_{\alpha\beta}$ and in the argument of $\mathbf{F}_{\mu\eta}^{(\alpha,\beta)}$. In addition, we assume that the mass of the solvent molecule is much less than that of the macromolecule, so that $m_0 \ll m_\alpha$.

In each of the equations of motion (Eqs. 9.4 and 9.7) the first two terms —those containing the time and space derivatives—are referred to as "acceleration terms." These terms are believed to be negligibly small provided that the particles are small, the friction coefficients are large, and the time changes of the velocity field are not too rapid. Order-of-magnitude arguments have been given by Fixman,[20] based on a characteristic time $m/\zeta \approx 10^{-13}$ sec for macromolecules. This leads us to the following assumption.

Assumption A. Acceleration terms in the equations of motion (Eqs. 9.4 and 9.7) are omitted.

In the last portion of this paper, we neglect polymer–polymer interactions by making the following simplification.

Assumption D. This is the assumption of a *dilute solution*, namely that the solute molecules are present in such low concentrations that macromolecule–macromolecule interactions can be entirely neglected and that $\mathbf{u}=\mathbf{u}_0$.

Finally, there are several additional assumptions that are needed primarily in order to make comparisons with earlier theories.

Assumption I. It is assumed that the macromolecular fluid is *incompressible* so that $(\partial/\partial\mathbf{r})\cdot\mathbf{u}=0$.

Assumption H. This is the assumption of *homogeneous flow*, namely that $(\partial/\partial\mathbf{r})\mathbf{u}$, ψ_α and $\tilde{\psi}_{\alpha\beta}$ are independent of position.

XI. APPROXIMATE "DIFFUSION EQUATION" FOR ψ_α IN CONFIGURATION SPACE

We are now ready to continue the discussion of the kinetic equations of Section IX. We consider a solution made up of several macromolecular solutes $\alpha = 1,2,3,\ldots$ and a single structureless solvent (labeled "0").

First, we introduce into the $\beta = 0$ contribution to term (6) in (9.7) the expression for $\overline{\mathbf{F}}_{\nu 0}^{(\alpha,0)}$ from (10.6). Then Eqs. 7.10 and 10.7 are used for $\mathbf{u}_{\alpha\nu}(\mathbf{r}, Q_\alpha)$. In this way the $\beta = 0$ contribution to term (6) becomes

$$-\sum_\nu m_{\alpha\nu}^{-1/2}\,\zeta_{\alpha\nu}\mathbf{b}_{\nu s}^{(\alpha)}\cdot\left[\mathbf{V}_{\alpha\nu}\right]^{(\alpha)}\psi_\alpha - \sum_\nu m_{\alpha\nu}^{-1/2}\,\zeta_{\alpha\nu}\mathbf{b}_{\nu s}^{(\alpha)}$$

$$\cdot\left\{\left[\frac{1}{m}\mathbf{p}_\alpha\right]^{(\alpha)} - \mathbf{R}_{\alpha\nu}\cdot\frac{\partial}{\partial\mathbf{r}}\int_0^1 \mathbf{u}_0(\mathbf{r}+\xi\mathbf{R}_{\alpha\nu})\,d\xi - \mathbf{u}_0\right\}\psi_\alpha$$

$$\equiv -\sum_\nu m_{\alpha\nu}^{-1/2}\,\zeta_{\alpha\nu}\mathbf{b}_{\nu s}^{(\alpha)}\cdot\left[\mathbf{V}_{\alpha\nu}\right]^{(\alpha)}\psi_\alpha - B_s^{(\alpha)} \qquad \text{F:(11.1)}$$

Then, with the help of (2.4), (2.5), (2.20), and (4.1), Eq. 9.7 can be written as

$$A_s^{(\alpha)} = -\psi_\alpha \sum_{tu} \mathcal{G}_{st}^{(\alpha)} G_{tu}^{(\alpha)}\left[P_{\alpha u}\right]^{(\alpha)} - B_s^{(\alpha)} \qquad \text{F:(11.2)}$$

where

$$\mathcal{G}_{st}^{(\alpha)} = \sum_\nu \frac{\zeta_{\alpha\nu}}{m_{\alpha\nu}}\mathbf{b}_{\nu s}^{(\alpha)}\cdot\mathbf{b}_{\nu t}^{(\alpha)} \qquad (11.3)$$

and $A_s^{(\alpha)}$ stands for terms (1) through (5) and the $\beta \neq 0$ contribution to term (6). Equation 11.2 may be solved for $[P_{\alpha s}]^{(\alpha)}$ to give

$$\left[P_{\alpha s}\right]^{(\alpha)}\psi_\alpha = -\sum_{tu} g_{su}^{(\alpha)}\mathcal{G}_{ut}^{(\alpha)}\left(A_t^{(\alpha)} + B_t^{(\alpha)}\right) \qquad \text{F:(11.4)}$$

where the $\mathcal{G}_{ut}^{(\alpha)}$ are the elements of the matrix inverse to $[g_{st}^{(\alpha)}]$, and

$$\begin{aligned}
A_t^{(\alpha)}+B_t^{(\alpha)} &= \overset{(1)}{\frac{\partial}{\partial t}\left(\left[P_{\alpha t}\right]^{(\alpha)}\psi_\alpha\right)} + \overset{(2)}{\frac{1}{m_\alpha}\left(\frac{\partial}{\partial \mathbf{r}}\cdot\left[\mathbf{p}_\alpha P_{\alpha t}\right]^{(\alpha)}\psi_\alpha\right)} \\
&\quad \overset{(3)}{+\sum_{uv}\frac{\partial}{\partial Q_{\alpha v}}\left(G_{uv}^{(\alpha)}\left[P_{\alpha u}P_{\alpha t}\right]^{(\alpha)}\psi_\alpha\right)} \\
&\quad \overset{(4)}{+\frac{1}{2}\psi_\alpha\sum_{uv}\left(\frac{\partial}{\partial Q_{\alpha t}}G_{uv}^{(\alpha)}\right)\left[P_{\alpha u}P_{\alpha v}\right]^{(\alpha)}} \\
&\quad \overset{(5)}{-\psi_\alpha\sum_{\nu\mu} m_{\alpha\nu}^{-1/2}\mathbf{b}_{\nu t}^{(\alpha)}\cdot\mathbf{F}_{\nu\mu}^{(\alpha)}} \\
&\quad +\sum_{\nu} m_{\alpha\nu}^{-1/2}\,\zeta_{\alpha\nu}\mathbf{b}_{\nu t}^{(\alpha)}\cdot\left\{\overset{(6b)}{\left(\left[\frac{1}{m_\alpha}\mathbf{p}_\alpha\right]^{(\alpha)}-\mathbf{u}_0\right)}\right. \\
&\quad \left.\overset{(6c)}{-\mathbf{R}_{\alpha\nu}\cdot\frac{\partial}{\partial \mathbf{r}}\int_0^1 \mathbf{u}_0(\mathbf{r}+\xi\mathbf{R}_{\alpha\nu})\,d\xi}\right\}\psi_\alpha \\
&\quad \overset{(6a)}{-\sum_{\substack{\beta\neq 0\\ \nu\mu}} v_{\alpha\beta}m_{\alpha\nu}^{-1/2}\,\psi_\alpha(\mathbf{r}+\mathbf{R}_{\alpha\nu})\mathbf{b}_{\nu t}^{(\alpha)}\cdot\int\overline{\mathbf{F}}_{\nu\mu}^{(\alpha,\beta)}\psi_\beta(\mathbf{r}+\mathbf{R}_{\alpha\nu})\,dQ_\beta} \\
&\equiv \text{term }(6b)+C_t^{(\alpha)} \qquad \text{F:(11.5)}
\end{aligned}$$

The numbers above the terms refer to the corresponding terms in (9.7).

Next we want to replace the expression $([\mathbf{p}_\alpha/m_\alpha]^{(\alpha)}-\mathbf{u}_0)$ in term (6*b*) by using the equation of motion for the center of mass, namely Eq. 9.4, since this will introduce additional terms involving $[P_{\alpha s}]^{(\alpha)}$. When Assumption F,

(Eq. 10.6) is introduced into (9.4), we get

$$\sum_{\nu} \zeta_{\alpha\nu} \left\{ \left[\frac{1}{m_\alpha} \mathbf{p}_\alpha + \mathbf{V}_{\alpha\nu} \right]^{(\alpha)} - \mathbf{u}_0(\mathbf{r}+\mathbf{R}_{\alpha\nu}) \right\} \psi_\alpha$$

$$= -\frac{\partial}{\partial t} [\mathbf{p}_\alpha]^{(\alpha)} \psi_\alpha - \frac{1}{m_\alpha} \frac{\partial}{\partial \mathbf{r}} \cdot [\mathbf{p}_\alpha \mathbf{p}_\alpha]^{(\alpha)} \psi_\alpha - \sum_{st} \frac{\partial}{\partial Q_{\alpha s}} G_{st}^{(\alpha)} [\mathbf{p}_\alpha P_{\alpha t}]^{(\alpha)} \psi_\alpha$$

$$+ \sum_{\beta \neq 0} \sum_{\nu\mu} v_{\alpha\beta} \psi_\alpha (\mathbf{r}+\mathbf{R}_{\alpha\nu}, Q_\alpha) \int \overline{\mathbf{F}}_{\nu\mu}^{(\alpha,\beta)} \psi_\beta (\mathbf{r}+\mathbf{R}_{\alpha\nu}, Q_\beta) \, dQ_\beta \qquad \text{F}:(11.6)$$

Then using Eq. 7.10 for $\mathbf{u}_0(\mathbf{r}+\mathbf{R}_{\alpha\nu})$ we get

$$\left(\frac{1}{m_\alpha} [\mathbf{p}_\alpha]^{(\alpha)} - \mathbf{u}_0 \right) \zeta_\alpha \psi_\alpha = -\sum_{\nu} \zeta_{\alpha\nu} [\mathbf{V}_{\alpha\nu}]^{(\alpha)} \psi_\alpha$$

$$+ \psi_\alpha \frac{\partial}{\partial \mathbf{r}} \cdot \sum_{\nu} \zeta_{\alpha\nu} \mathbf{R}_{\alpha\nu} \int_0^1 \mathbf{u}_0 (\mathbf{r}+\xi \mathbf{R}_{\alpha\nu}) \, d\xi$$

$$- \frac{\partial}{\partial t} \left([\mathbf{p}_\alpha]^{(\alpha)} \psi_\alpha \right) - \frac{1}{m_\alpha} \frac{\partial}{\partial \mathbf{r}} \cdot \left([\mathbf{p}_\alpha \mathbf{p}_\alpha]^{(\alpha)} \psi_\alpha \right)$$

$$- \sum_{st} \frac{\partial}{\partial Q_{\alpha s}} \left(G_{st}^{(\alpha)} [\mathbf{p}_\alpha P_{\alpha t}]^{(\alpha)} \psi_\alpha \right)$$

$$+ \sum_{\substack{\beta \neq 0 \\ \nu\mu}} v_{\alpha\beta} \psi_\alpha (\mathbf{r}+\mathbf{R}_{\alpha\nu}, Q_\alpha)$$

$$\times \int \overline{\mathbf{F}}_{\nu\mu}^{(\alpha,\beta)} \psi_\beta (\mathbf{r}+\mathbf{R}_{\alpha\nu}, Q_\beta) \, dQ_\beta \qquad \text{F}:(11.7)$$

in which $\zeta_\alpha = \Sigma \zeta_{\alpha\nu}$.

Next divide Eq. 11.6 by ζ_α and form the scalar product with $\Sigma_\nu m_{\alpha\nu}^{-1/2} \zeta_{\alpha\nu} \mathbf{b}_{\nu t}^{(\alpha)}$. Then use Eqs. 2.16 and 4.1 to transform the second term:

$$\sum_{\nu} m_{\alpha\nu}^{-1/2} \zeta_{\alpha\nu} \mathbf{b}_{\nu t}^{(\alpha)} \cdot \left(\left[\frac{1}{m_\alpha} \mathbf{p}_\alpha \right]^{(\alpha)} - \mathbf{u}_0 \right) \psi_\alpha$$

$$= \psi_\alpha \sum_{uv} \left(\tilde{g}_{tu}^{(\alpha)} - g_{tu}^{(\alpha)} \right) G_{uv}^{(\alpha)} [P_{\alpha v}]^{(\alpha)} + D_t^{(\alpha)} \qquad \text{F}:(11.8)$$

where

$$D_t^{(\alpha)} = \frac{1}{\zeta_\alpha} \sum_\nu m_{\alpha\nu}^{-1/2} \zeta_{\alpha\nu} \mathbf{b}_{\nu t}^{(\alpha)} \cdot \Bigg[\overset{(7)}{\psi_\alpha \frac{\partial}{\partial \mathbf{r}} \cdot \sum_\mu \zeta_{\alpha\mu} \mathbf{R}_{\alpha\mu} \int_0^1 \mathbf{u}_0(\mathbf{r} + \xi \mathbf{R}_{\alpha\mu})\, d\xi}$$

$$\overset{(8)}{- \frac{\partial}{\partial t}([\mathbf{p}_\alpha]^{(\alpha)} \psi_\alpha)} \overset{(9)}{- \frac{1}{m_\alpha} \frac{\partial}{\partial \mathbf{r}} \cdot ([\mathbf{p}_\alpha \mathbf{p}_\alpha]^{(\alpha)} \psi_\alpha)} \overset{(10)}{- \sum_{vu} \frac{\partial}{\partial Q_{\alpha v}} (G_{vu}^{(\alpha)} [\mathbf{p}_\alpha P_{\alpha u}]^{(\alpha)} \psi_\alpha)}$$

$$\overset{(11)}{+ \sum_{\substack{\beta \neq 0 \\ \mu\eta}} v_{\alpha\beta} \psi_\alpha(\mathbf{r} + \mathbf{R}_{\alpha\mu}, Q_\alpha) \int \overline{\mathbf{F}}_{\mu\eta}^{(\alpha,\beta)} \psi_\beta(\mathbf{r} + \mathbf{R}_{\alpha\mu}, Q_\beta)\, dQ_\beta} \Bigg] \qquad \text{F}:(11.9)$$

Then substitution of (11.5) and (11.8) into (11.4) gives

$$[P_{\alpha s}]^{(\alpha)} \psi_\alpha = - \sum_{tu} g_{su}^{(\alpha)} \mathcal{G}_{ut}^{(\alpha)} \Big\{ C_t^{(\alpha)} + D_t^{(\alpha)} + \psi_\alpha \sum_{uv} (\tilde{g}_{tu}^{(\alpha)} - \mathcal{g}_{tu}^{(\alpha)}) G_{uv}^{(\alpha)} [P_{\alpha v}]^{(\alpha)} \Big\} \qquad \text{F}:(11.10)$$

This may be solved for $[P_{\alpha s}]^{(\alpha)} \psi_\alpha$ to give

$$[P_{\alpha s}]^{(\alpha)} \psi_\alpha = \sum_{tu} g_{su}^{(\alpha)} \tilde{G}_{ut}^{(\alpha)} E_t^{(\alpha)} \qquad \text{F}:(11.11)$$

where $E_t^{(\alpha)} = -(C_t^{(\alpha)} + D_t^{(\alpha)})$ is given by

$$E_t^{(\alpha)} = \overset{(5)}{\psi_\alpha \sum_{\nu\mu} m_{\alpha\nu}^{-1/2} \mathbf{b}_{\nu t}^{(\alpha)} \cdot \mathbf{F}_{\nu\mu}^{(\alpha)}} \overset{(3)}{- \sum_{uv} \frac{\partial}{\partial Q_{\alpha v}} (G_{uv}^{(\alpha)} [P_{\alpha u} P_{\alpha t}]^{(\alpha)} \psi_\alpha)}$$

$$\overset{(4)}{- \frac{1}{2} \psi_\alpha \sum_{uv} \left(\frac{\partial}{\partial Q_{\alpha t}} G_{uv}^{(\alpha)} \right) [P_{\alpha u} P_{\alpha v}]^{(\alpha)}} \overset{(6c)+(7)}{+ \psi_\alpha \frac{\partial}{\partial \mathbf{r}} \cdot \sum_\nu \zeta_{\alpha\nu}^{1/2} \mathbf{R}_{\alpha\nu} \tilde{\mathbf{b}}_{\nu t}^{(\alpha)}}$$

$$\cdot\int_0^1 \mathbf{u}_0(\mathbf{r}+\xi\mathbf{R}_{\alpha\nu})\,d\xi - \overset{(1)}{\frac{\partial}{\partial t}[P_{\alpha t}]^{(\alpha)}\psi_\alpha} - \overset{(2)}{\frac{1}{m_\alpha}\frac{\partial}{\partial \mathbf{r}}\cdot[\mathbf{p}_\alpha P_{\alpha t}]^{(\alpha)}\psi_\alpha}$$

$$\overset{(6d)+(11)}{+\sum_{\substack{\beta\neq 0\\ \nu\mu}} v_{\alpha\beta}\zeta_{\alpha\nu}^{-1/2}\,\psi_\alpha(\mathbf{r}+\mathbf{R}_{\alpha\nu},Q_\alpha)\tilde{\mathbf{b}}_{\nu t}^{(\alpha)}\cdot\int \overline{\mathbf{F}}_{\nu\mu}^{(\alpha,\beta)}\psi_\beta(\mathbf{r}+\mathbf{R}_{\alpha\nu},Q_\beta)\,dQ_\beta}$$

$$+\left(\frac{1}{\zeta_\alpha}\sum_\nu m_{\alpha\nu}^{-1/2}\,\zeta_{\alpha\nu}\mathbf{b}_{\nu t}^{(\alpha)}\right)\cdot\left\{\overset{(8)}{\frac{\partial}{\partial t}([\mathbf{p}_\alpha]^{(\alpha)}\psi_\alpha)}+\overset{(9)}{\frac{1}{m_\alpha}\left(\frac{\partial}{\partial \mathbf{r}}\cdot[\mathbf{p}_\alpha\mathbf{p}_\alpha]^{(\alpha)}\psi_\alpha\right)}\right.$$

$$\left.\overset{(10)}{+\sum_{uv}\frac{\partial}{\partial Q_{\alpha v}}\left(G_{vu}^{(\alpha)}[\mathbf{p}_\alpha P_{\alpha u}]^{(\alpha)}\psi_\alpha\right)}\right\} \qquad \text{F}:(11.12)$$

In combining terms ($6c$) and (7), and ($6d$) and (11), Eq. 2.6 was used, and this gives rise to the base vectors $\tilde{\mathbf{b}}_t^{(\alpha)}$. Note that if the $\zeta_{\alpha\nu}=C_\alpha m_{\alpha\nu}$ (Assumption P), then terms (8) through (10) vanish and $\tilde{\mathbf{b}}_{\nu t}^{(\alpha)}$ becomes $C_\alpha^{1/2}\mathbf{b}_{\nu s}^{(\alpha)}$ in two places.

Substituting Eq. 11.11 into the equation of continuity in (9.2) gives

$$\frac{\partial}{\partial t}\psi_\alpha+\frac{1}{m_\alpha}\left(\frac{\partial}{\partial \mathbf{r}}\cdot[\mathbf{p}_\alpha]^{(\alpha)}\psi_\alpha\right)+\sum_{st}\frac{\partial}{\partial Q_{\alpha s}}\left(\tilde{G}_{st}^{(\alpha)}E_t^{(\alpha)}\right)=0 \qquad \text{F}:(11.13)$$

This is the "diffusion equation" for ψ_α. Up to this point only Assumption F has been made: the introduction of the friction factor for the mass points and the assumption that the solvent concentration n_0 is independent of position.

The quantity $E_t^{(\alpha)}$ may be simplified by introducing a few additional assumptions:

Assumption D removes term ($6d$)+(11).

Assumption H allows us to simplify term ($6c$)+(7) by replacing $\int_0^1\mathbf{u}_0(\mathbf{r}+\xi\mathbf{R}_{\alpha\nu})\,d\xi$ by $\mathbf{u}_0(\mathbf{r})$.

Assumption A allows us to omit terms (1), (2), (8), and (9).

Assumption E allows terms (3) and (4) to be simplified and combined into $-kTg_\alpha^{1/2}(\partial/\partial Q_{\alpha t})(g_\alpha^{-1/2}\psi_\alpha)$. In addition, term (10) drops out.

Assumptions H, E, and I cause the second term in (11.13) to vanish. With these assumptions and with (2.7), Eq. 11.13 becomes finally:

$$\overset{(a)}{\frac{\partial\psi_\alpha}{\partial t}} + \sum_{st}\frac{\partial}{\partial Q_{\alpha s}}\left\{\tilde{G}_{st}^{(\alpha)}\left[\overset{(c)}{\psi_\alpha\sum_{\nu\mu} m_{\alpha\nu}^{-1/2}\,\mathbf{b}_{\nu t}^{(\alpha)}\cdot\mathbf{F}_{\nu\mu}^{(\alpha)}} - \overset{(d)}{kTg_\alpha^{1/2}\frac{\partial}{\partial Q_{\alpha t}}(g_\alpha^{-1/2}\psi_\alpha)}\right.\right.$$

$$\left.\left. + \overset{(b)}{\psi_\alpha\sum_\nu \zeta_{\alpha\nu}^{1/2}\,\tilde{\mathbf{b}}_{\nu t}^{(\alpha)}\mathbf{R}_{\alpha\nu}:\frac{\partial}{\partial\mathbf{r}}\mathbf{u}}\right]\right\} = 0 \qquad \text{FDHAEI}:(11.14)$$

A number of comments need to be made regarding this important result:

(1) A diffusion equation similar to (11.14) was given earlier by Riseman and Kirkwood (see Ref. 28, Eq. 19) for the special case that all masses and friction coefficients are equal. Although their equation was specifically intended for a linear bead–rod model with fixed-bond angles, the result is more general with proper interpretation. In comparing Eq. 11.14 with the Riseman–Kirkwood equation, it must be noted that their $fg^{1/2}$ corresponds to our ψ_α.

(2) For steady-state flows with $(\partial/\partial\mathbf{r})\mathbf{u}$ equal to its transpose (i.e., zero vorticity), Eq. 11.14 has the solution:

$$\psi_\alpha = \text{const}\; g_\alpha^{1/2}\exp\left(\frac{\frac{1}{2}\Sigma_\nu\zeta_{\alpha\nu}\tilde{\mathbf{R}}_{\alpha\nu}\tilde{\mathbf{R}}_{\alpha\nu}:(\partial/\partial\mathbf{r})\mathbf{u}-\phi_\alpha}{kT}\right) \qquad \text{FDHAEI}:(11.15)$$

Steady-State, Irrotational Flow

where $\phi_\alpha = \frac{1}{2}\Sigma_{\nu\mu}\phi_{\alpha\nu,\alpha\mu}$. This expression for ψ_α causes the expression in brackets in (11.14) to vanish. Special cases of this equation have been given by Kramers[2] in 1944, and by Bird, Johnson, and Curtiss[5] in 1969. Hence for such flows the configuration space distribution function can be written down directly. Use has recently been made of this fact by Hassager[7] in his study of flexible macromolecules.

(3) The solution of the diffusion equation for rotational flows is clearly a formidable problem. Even for such simple systems as rigid dumbbells or chains of beads and rods the solution of this equation

presents difficulties. Usually it has been possible to obtain solutions by perturbation methods for flows that depart only slightly from the equilibrium state.[6]

(4) Note that Eq. 11.14 is valid for bead–rod–spring models with beads of differing masses and friction coefficients. This seems to be the first time that a diffusion equation for ψ_α has been presented for models of this degree of generality.

(5) Note that in 11.14, terms $(c)+(d)$ inside the bracket may be rewritten as

$$\psi_\alpha \Sigma_{\nu\mu} m_{\alpha\nu}^{-1/2}\, \mathbf{b}_{\nu t}^{(\alpha)} \cdot \mathbf{F}_{\nu\mu}^{(\alpha)} - kT g_\alpha^{1/2} \frac{\partial}{\partial Q_{\alpha t}} \left(g_\alpha^{-1/2} \psi_\alpha \right) = -kT \psi_\alpha \frac{\partial}{\partial Q_{\alpha t}} \ln \frac{\psi_\alpha}{\psi_{\alpha,\text{eq.}}} \tag{11.16}$$

where $\psi_{\alpha,\text{eq.}}$ is the equilibrium distribution function given by (11.15) with $(\partial/\partial\mathbf{r})\mathbf{u}=0$.

XII. AN EXPRESSION FOR THE MOMENTUM-FLUX TENSOR IN TERMS OF THE CONFIGURATION-SPACE DISTRIBUTION FUNCTION

In Table III we summarized the various contributions to the momentum-flux tensor, $\mathsf{p}^{(k)}$, $\mathsf{p}^{(\mathrm{I})}$, and $\mathsf{p}^{(\mathrm{II})}$, which are given in (7.4), (7.12), and (7.13), respectively, in terms of the distribution functions f_α and $f_{\alpha\beta}$ in phase space. In this section we derive several expressions for the momentum-flux tensor in terms of the configuration-space distribution functions. The expression given in (12.8) contains only the assumption that the drag force can be expressed in terms of a drag coefficient $\zeta_{\alpha\nu}$ (Assumption F) and the short-range force assumption (Assumption S).

We first define $\mathsf{p}_{\text{int}}^{(k)}$ and $\mathsf{p}_{\text{tr}}^{(k)}$ by (see Eq. 7.4 for the definition of $\mathsf{p}^{(k)}$):

$$\mathsf{p}^{(k)} = \mathsf{p}_{\text{tr}}^{(k)} + \mathsf{p}_{\text{int}}^{(k)} \tag{12.1}$$

and

$$\mathsf{p}_{\text{int}}^{(k)} = \sum_{\alpha\nu} m_{\alpha\nu} \int \int_0^1 \left[\mathbf{V}_{\alpha\nu} \mathbf{V}_{\alpha\nu} \right]^{(\alpha)} \psi_\alpha(\mathbf{r} - \xi \mathbf{R}_{\alpha\nu}, Q_\alpha)\, d\xi\, dQ_\alpha \tag{12.2}$$

Then using Eqs. 4.1, 7.6, and 7.12 we get

$$\mathsf{p}_{\text{int}}^{(k)}+\mathsf{p}^{(\text{I})}=\sum_{\substack{\alpha\nu\mu\eta\\ st}}\left(\frac{m_{\alpha\nu}}{m_{\alpha\mu}}\right)^{1/2}\int\int_0^1 G_{st}^{(\alpha)}\mathbf{R}_{\alpha\nu}\mathbf{b}_{\nu s}^{(\alpha)}\mathbf{b}_{\mu t}^{(\alpha)}$$

$$\cdot\mathbf{F}_{\mu\eta}^{(\alpha)}\psi_\alpha(\mathbf{r}-\xi\mathbf{R}_{\alpha\nu},Q_\alpha)\,d\xi\,dQ_\alpha+\sum_{\substack{\alpha\nu st\\ uv}}m_{\alpha\nu}^{1/2}\int\int_0^1\left\{G_{tv}^{(\alpha)}\left(\frac{\partial}{\partial Q_{\alpha t}}G_{su}^{(\alpha)}\mathbf{R}_{\alpha\nu}\mathbf{b}_{\nu s}^{(\alpha)}\right)\right.$$

$$\left.-\frac{1}{2}G_{st}^{(\alpha)}\mathbf{R}_{\alpha\nu}\mathbf{b}_{\nu s}\left(\frac{\partial}{\partial Q_{\alpha t}}G_{uv}^{(\alpha)}\right)\right\}[P_{\alpha u}P_{\alpha v}]^{(\alpha)}\psi_\alpha(\mathbf{r}-\xi\mathbf{R}_{\alpha\nu},Q_\alpha)\,d\xi\,dQ_\alpha \tag{12.3}$$

Next, the use of Approximations S and F gives for $\mathsf{p}^{(\text{II})}$ from (7.7), (7.13), and (9.5):

$$\mathsf{p}^{(\text{II})}=\mathsf{p}_0^{(\text{II})}-\sum_{\substack{\alpha\nu\mu\\ st}}\zeta_{\alpha\mu}\left(\frac{m_{\alpha\nu}}{m_{\alpha\mu}}\right)^{1/2}\int\int_0^1 G_{st}^{(\alpha)}\mathbf{R}_{\alpha\nu}\mathbf{b}_{\nu s}^{(\alpha)}\mathbf{b}_{\mu t}^{(\alpha)}$$

$$\cdot\left\{\left[\frac{1}{m_\alpha}\mathbf{p}_\alpha+\mathbf{V}_{\alpha\mu}\right]^{(\alpha)}-\mathbf{u}_0(\mathbf{r}-\xi\mathbf{R}_{\alpha\nu}+\mathbf{R}_{\alpha\mu})\right\}\psi_\alpha(\mathbf{r}-\xi\mathbf{R}_{\alpha\nu},Q_\alpha)\,d\xi\,dQ_\alpha$$

$$+\sum_{\alpha\nu}\zeta_{\alpha\nu}\int\int_0^1\mathbf{R}_{\alpha\nu}\left\{\left[\frac{1}{m_\alpha}\mathbf{p}_\alpha+\mathbf{V}_{\alpha\nu}\right]^{(\alpha)}\right.$$

$$\left.-\mathbf{u}_0(\mathbf{r}+(1-\xi)\mathbf{R}_{\alpha\nu})\right\}\psi_\alpha(\mathbf{r}-\xi\mathbf{R}_{\alpha\nu},Q_\alpha)\,d\xi\,dQ_\alpha$$

$$-\sum_{\alpha\nu\mu}\zeta_{\alpha\mu}\frac{m_{\alpha\nu}}{m_\alpha}\int\int_0^1\mathbf{R}_{\alpha\nu}\left\{\left[\frac{1}{m_\alpha}\mathbf{p}_\alpha+\mathbf{V}_{\alpha\mu}\right]^{(\alpha)}\right.$$

$$\left.-\mathbf{u}_0(\mathbf{r}-\xi\mathbf{R}_{\alpha\nu}+\mathbf{R}_{\alpha\mu})\right\}\psi_\alpha(\mathbf{r}-\xi\mathbf{R}_{\alpha\nu},Q_\alpha)\,d\xi\,dQ_\alpha$$

$$+\sum_{\substack{\alpha\beta\nu\\ \alpha\neq0\\ \beta\neq0}}\int\int_0^1\left(\mathbf{R}_{\alpha\nu}-\frac{m_\beta}{m_\alpha+m_\beta}\mathbf{R}_{\alpha\beta}\right)\left\{\frac{m_{\alpha\nu}}{m_\alpha}\mathbf{F}^{(\alpha,\beta)}\right.$$

$$\left.+\sum_{st\mu\eta}\left(\frac{m_{\alpha\nu}}{m_{\alpha\mu}}\right)^{1/2}G_{st}\mathbf{b}_{\nu s}\mathbf{b}_{\mu t}\cdot\mathbf{F}_{\mu\eta}^{(\alpha,\beta)}\right\}$$

$$\times \tilde{\psi}_{\alpha\beta}\left(\mathbf{r}-\xi\mathbf{R}_{\alpha\nu}+\frac{\xi m_\beta}{m_\alpha+m_\beta}\mathbf{R}_{\alpha\beta},\mathbf{R}_{\alpha\beta},Q_\alpha,Q_\beta,t\right)d\xi\, d\mathbf{R}_{\alpha\beta}\, dQ_\alpha\, dQ_\beta$$

FS: (12.4)

where $\mathbf{p}_0^{(\mathrm{II})}$ arises from the term in (7.7) in which $\alpha=0$ and $\beta=0$, representing the solvent–solvent interaction. We now want to eliminate the velocity averages appearing in (12.4) in favor of averages over the momentum coordinates. We first use Eq. (7.10) to rewrite the terms containing $\mathbf{u}_0$ as follows:

$$\mathbf{p}^{(\mathrm{II})}=\mathbf{p}_0^{(\mathrm{II})}-\sum_{\substack{\alpha\nu\mu\\ st}}\zeta_{\alpha\mu}\left(\frac{m_{\alpha\nu}}{m_{\alpha\mu}}\right)^{1/2}\int\int_0^1 G_{st}^{(\alpha)}\mathbf{R}_{\alpha\nu}\mathbf{b}_{\nu s}^{(\alpha)}\mathbf{b}_{\mu t}^{(\alpha)}\cdot\left\{\left[\frac{1}{m_\alpha}\mathbf{p}_\alpha-\mathbf{u}_0+\mathbf{V}_{\alpha\mu}\right]^{(\alpha)}\right.$$

$$\left.-\mathbf{R}_{\alpha\mu}\cdot\int_0^1\frac{\partial}{\partial\mathbf{r}}\mathbf{u}_0(\mathbf{r}-\xi\mathbf{R}_{\alpha\nu}+\xi'\mathbf{R}_{\alpha\mu})\,d\xi'\right\}\psi_\alpha(\mathbf{r}-\xi\mathbf{R}_{\alpha\nu},Q_\alpha)\,d\xi\, dQ_\alpha$$

$$+\sum_{\alpha\nu}\zeta_{\alpha\mu}\int\int_0^1\mathbf{R}_{\alpha\nu}\left\{\left[\frac{1}{m_\alpha}\mathbf{p}_\alpha-\mathbf{u}_0+\mathbf{V}_{\alpha\mu}\right]^{(\alpha)}\right.$$

$$\left.-\mathbf{R}_{\alpha\mu}\cdot\int_0^1\frac{\partial}{\partial\mathbf{r}}\mathbf{u}_0(\mathbf{r}-\xi\mathbf{R}_{\alpha\nu}+\xi'\mathbf{R}_{\alpha\mu})\,d\xi'\right\}\psi_\alpha(\mathbf{r}-\xi\mathbf{R}_{\alpha\nu},Q_\alpha)\,d\xi\, dQ_\alpha$$

$$-\sum_{\alpha\nu\mu}\zeta_{\alpha\mu}\frac{m_{\alpha\nu}}{m_\alpha}\int\int_0^1\mathbf{R}_{\alpha\nu}\left\{\left[\frac{1}{m_\alpha}\mathbf{p}_\alpha-\mathbf{u}_0+\mathbf{V}_{\alpha\mu}\right]^{(\alpha)}\right.$$

$$\left.-\mathbf{R}_{\alpha\mu}\cdot\int_0^1\frac{\partial}{\partial\mathbf{r}}\mathbf{u}_0(\mathbf{r}-\xi\mathbf{R}_{\alpha\nu}+\xi'\mathbf{R}_{\alpha\mu})\,d\xi'\right\}\psi_\alpha(\mathbf{r}-\xi\mathbf{R}_{\alpha\nu},Q_\alpha)\,d\xi\, dQ_\alpha$$

$$+\sum_{\substack{\alpha\beta\nu\\ \alpha\neq0\\ \beta\neq0}}\int\int_0^1\left(\mathbf{R}_{\alpha\nu}-\frac{m_\beta}{m_\alpha+m_\beta}\mathbf{R}_{\alpha\beta}\right)\left\{\frac{m_{\alpha\nu}}{m_\alpha}\mathbf{F}^{(\alpha,\beta)}\right.$$

$$\left.+\sum_{\substack{st\\ \mu\eta}}\left(\frac{m_{\alpha\nu}}{m_{\alpha\mu}}\right)^{1/2}G_{st}^{(\alpha)}\mathbf{b}_{\nu s}^{(\alpha)}\mathbf{b}_{\mu t}^{(\alpha)}\cdot\mathbf{F}_{\mu\eta}^{(\alpha,\beta)}\right\}$$

$$\times\tilde{\psi}_{\alpha\beta}\left(\mathbf{r}-\xi\mathbf{R}_{\alpha\nu}+\frac{\xi m_\beta}{m_\alpha+m_\beta}\mathbf{R}_{\alpha\beta},\mathbf{R}_{\alpha\beta},Q_\alpha,Q_\beta,t\right)d\xi\, d\mathbf{R}_{\alpha\beta}\, dQ_\alpha\, dQ_\beta$$

FS: (12.5)

Now let us note that Eq. 11.11 yields the following expression for $[\mathbf{V}_{\alpha\nu}]^{(\alpha)}$:

$$[\mathbf{V}_{\alpha\nu}]^{(\alpha)}\psi_\alpha = \zeta_{\alpha\nu}^{-1/2}\sum_{st}\tilde{G}_{st}^{(\alpha)}E_t^{(\alpha)}\tilde{\mathbf{b}}_{\nu s}^{(\alpha)} + \frac{1}{\zeta_\alpha}\sum_\mu \zeta_{\alpha\mu}[\mathbf{V}_{\alpha\mu}]^{(\alpha)} \qquad \text{F:(12.6)}$$

where $E_t^{(\alpha)}(\mathbf{r}_\alpha, Q_\alpha)$ is given by (11.12). We now want to insert this expression for $[\mathbf{V}_{\alpha\nu}]^{(\alpha)}$ into (12.5). We then need to use Eq. 12.6 written for $\psi_\alpha(\mathbf{r}-\xi\mathbf{R}_{\alpha\nu}, Q_\alpha)$ rather than for $\psi_\alpha(\mathbf{r}, Q_\alpha)$. In so doing, however, note that in the expression for $E_t^{(\alpha)}$ the differentiations of the ψ_α with respect to the $Q_{\alpha s}$ are understood to operate on the Q_α arguments of ψ_α only, and not on the $\mathbf{r}-\xi\mathbf{R}_{\alpha\nu}$ argument. These comments will also apply to (12.7) and (12.8). With this in mind we find,

$$\mathsf{p}^{(\mathrm{II})} = \mathsf{p}_0^{(\mathrm{II})} - \sum_{\substack{\alpha\nu\\ st}} m_{\alpha\nu}^{1/2}\int\int_0^1 G_{st}^{(\alpha)}\mathbf{R}_{\alpha\nu}\mathbf{b}_{\nu s}^{(\alpha)}E_t^{(\alpha)}(\mathbf{r}-\xi\mathbf{R}_{\alpha\nu}, Q_\alpha)\,d\xi\,dQ_\alpha$$

$$-\sum_{\substack{\alpha\nu\mu\\ st}}\zeta_{\alpha\mu}\left(\frac{m_{\alpha\nu}}{m_{\alpha\mu}}\right)^{1/2}\int\int_0^1 G_{st}^{(\alpha)}\mathbf{R}_{\alpha\nu}\mathbf{b}_{\nu s}^{(\alpha)}\mathbf{b}_{\mu t}^{(\alpha)}\cdot\left[\frac{1}{m_\alpha}\mathbf{p}_\alpha - \mathbf{u}_0 + \frac{1}{\zeta_\alpha}\sum_\eta \zeta_{\alpha\eta}\mathbf{V}_{\alpha\eta}\right]^{(\alpha)}$$

$$\times\psi_\alpha(\mathbf{r}-\xi\mathbf{R}_{\alpha\nu}, Q_\alpha)\,d\xi\,dQ_\alpha + \sum_{\substack{\alpha\nu\mu\\ st}}\zeta_{\alpha\mu}\left(\frac{m_{\alpha\nu}}{m_{\alpha\mu}}\right)^{1/2}\int\int_0^1\int_0^1 G_{st}^{(\alpha)}\mathbf{R}_{\alpha\nu}\mathbf{b}_{\nu s}^{(\alpha)}\mathbf{b}_{\mu t}^{(\alpha)}\mathbf{R}_{\alpha\mu}$$

$$:\psi_\alpha(\mathbf{r}-\xi\mathbf{R}_{\alpha\nu}, Q_\alpha)\frac{\partial}{\partial\mathbf{r}}\mathbf{u}_0(\mathbf{r}-\xi\mathbf{R}_{\alpha\nu}+\xi'\mathbf{R}_{\alpha\mu})\,d\xi'\,d\xi\,dQ_\alpha$$

$$+\sum_{\substack{\alpha\nu\\ st}}\zeta_{\alpha\nu}^{1/2}\int\int_0^1\tilde{G}_{st}^{(\alpha)}\mathbf{R}_{\alpha\nu}\tilde{\mathbf{b}}_{\nu s}^{(\alpha)}E_t^{(\alpha)}(\mathbf{r}-\xi\mathbf{R}_{\alpha\nu}, Q_\alpha)\,d\xi\,dQ_\alpha$$

$$+\sum_{\alpha\nu}\zeta_{\alpha\nu}\int\int_0^1\mathbf{R}_{\alpha\nu}\left[\frac{1}{m_\alpha}\mathbf{p}_\alpha - \mathbf{u}_0 + \frac{1}{\zeta_\alpha}\sum_\mu\zeta_{\alpha\mu}\mathbf{V}_{\alpha\mu}\right]^{(\alpha)}\psi_\alpha(\mathbf{r}-\xi\mathbf{R}_{\alpha\nu}, Q_\alpha)\,d\xi\,dQ_\alpha$$

$$-\sum_{\alpha\nu}\zeta_{\alpha\nu}\int\int_0^1\int_0^1\psi_\alpha(\mathbf{r}-\xi\mathbf{R}_{\alpha\nu}, Q_\alpha)\mathbf{R}_{\alpha\nu}\mathbf{R}_{\alpha\nu}\cdot\frac{\partial}{\partial\mathbf{r}}\mathbf{u}_0(\mathbf{r}-\xi\mathbf{R}_{\alpha\nu}+\xi'\mathbf{R}_{\alpha\nu})\,d\xi'\,d\xi\,dQ_\alpha$$

$$-\sum_{\alpha\nu}\zeta_\alpha\frac{m_{\alpha\nu}}{m_\alpha}\int\int_0^1\mathbf{R}_{\alpha\nu}\left[\frac{1}{m_\alpha}\mathbf{p}_\alpha - \mathbf{u}_0 + \frac{1}{\zeta_\alpha}\sum_\mu\zeta_{\alpha\mu}\mathbf{V}_{\alpha\mu}\right]^{(\alpha)}\psi_\alpha(\mathbf{r}-\mathbf{R}_{\alpha\nu}, Q_\alpha)\,d\xi\,dQ_\alpha$$

$$+\sum_{\alpha\nu\mu}\zeta_{\alpha\mu}\frac{m_{\alpha\nu}}{m_\alpha}\int\int_0^1\int_0^1\psi_\alpha(\mathbf{r}-\xi\mathbf{R}_{\alpha\nu},Q_\alpha)\mathbf{R}_{\alpha\nu}\mathbf{R}_{\alpha\mu}$$

$$\cdot\frac{\partial}{\partial\mathbf{r}}\mathbf{u}_0(\mathbf{r}-\xi\mathbf{R}_{\alpha\nu}+\xi'\mathbf{R}_{\alpha\mu})\,d\xi'\,d\xi\,dQ_\alpha$$

$$+\sum_{\substack{\alpha\beta\nu\\ \alpha\neq0\\ \beta\neq0}}\int\int_0^1\left(\mathbf{R}_{\alpha\nu}-\frac{m_\beta}{m_\alpha+m_\beta}\mathbf{R}_{\alpha\beta}\right)\left\{\frac{m_{\alpha\nu}}{m_\alpha}\mathbf{F}^{(\alpha,\beta)}\right.$$

$$\left.+\sum_{\substack{st\\ \mu\eta}}\left(\frac{m_{\alpha\nu}}{m_{\alpha\mu}}\right)^{1/2}G_{st}^{(\alpha)}\mathbf{b}_{\nu s}^{(\alpha)}\mathbf{b}_{\mu t}^{(\alpha)}\cdot\mathbf{F}_{\mu\eta}^{(\alpha,\beta)}\right\}$$

$$\times\tilde{\psi}_{\alpha\beta}\left(\mathbf{r}-\xi\mathbf{R}_{\alpha\nu}+\frac{\xi m_\beta}{m_\alpha+m_\beta}\mathbf{R}_{\alpha\beta},\mathbf{R}_{\alpha\beta},Q_\alpha,Q_\beta,t\right)d\xi\,d\mathbf{R}_{\alpha\beta}\,dQ_\alpha\,dQ_\beta$$

FS: (12.7)

Finally, we substitute the expression for $E_t^{(\alpha)}(\mathbf{r}-\xi\mathbf{R}_{\alpha\nu},Q_\alpha)$ [from (11.12)] and the expression for

$$\left[\frac{1}{m_\alpha}\mathbf{p}_\alpha-\mathbf{u}_0+\frac{1}{\zeta_\alpha}\sum_\mu\zeta_{\alpha\mu}\mathbf{V}_{\alpha\mu}\right]^{(\alpha)}\psi_\alpha(\mathbf{r}-\xi\mathbf{R}_{\alpha\nu},Q_\alpha)$$

[from (11.7)] into (12.7). When this is done, and the result is combined with (12.3) for $\mathsf{p}_{\text{int}}^{(k)}+\mathsf{p}^{(\text{I})}$, we find for p:

$$\mathsf{p}=\mathsf{p}_{\text{tr}}^{(k)}+\mathsf{p}_0^{(\text{II})}+\sum_{\alpha\nu st}\zeta_{\alpha\nu}^{1/2}\int\int_0^1\tilde{G}_{st}^{(\alpha)}\mathbf{R}_{\alpha\nu}\tilde{\mathbf{b}}_{\nu s}^{(\alpha)}\left\{\psi_\alpha(\mathbf{r}-\xi\mathbf{R}_{\alpha\nu},Q_\alpha)\right.$$

$$\times\sum_{\mu\eta}m_{\alpha\mu}^{-1/2}\,\mathbf{b}_{\mu t}^{(\alpha)}\cdot\mathbf{F}_{\mu\eta}-\sum_{\mu v}\frac{\partial}{\partial Q_{\alpha v}}\left(G_{uv}^{(\alpha)}[P_{\alpha u}P_{\alpha t}]^{(\alpha)}\psi_\alpha(\mathbf{r}-\xi\mathbf{R}_{\alpha\nu},Q_\alpha)\right)$$

$$-\frac{1}{2}\psi_\alpha(\mathbf{r}-\xi\mathbf{R}_{\alpha\nu},Q_\alpha)\sum_{uv}\left(\frac{\partial}{\partial Q_{\alpha t}}G_{uv}^{(\alpha)}\right)[P_{\alpha u}P_{\alpha v}]^{(\alpha)}+\sum_\mu\zeta_{\alpha\mu}^{1/2}\,\tilde{\mathbf{b}}_{\mu t}^{(\alpha)}\mathbf{R}_{\alpha\mu}$$

$$\left.\cdot\psi_\alpha(\mathbf{r}-\xi\mathbf{R}_{\alpha\nu},Q_\alpha)\int_0^1\frac{\partial}{\partial\mathbf{r}}\mathbf{u}_0(\mathbf{r}+\xi'\mathbf{R}_{\alpha\mu}-\xi\mathbf{R}_{\alpha\nu})d\xi'\right\}d\xi\,dQ_\alpha$$

$$-\sum_{\alpha\nu\mu}\left(\delta_{\nu\mu}\zeta_{\alpha\nu}-\frac{\zeta_{\alpha\nu}\zeta_{\alpha\mu}}{\zeta_\alpha}\right)\int\int_0^1\int_0^1\psi_\alpha(\mathbf{r}-\xi\mathbf{R}_{\alpha\nu},Q_\alpha)\mathbf{R}_{\alpha\nu}\mathbf{R}_{\alpha\mu}$$

$$\cdot\frac{\partial}{\partial \mathbf{r}}\mathbf{u}_0(\mathbf{r}+\xi'\mathbf{R}_{\alpha\mu}-\xi\mathbf{R}_{\alpha\nu})\,d\xi'\,d\xi\,dQ_\alpha+\sum_{\alpha\neq 0}\sum_{\beta\neq 0}\sum_{\nu}$$

$$\int\int_0^1\left(\mathbf{R}_{\alpha\nu}-\frac{m_\beta}{m_\alpha+m_\beta}\mathbf{R}_{\alpha\beta}\right)\left\{\frac{m_{\alpha\nu}}{m_\alpha}\mathbf{F}^{(\alpha,\beta)}+\sum_{st\mu\eta}\left(\frac{m_{\alpha\nu}}{m_{\alpha\mu}}\right)^{1/2}G_{st}^{(\alpha)}\mathbf{b}_{\nu s}^{(\alpha)}\mathbf{b}_{\mu t}^{(\alpha)}\cdot\mathbf{F}_{\mu\eta}^{(\alpha,\beta)}\right\}$$

$$\times\tilde{\psi}_{\alpha\beta}\left(\mathbf{r}-\xi\mathbf{R}_{\alpha\nu}+\frac{\xi m_\beta}{m_\alpha+m_\beta}\mathbf{R}_{\alpha\beta},\mathbf{R}_{\alpha\beta},Q_\alpha,Q_\beta,t\right)d\xi\,d\mathbf{R}_{\alpha\beta}\,dQ_\alpha\,dQ_\beta$$

$$+\sum_{\substack{\alpha\nu\\ st}}m_{\alpha\nu}^{1/2}\int\int_0^1\mathbf{R}_{\alpha\nu}\left[G_{st}^{(\alpha)}\mathbf{b}_{\nu s}^{(\alpha)}-\left(\frac{\zeta_{\alpha\nu}}{m_{\alpha\nu}}\right)^{1/2}\tilde{G}_{st}^{(\alpha)}\tilde{\mathbf{b}}_{\nu s}^{(\alpha)}\right]$$

$$\times\left[\frac{\partial}{\partial t}\left([P_{\alpha t}]^{(\alpha)}\psi_\alpha(\mathbf{r}-\xi\mathbf{R}_{\alpha\nu},Q_\alpha)\right)\right.$$

$$\left.+\frac{1}{m_\alpha}\frac{\partial}{\partial \mathbf{r}}\cdot\left([\mathbf{p}_\alpha P_{\alpha t}]^{(\alpha)}\psi_\alpha(\mathbf{r}-\xi\mathbf{R}_{\alpha\nu},Q_\alpha)\right)\right]d\xi\,dQ_\alpha$$

$$+\sum_{\alpha\nu}\int\int_0^1\mathbf{R}_{\alpha\nu}\left\{\left(\frac{m_{\alpha\nu}}{m_\alpha}-\frac{\zeta_{\alpha\nu}}{\zeta_\alpha}\right)\mathsf{U}+\Sigma_{st}\tilde{G}_{st}^{(\alpha)}\tilde{\mathbf{b}}_{\nu s}^{(\alpha)}\left(\left(\frac{\zeta_{\alpha\nu}}{m_{\alpha\nu}}\right)^{1/2}\mathbf{b}_{\nu t}^{(\alpha)}-\tilde{\mathbf{b}}_{\nu t}^{(\alpha)}\right)\right.$$

$$\cdot\left\{\frac{\partial}{\partial t}[\mathbf{p}_\alpha]^{(\alpha)}\psi_\alpha(\mathbf{r}-\xi\mathbf{R}_{\alpha\nu},Q_\alpha)+\frac{1}{m_\alpha}\frac{\partial}{\partial \mathbf{r}}\cdot[\mathbf{p}_\alpha\mathbf{p}_\alpha]^{(\alpha)}\psi_\alpha(\mathbf{r}-\xi\mathbf{R}_{\alpha\nu},Q_\alpha)\right.$$

$$\left.+\sum_{st}\frac{\partial}{\partial Q_{\alpha s}}\left(G_{st}^{(\alpha)}[\mathbf{p}_\alpha P_{\alpha t}]^{(\alpha)}\psi_\alpha(\mathbf{r}-\xi R_{\alpha\nu},Q_\alpha)\right)\right\}d\xi\,dQ_\alpha+\sum_{\beta\neq 0}\sum_{\alpha\mu\eta\nu}v_{\alpha\beta}\int\int_0^1\mathbf{R}_{\alpha\nu}$$

$$\times\left\{\left(\frac{\zeta_{\alpha\nu}}{\zeta_\alpha}-\frac{m_{\alpha\nu}}{m_\alpha}\right)\mathsf{U}-\left(\frac{m_{\alpha\nu}}{m_{\alpha\mu}}\right)^{1/2}\sum_{st}\left[G_{st}^{(\alpha)}\mathbf{b}_{\nu s}^{(\alpha)}\mathbf{b}_{\mu t}^{(\alpha)}-\tilde{G}_{st}^{(\alpha)}\tilde{\mathbf{b}}_{\nu s}^{(\alpha)}\tilde{\mathbf{b}}_{\mu t}^{(\alpha)}\right]\right\}$$

$$\left.\cdot\psi_\alpha(\mathbf{r}+\mathbf{R}_{\alpha\mu}-\xi\mathbf{R}_{\alpha\nu},Q_\alpha)\int\overline{\mathbf{F}}_{\mu\eta}^{(\alpha,\beta)}\psi_\beta(\mathbf{r}+\mathbf{R}_{\alpha\mu}-\xi\mathbf{R}_{\alpha\nu},Q_\beta)\,dQ_\beta\right\}d\xi\,dQ_\alpha$$

FS: (12.8))

Equation 12.8 is the principal result of this section. It gives the momentum-flux tensor in terms of the configuration-space distribution function ψ_α. It does, however, contain the $[\ ]^{(\alpha)}$ quantities that are integrals involving the phase-space distribution function f_α.

The expression for p can now be simplified by introducing several additional assumptions:

Assumption D causes the terms containing $\overline{\mathbf{F}}_{\mu\eta}^{(\alpha,\beta)}$ to drop out.

Assumption H causes the terms containing $\mathbf{u}_0$ to become linear in $(\partial/\partial\mathbf{r})\mathbf{u}_0$.

Assumption A allows the acceleration terms to be omitted.

Assumption E leads to explicit evaluation of the Brownian motion terms containing $[P_{\alpha s}P_{\alpha t}]^{(\alpha)}$. This approximation also causes the term containing $[\mathbf{p}_\alpha P_{\alpha t}]^{(\alpha)}$ to drop out.

Finally, we obtain the following:

$$
\begin{aligned}
\mathsf{p}=\mathsf{p}_0&+\Big(\sum_{\alpha\neq 0} n_\alpha\Big)kT\mathsf{U}-\sum_{\alpha\nu}\zeta_{\alpha\nu}\int\tilde{\mathbf{R}}_{\alpha\nu}\tilde{\mathbf{R}}_{\alpha\nu}\psi_\alpha\,dQ_\alpha\cdot\frac{\partial}{\partial\mathbf{r}}\mathbf{u}\\
&+\sum_{\substack{\alpha\nu\mu\\ st}}(\zeta_{\alpha\nu}\zeta_{\alpha\mu})^{1/2}\int\tilde{G}_{st}^{(\alpha)}\tilde{\mathbf{R}}_{\alpha\nu}\tilde{\mathbf{b}}_{\nu s}^{(\alpha)}\tilde{\mathbf{b}}_{\mu t}^{(\alpha)}\tilde{\mathbf{R}}_{\alpha\mu}\psi_\alpha\,dQ_\alpha:\frac{\partial}{\partial\mathbf{r}}\mathbf{u}\\
&-kT\sum_{\alpha\nu st}\zeta_{\alpha\nu}^{1/2}\int g_\alpha^{1/2}\tilde{G}_{st}^{(\alpha)}\tilde{\mathbf{R}}_{\alpha\nu}\tilde{\mathbf{b}}_{\nu s}^{(\alpha)}\left(\frac{\partial}{\partial Q_{\alpha t}}g_\alpha^{-1/2}\psi_\alpha\right)dQ_\alpha\\
&+\sum_{\substack{\alpha\nu st\\ \mu\eta}}\left(\frac{\zeta_{\alpha\nu}}{\zeta_{\alpha\mu}}\right)^{1/2}\int\tilde{G}_{st}^{(\alpha)}\tilde{\mathbf{R}}_{\alpha\nu}\tilde{\mathbf{b}}_{\nu s}^{(\alpha)}\tilde{\mathbf{b}}_{\mu t}^{(\alpha)}\cdot\mathbf{F}_{\mu\eta}^{(\alpha)}\psi_\alpha\,dQ_\alpha \qquad \text{FSDHAE}:(12.9)
\end{aligned}
$$

where $\mathsf{p}_0=\mathsf{p}_0^{(\mathrm{II})}+n_0kT\mathsf{U}$ is the solvent contribution to p.

Equation 12.9 seems to be new. Expressions for the momentum flux given previously have not been as complete as this expression, which is valid for models in which all the $m_{\alpha\nu}$ and $\zeta_{\alpha\nu}$ are different. Several points need to be made about the result.

(1) It is not possible by casual inspection to determine the symmetry of p from (12.9); it is shown in Section XIV that this expression for p is symmetric.

(2) This expression for p is not easily compared with other formulas given in the literature. Alternative forms of p are given in Sections XIII and XIV that establish a connection with earlier formulas of Kramers, Kirkwood, Giesekus, and others.

(3) It is not easy to interpret the various terms in (12.9). Some feeling for what the terms contain can be obtained by examining the way in which the terms simplify for two well-known models (see Table IV).

TABLE IV. Equation (12.9) for Two Macromolecular Models

Term on right-hand side of (12.9)		Rouse model (N identical beads connected by N − 1 springs with universal joints)	Rigid-dumbbell model (two identical beads connected by a rigid rod)
2	Translational motion	$n_\alpha kT\mathbf{U}$	$n_\alpha kT\mathbf{U}$
3 and 4	Hydrodynamic forces	0	$-\frac{\zeta L^2}{2}\int(\frac{\partial}{\partial \mathbf{r}}\mathbf{u}):\check{\boldsymbol{\delta}}_R\check{\boldsymbol{\delta}}_R\check{\boldsymbol{\delta}}_R\check{\boldsymbol{\delta}}_R\psi_\alpha\, dQ_\alpha$
5	Brownnian motion forces	$n_\alpha kT(N-1)\mathbf{U}$	$-3kT\int\check{\boldsymbol{\delta}}_R\check{\boldsymbol{\delta}}_R\psi_\alpha\, dQ_\alpha+n_\alpha kT\mathbf{U}$
6	Spring forces	$-\sum_{a=1}^{N-1}\int\mathbf{F}_a^{(c)}\mathbf{Q}_a\psi_\alpha\, dQ_\alpha$	0

Notes:

(1) $\mathbf{Q}_a$ in the Rouse model stands for $\mathbf{R}_{a+1}-\mathbf{R}_a$, and $\mathbf{F}_a^{(c)}$, the tension in the ath spring, is collinear with $\mathbf{Q}_a$.

(2) $\int\cdots\psi_\alpha\, dQ_\alpha$ in both models indicates the appropriate integration over configuration space.

(3) $\check{\boldsymbol{\delta}}_R$ in the rigid-dumbbell model is the unit vector from bead 1 to bead 2.

(4) The rigid-dumbbell results should be compared with Eq. 4-22 in Ref. 6.

(4) Note that terms (4) and (5) can be combined to give

$$-kT\sum_{\substack{\alpha\nu\\ st}}\zeta_{\alpha\nu}^{1/2}\int\psi_{\alpha,\text{eq.}}\tilde{G}_{st}^{(\alpha)}\tilde{\mathbf{R}}_{\alpha\nu}\tilde{\mathbf{b}}_{\nu s}^{(\alpha)}\frac{\partial}{\partial Q_{\alpha t}}\left(\frac{\psi_\alpha}{\psi_{\alpha,\text{eq.}}}\right)dQ_\alpha \tag{12.10}$$

where $\psi_{\alpha,\text{eq.}}$ is given by (11.15) with the $(\partial/\partial\mathbf{r})\mathbf{u}$ set equal to zero.

XIII. THE MOMENTUM–FLUX TENSOR EXPRESSIONS OF KRAMERS, KIRKWOOD, KOTAKA, AND FIXMAN

The momentum-flux tensor given in (12.9) bears little resemblance to those given by previous investigators. In this section we point out how our result is related to those published earlier.

First, we note that in (12.4) the terms in braces involving differences of velocities are in fact very closely related to the expression for $\overline{\mathbf{F}}_{\mu 0}^{(\alpha,0)}\psi_\alpha(\mathbf{r}+\mathbf{R}_{\alpha\mu}-\xi\mathbf{R}_{\alpha\nu}, Q_\alpha)$ obtained by combining Eqs. 10.6 and 10.7 and using a displaced argument. Then the expression for $\mathsf{p}^{(\mathrm{II})}$ can be rewritten as

$$\mathbf{p}^{(\text{II})}=\mathbf{p}_0^{(\text{II})}+\sum_{\alpha\mu\nu st} n_0 v_{\alpha 0}\left(\frac{m_{\alpha\nu}}{m_{\alpha\mu}}\right)^{1/2}\int\int_0^1 G_{st}^{(\alpha)}\mathbf{R}_{\alpha\nu}\mathbf{b}_{\nu s}^{(\alpha)}\mathbf{b}_{\mu t}^{(\alpha)}$$

$$\cdot\overline{\mathbf{F}}_{\mu 0}^{(\alpha 0)}\psi_\alpha(\mathbf{r}+\mathbf{R}_{\alpha\mu}-\xi\mathbf{R}_{\alpha\nu},Q_\alpha)\,d\xi\,dQ_\alpha-\sum_{\alpha\nu\mu} n_0 v_{\alpha 0}\left(\delta_{\mu\nu}-\frac{m_{\alpha\nu}}{m_\alpha}\right)$$

$$\times\int\int_0^1 \mathbf{R}_{\alpha\nu}\overline{\mathbf{F}}_{\mu 0}^{(\alpha,0)}\psi_\alpha(\mathbf{r}+\mathbf{R}_{\alpha\mu}-\xi\mathbf{R}_{\alpha\nu},Q_\alpha)\,d\xi\,dQ_\alpha$$

$$+\sum_{\substack{\alpha\beta\nu\\ \alpha\neq 0\\ \beta\neq 0}}\int\int_0^1\left(\mathbf{R}_{\alpha\nu}-\frac{m_\beta}{m_\alpha+m_\beta}\mathbf{R}_{\alpha\beta}\right)$$

$$\times\left[\frac{m_{\alpha\nu}}{m_\alpha}\mathbf{F}^{(\alpha,\beta)}+\sum_{st\mu\eta}\left(\frac{m_{\alpha\nu}}{m_{\alpha\mu}}\right)^{1/2}G_{st}^{(\alpha)}\mathbf{b}_{\nu s}^{(\alpha)}\mathbf{b}_{\mu t}^{(\alpha)}\cdot\mathbf{F}_{\mu\eta}^{(\alpha\beta)}\right]$$

$$\times\tilde{\psi}_{\alpha\beta}\left(\mathbf{r}-\xi\mathbf{R}_{\alpha\nu}+\frac{\xi m_\beta}{m_\alpha+m_\beta}\mathbf{R}_{\alpha\beta},\mathbf{R}_{\alpha\beta},Q_\alpha,Q_\beta,t\right)d\xi\,d\mathbf{R}_{\alpha\beta}\,dQ_\alpha\,dQ_\beta \qquad \text{S:(13.1)}$$

Next we turn to $\mathbf{p}^{(\text{I})}$ given by (4.21) and (7.12); in this expression we eliminate the quantity $\sum_{\mu\eta} m_{\alpha\mu}^{-1/2}\,\mathbf{b}_{\mu t}^{(\alpha)}\cdot\mathbf{F}_{\mu\eta}^{(\alpha)}\,\psi_\alpha$ by using Eq. 9.7. When this is done there is some cancellation and combination of terms, and $\mathbf{p}^{(\text{I})}$ becomes

$$\mathbf{p}^{(\text{I})}=\sum_{\alpha\nu st} m_{\alpha\nu}^{1/2}\int\int_0^1 G_{st}^{(\alpha)}\mathbf{R}_{\alpha\nu}\mathbf{b}_{\nu s}^{(\alpha)}\left(\frac{\partial}{\partial t}\left([P_{\alpha t}]^{(\alpha)}\psi_\alpha(\mathbf{r}-\xi\mathbf{R}_{\alpha\nu},Q_\alpha)\right)\right.$$

$$\left.+\frac{1}{m_\alpha}\left(\frac{\partial}{\partial\mathbf{r}}\cdot[\mathbf{p}_\alpha P_{\alpha t}]^{(\alpha)}\psi_\alpha(\mathbf{r}-\xi\mathbf{R}_{\alpha\nu},Q_\alpha)\right)\right)d\xi\,dQ_\alpha+\sum_{\alpha\nu stuv} m_{\alpha\nu}^{1/2}\int\int_0^1\mathbf{R}_{\alpha\nu}\frac{\partial}{\partial Q_{\alpha t}}$$

$$\times\left(G_{su}^{(\alpha)}G_{tv}^{(\alpha)}\mathbf{b}_{\nu s}^{(\alpha)}[P_{\alpha u}P_{\alpha v}]^{(\alpha)}\psi_\alpha(\mathbf{r}-\xi\mathbf{R}_{\alpha\nu},Q_\alpha)\right)d\xi\,dQ_\alpha-\sum_{\substack{\alpha\nu\mu\\ st}} n_0 v_{\alpha 0}\left(\frac{m_{\alpha\nu}}{m_{\alpha\mu}}\right)^{1/2}$$

$$\times\int\int_0^1 G_{st}^{(\alpha)}\mathbf{R}_{\alpha\nu}\mathbf{b}_{\nu s}^{(\alpha)}\mathbf{b}_{\nu t}^{(\alpha)}\cdot\overline{\mathbf{F}}_{\mu 0}^{(\alpha,0)}\psi_\alpha(\mathbf{r}+\mathbf{R}_{\alpha\mu}-\xi\mathbf{R}_{\alpha\nu},Q_\alpha)\,d\xi\,dQ_\alpha$$

$$-\sum_{\substack{\alpha\beta st\\ \nu\mu\eta\\ \alpha\neq 0\\ \beta\neq 0}}\left(\frac{m_{\alpha\nu}}{m_{\alpha\mu}}\right)^{1/2}\int\int_0^1 G_{st}^{(\alpha)}\mathbf{R}_{\alpha\nu}\mathbf{b}_{\nu s}^{(\alpha)}\mathbf{b}_{\nu t}^{(\alpha)}\cdot\mathbf{F}_{\mu\eta}^{(\alpha,\beta)}$$

$$\times\tilde{\psi}_{\alpha\beta}\left(\mathbf{r}-\xi\mathbf{R}_{\alpha\nu}+\frac{m_\beta}{m_\alpha+m_\beta}\mathbf{R}_{\alpha\beta},\mathbf{R}_{\alpha\beta},Q_\alpha,Q_\beta,t\right)d\xi\,d\mathbf{R}_{\alpha\beta}\,dQ_\alpha\,dQ_\beta \qquad (13.2)$$

In the second term a notational difficulty arises similar to that encountered in (12.7) in that the differentiation with respect to $Q_{\alpha t}$ does not apply to the $\mathbf{R}_{\alpha\nu}$ in the arguments of $[P_{\alpha u}P_{\alpha v}]^{(\alpha)}$ and ψ_{α}. Note that the argument of $\tilde{\psi}_{\alpha\beta}$ in (13.2) is not quite the same as that of $\psi_{\alpha\beta}$ in (13.1). The $\partial/\partial Q_{\alpha t}$ term can be integrated by parts, and then Eq. 4.1 can be used to simplify it. Combination of (7.4) for $\mathbf{p}^{(k)}$ and 13.1 and 13.2 for $\mathbf{p}^{(\mathrm{I})}$ and $\mathbf{p}^{(\mathrm{II})}$ then gives

$$\mathsf{p}=\mathsf{p}^{(k)}+\mathsf{p}^{(\mathrm{I})}+\mathsf{p}^{(\mathrm{II})}$$

$$\overset{(1)}{=\mathsf{p}_0^{(\mathrm{II})}+\sum_{\alpha\nu} m_{\alpha\nu}\int\left[\left(\frac{\mathbf{p}_\alpha}{m_\alpha}+\mathbf{V}_{\alpha\nu}-\mathbf{u}\right)\left(\frac{\mathbf{p}_\alpha}{m_\alpha}+\mathbf{V}_{\alpha\nu}-\mathbf{u}\right)\right]^{(\alpha)}\psi_\alpha(\mathbf{r}-\mathbf{R}_{\alpha\nu},Q_\alpha)\,dQ_\alpha}$$

$$\overset{(2)}{-\sum_{\alpha\nu} m_{\alpha\nu}\int\int_0^1[\mathbf{V}_{\alpha\nu}\mathbf{V}_{\alpha\nu}]^{(\alpha)}\psi_\alpha(\mathbf{r}-\xi\mathbf{R}_{\alpha\nu},Q_\alpha)\,d\xi\,dQ_\alpha}\;\overset{(3)}{-\sum_{\alpha\nu\mu} n_0 v_{\alpha0}\left(\delta_{\mu\nu}-\frac{m_{\alpha\nu}}{m_\alpha}\right)}$$

$$\times\int\int_0^1\mathbf{R}_{\alpha\nu}\overline{\mathbf{F}}_\mu^{(\alpha0)}\psi_\alpha(\mathbf{r}+\mathbf{R}_{\alpha\mu}-\xi\mathbf{R}_{\alpha\nu},Q_\alpha)\,d\xi\,dQ_\alpha\;\overset{(4)}{+\sum_{\alpha\nu st} m_{\alpha\nu}^{1/2}\int\int_0^1\mathbf{R}_{\alpha\nu}G_{st}\mathbf{b}_{\nu s}}$$

$$\times\left[\frac{\partial}{\partial t}\left([P_{\alpha t}]^{(\alpha)}\psi_\alpha(\mathbf{r}-\xi\mathbf{R}_{\alpha\nu},Q_\alpha)\right)\right.$$

$$\left.+\frac{1}{m_\alpha}\left(\frac{\partial}{\partial\mathbf{r}}\cdot[\mathbf{p}_\alpha P_{\alpha t}]^{(\alpha)}\psi_\alpha(\mathbf{r}-\xi\mathbf{R}_{\alpha\nu},Q_\alpha)\right)\right]d\xi\,dQ_\alpha$$

$$\overset{(5)}{-\sum_{\substack{\alpha\beta st\\ \nu\mu\eta\\ \alpha\neq0\\ \beta\neq0}}\left(\frac{m_{\alpha\nu}}{m_{\alpha\mu}}\right)^{1/2}\int\int_0^1 G_{st}\mathbf{R}_{\alpha\nu}\mathbf{b}_{\nu s}\mathbf{b}_{\mu t}\cdot\mathbf{F}_{\mu\eta}^{(\alpha\beta)}}$$

$$\times\tilde{\psi}_{\alpha\beta}\left(\mathbf{r}-\xi\mathbf{R}_{\alpha\nu}+\frac{m_\beta}{m_\alpha+m_\beta}\mathbf{R}_{\alpha\beta},\mathbf{R}_{\alpha\beta},Q_\alpha,Q_\beta,t\right)d\xi\,d\mathbf{R}_{\alpha\beta}\,dQ_\alpha\,dQ_\beta$$

(6)

$$+\sum_{\substack{\alpha\beta\nu \\ \alpha\neq 0 \\ \beta\neq 0}} \int\int_0^1 \left(\mathbf{R}_{\alpha\nu} - \frac{m_\beta}{m_\alpha + m_\beta}\mathbf{R}_{\alpha\beta}\right)\left[\frac{m_{\alpha\nu}}{m_\alpha}\mathbf{F}^{(\alpha,\beta)}\right.$$

$$\left. + \sum_{st\mu\eta}\left(\frac{m_{\alpha\nu}}{m_{\alpha\mu}}\right)^{1/2} G_{st}^{(\alpha)}\mathbf{b}_{\nu s}^{(\alpha)}\mathbf{b}_{\mu t}^{(\alpha)}\cdot\mathbf{F}_{\mu\eta}^{(\alpha\beta)}\right]$$

$$\times\tilde{\psi}_{\alpha\beta}\left(\mathbf{r} - \xi\mathbf{R}_{\alpha\nu} + \frac{\xi m_\beta}{m_\alpha + m_\beta}\mathbf{R}_{\alpha\beta}, \mathbf{R}_{\alpha\beta}, Q_\alpha, Q_\beta, t\right) d\xi\, d\mathbf{R}_{\alpha\beta}\, dQ_\alpha\, dQ_\beta$$

FS: (13.3)

Up to this point only the assumptions of short-range forces and frictional drag have been made. We now make the following simplifications:

Assumption A allows term (4), the acceleration terms, to be omitted.

Assumption H has the following consequences: Term (2) cancels the $\mathbf{V}_{\alpha\nu}\mathbf{V}_{\alpha\nu}$ part of term (1); the $m_{\alpha\nu}/m_\alpha$ part of term (3) vanishes by (2.9); term (5) cancels the $\mathbf{R}_{\alpha\nu}\mathbf{b}_{\nu s}^{(\alpha)}\mathbf{b}_{\mu t}^{(\alpha)}\cdot\mathbf{F}_{\mu\eta}^{(\alpha,\beta)}$ part of term (6); the $\mathbf{R}_{\alpha\beta}\mathbf{b}_{\nu s}^{(\alpha)}\mathbf{b}_{\mu t}^{(\alpha)}\cdot\mathbf{F}_{\mu\eta}^{(\alpha,\beta)}$ term in term (6) drops out by (2.11).

Assumption E enables the remaining portion of term (1) to be simplified to $(\Sigma_\alpha n_\alpha)kT\mathsf{U}$.

With these assumptions the momentum-flux tensor then becomes considerably simpler:

$$\mathsf{p} = \mathsf{p}_0 + \left(\sum_{\alpha\neq 0} n_\alpha\right)kT\mathsf{U} - \sum_{\alpha\nu} n_0 v_{\alpha 0}\int \mathbf{R}_{\alpha\nu}\overline{\mathbf{F}}_{\nu 0}^{(\alpha,0)}\psi_\alpha(Q_\alpha, t)\, dQ_\alpha$$

$$-\frac{1}{2}\sum_{\substack{\alpha\beta \\ \alpha\neq 0 \\ \beta\neq 0}} \int \mathbf{R}_{\alpha\beta}\mathbf{F}^{(\alpha,\beta)}\tilde{\psi}_{\alpha\beta}(\mathbf{R}_{\alpha\beta}, Q_\alpha, Q_\beta, t)\, d\mathbf{R}_{\alpha\beta}\, dQ_\alpha\, dQ_\beta \qquad \text{FSAHE: (13.4)}$$

When the momentum-flux tensor is written in this form, the connection with previous theories becomes clear. The term containing $\mathbf{R}_{\alpha\nu}\overline{\mathbf{F}}_{\nu 0}^{(\alpha,0)}$ is that obtained initially by Kramers[2] (for the *xy*-component only, and for a flexible linear chain with all beads identical) and used as the starting point by Kirkwood and his collaborators,[8] by Zimm,[12] and others; the same term, for any component, was used by Kotaka,[9] Williams,[13] Giesekus,[17] and others. The Kramers–Kirkwood–Kotaka expression has been universally adopted for dilute solution calculations, without questioning its applicability outside the range implied by the original derivation (steady state, potential flow of solutions of freely-jointed linear chain models with identical beads). Only Booij and van Wiechen[30] have expressed doubts about the Kramers expression. The term containing $\mathbf{R}_{\alpha\beta}\mathbf{F}^{(\alpha,\beta)}$ was proposed by Fixman[20] for systems containing one solute; calculations based on Fixman's polymer–polymer interaction term have been made by Williams.[22]

The derivation that we have given does then give credence to the earlier results obtained by configuration-space arguments. In addition, we can see exactly what approximations are contained in these results (i.e., Eq. 13.4), and Eq. 13.3 provides the basis for making a systematic examination of the errors involved in Approximations A, H, and E.

In using the Kramers–Kirkwood–Kotaka formula for dilute solutions, one customarily replaces the hydrodynamic drag force $\overline{\mathbf{F}}_{\nu}^{\alpha 0)}$ by the sum of the Brownian motion and connector forces with the use of an equation of motion, such as Eq. 9.7, with the acceleration terms omitted and the bracket quantities evaluated using the equilibration approximation (i.e., using Assumptions A and E).

XIV. COMMENTS ON THE SYMMETRY OF THE MOMENTUM–FLUX TENSOR

From Table IV it is evident that the momentum flux is symmetric for the two models under consideration. However, it is not at all straightforward to see that the stress tensor as given in (12.9) is symmetric for all models. Let us examine the antisymmetric part of each term in this equation; we make use of many results from Section III.

The integrand of the second term on the right-hand side can be rewritten as follows, using the notation of (3.33):

$$-\Sigma_{\nu}\zeta_{\alpha\nu}\tilde{\mathbf{R}}_{\alpha\nu}\tilde{\mathbf{R}}_{\alpha\nu}\cdot\frac{\partial}{\partial\mathbf{r}}\mathbf{u}=-\Sigma_{mnpq}K_{mn}^{(\alpha)}S_{mq}S_{np}\boldsymbol{\delta}_{p}\boldsymbol{\delta}_{q}\cdot\frac{\partial}{\partial\mathbf{r}}\mathbf{u} \tag{14.1}$$

and this has as its antisymmetric part

$$-\frac{1}{2}\Sigma_{mnpq}K_{mn}^{(\alpha)}S_{mq}S_{np}\left[\boldsymbol{\delta}_{p}\boldsymbol{\delta}_{q}\cdot\frac{\partial}{\partial\mathbf{r}}\mathbf{u}-\left(\frac{\partial}{\partial\mathbf{r}}\mathbf{u}\right)^{\dagger}\cdot\boldsymbol{\delta}_{q}\boldsymbol{\delta}_{p}\right] \tag{14.2}$$

The integrand of the third term contains the tensorial contribution

$$\Sigma_{\nu s}\zeta_{\alpha\nu}^{1/2}\,\tilde{G}_{st}^{(\alpha)}\tilde{\mathbf{R}}_{\alpha\nu}\tilde{\mathbf{b}}_{\nu s}^{(\alpha)} \tag{14.3}$$

which has as its antisymmetric part:

$$\frac{1}{2}\Sigma_{mnpqu}\Lambda_{tu}^{-1}\epsilon_{nmu}S_{mp}S_{nq}\boldsymbol{\delta}_q\boldsymbol{\delta}_p \tag{14.4}$$

To prove this Eqs. 3.29 and 3.31 are used. Then using Eqs. 3.27 and 3.32, we see that the antisymmetric part of

$$\Sigma_{\nu\mu st}(\zeta_{\alpha\nu}\zeta_{\alpha\mu})^{1/2}\tilde{G}_{st}^{(\alpha)}\tilde{\mathbf{R}}_{\alpha\nu}\tilde{\mathbf{b}}_{\nu s}^{(\alpha)}\tilde{\mathbf{b}}_{\mu t}^{(\alpha)}\tilde{\mathbf{R}}_{\alpha\mu}:\frac{\partial}{\partial \mathbf{r}}\mathbf{u} \tag{14.5}$$

is

$$\frac{1}{2}\Sigma_{mnpqr}K_{mn}^{(\alpha)}S_{mq}S_{nr}(\boldsymbol{\delta}_q\boldsymbol{\delta}_p-\boldsymbol{\delta}_p\boldsymbol{\delta}_q)\left(\boldsymbol{\delta}_p\boldsymbol{\delta}_r:\frac{\partial}{\partial \mathbf{r}}\mathbf{u}\right)$$

$$=\frac{1}{2}\Sigma_{mnpq}K_{mn}^{(\alpha)}S_{mp}S_{nq}\left[\boldsymbol{\delta}_q\boldsymbol{\delta}_p\cdot\frac{\partial}{\partial \mathbf{r}}\mathbf{u}-\left(\frac{\partial}{\partial \mathbf{r}}\mathbf{u}\right)^{\dagger}\cdot\boldsymbol{\delta}_q\boldsymbol{\delta}_p\right] \tag{14.6}$$

This exactly cancels the antisymmetric part of the second term as given by (14.2).

The fourth term of (12.9) may be integrated by parts. Then using Eqs. 14.3 and 14.4 we may obtain the antisymmetric part of this term as

$$\frac{1}{2}kT\sum_{\substack{\alpha t\\ mnpqu}}\epsilon_{umn}\boldsymbol{\delta}_p\boldsymbol{\delta}_q\int g_\alpha^{-1/2}\psi_\alpha\left(\frac{\partial}{\partial Q_{\alpha t}}g_\alpha^{1/2}\Lambda_{tu}^{-1}S_{mp}S_{nq}\right)dQ_\alpha \tag{14.7}$$

Two cases have to be considered: For $t \geqslant 4$, Λ_{tu}^{-1} is zero since u is restricted here to values of three or less; for $t \leqslant 3$, the derivative with respect to $Q_{\alpha t}$ is regarded as the derivative of a product of $g_\alpha^{1/2}\Lambda_{tu}^{-1}$ and $S_{mp}S_{nq}$. The sum of derivatives $\Sigma_t\, \partial[g_\alpha^{1/2}\Lambda_{tu}^{-1}]/\partial Q_{\alpha t}$ can be shown to be zero using Eqs. 3.14 and 3.21. Then $\Sigma_{umnt}\epsilon_{umn}\Lambda_{tu}^{-1}\,\partial[S_{mp}S_{nq}]/\partial Q_{\alpha t}$ can be shown to be zero using Eq. 3.12 and the orthogonality relations for the S_{mn} mentioned after (3.2).

Finally, we come to the fifth term in (12.9), which involves the forces. In order to examine the symmetry properties of this term, we first rewrite the forces thus:

$$\mathbf{F}_{\mu\eta}^{(\alpha)}=\left(\frac{F_{\mu\eta}^{(\alpha)}}{R_{\mu\eta}^{(\alpha)}}\right)(\mathbf{R}_{\alpha\eta}-\mathbf{R}_{\alpha\mu}) \tag{14.8}$$

Here $F_{\mu\eta}^{(\alpha)}$ is the magnitude of the force and

$$R_{\mu\eta}^{(\alpha)}=|\mathbf{R}_{\alpha\eta}-\mathbf{R}_{\alpha\mu}|=\left[\Sigma_n(R_{\alpha\eta n}-R_{\alpha\mu n})^2\right]^{1/2} \tag{14.9}$$

is the distance between the mass points. One then finds using Eqs. 2.6, 3.19, and 4.14 that

$$\Sigma_{\mu\eta}\zeta_{\alpha\mu}^{-1/2}\tilde{\mathbf{b}}_{\mu t}^{(\alpha)}\cdot\mathbf{F}_{\mu\eta}^{(\alpha)}=\Sigma_{\mu\eta}m_{\alpha\mu}^{-1/2}\tilde{\mathbf{b}}_{\mu t}^{(\alpha)}\cdot\mathbf{F}_{\mu\eta}^{(\alpha)}$$

$$=\Sigma_{\mu\eta sm}m_{\alpha\mu}^{-1/2}\left(\frac{F_{\mu\eta}^{(\alpha)}}{R_{\mu\eta}^{(\alpha)}}\right)\Lambda_{st}b_{(\alpha)}^{\mu sm}(R_{\alpha\eta m}-R_{\alpha\mu m}) \quad (14.10)$$

Then use of (3.17) gives

$$\Sigma_{\mu\eta}\zeta_{\alpha\mu}^{-1/2}\,\tilde{\mathbf{b}}_{\mu t}^{(\alpha)}\cdot\mathbf{F}_{\mu\eta}^{(\alpha)}=0 \qquad (t\leqslant 3) \quad (14.11)$$

$$\Sigma_{\mu\eta}\zeta_{\alpha\mu}^{-1/2}\,\tilde{\mathbf{b}}_{\mu t}^{(\alpha)}\cdot\mathbf{F}_{\mu\eta}^{(\alpha)}=\Sigma_{\mu\eta m}\left(\frac{F_{\mu\eta}^{(\alpha)}}{R_{\mu\eta}^{(\alpha)}}\right)(R_{\alpha\eta m}-R_{\alpha\mu m})\left(\frac{\partial R_{\alpha\mu m}}{\partial Q_{\alpha t}}\right)$$

$$=-\frac{1}{2}\Sigma_{\mu\eta}\mathbf{F}_{\mu\eta}^{(\alpha)}\cdot\left(\frac{\partial\mathbf{R}_{\mu\eta}^{(\alpha)}}{\partial Q_{\alpha t}}\right)$$

$$=\frac{\partial\phi_\alpha}{\partial Q_{\alpha t}} \qquad (t\geqslant 4) \quad (14.12)$$

where ϕ_α is the quantity defined just after (10.2). These results are reasonable since $\mathbf{b}_{\nu t}^{(\alpha)}\cdot\mathbf{F}_{\nu\mu}^{(\alpha)}$ represents the projection of an interparticle force (e.g., a spring force) in the tth direction; clearly there would be no components in the directions of the Euler-angle coordinates since the spring forces are independent of the orientation in space.

Now when we use the result in (14.12) along with the information from (14.3) and (14.4), it can be seen that the antisymmetric part of the second term in (12.9) is zero. We have thus proven that the antisymmetric part of p vanishes, when the approximations inherent in (12.9) are made.

XV. AN ALTERNATE FORM FOR THE MOMENTUM-FLUX TENSOR: A GENERALIZATION OF THE GIESEKUS EQUATION

Once one knows that the momentum-flux tensor is symmetric, it is possible to rearrange Eq. 12.9 with the help of the diffusion equation (Eq. 11.14) into a much simpler form, suggested by Giesekus.[17]

We multiply Eq. 11.14 by $\Sigma_\eta\zeta_{\alpha\eta}\tilde{\mathbf{R}}_{\alpha\eta}\tilde{\mathbf{R}}_{\alpha\eta}$ and integrate over all the configuration space associated with species α; the various terms may then be rearranged as follows:

$$(a)\quad \int \Sigma_\eta \zeta_{\alpha\eta} \tilde{\mathbf{R}}_{\alpha\eta} \tilde{\mathbf{R}}_{\alpha\eta} \frac{\partial}{\partial t} \psi_\alpha \, dQ_\alpha = \frac{d}{dt} \Sigma_\eta \int \zeta_{\alpha\eta} \tilde{\mathbf{R}}_{\alpha\eta} \tilde{\mathbf{R}}_{\alpha\eta} \psi_\alpha \, dQ_\alpha \tag{15.1}$$

$$(b)\quad \int \Sigma_\eta \zeta_{\alpha\eta} \tilde{\mathbf{R}}_{\alpha\eta} \tilde{\mathbf{R}}_{\alpha\eta} \Sigma_{st} \frac{\partial}{\partial Q_{\alpha s}} \left(\tilde{G}_{st}^{(\alpha)} \psi_\alpha \Sigma_\nu \zeta_{\alpha\nu}^{1/2} \tilde{\mathbf{b}}_{\nu t}^{(\alpha)} \tilde{\mathbf{R}}_{\alpha\nu} : \frac{\partial}{\partial \mathbf{r}} \mathbf{u} \right) dQ_\alpha$$

$$= -\Sigma_{\nu\eta st} (\zeta_{\alpha\eta} \zeta_{\alpha\nu})^{1/2} \int \tilde{G}_{st}^{(\alpha)} \psi_\alpha (\tilde{\mathbf{b}}_{\eta s}^{(\alpha)} \tilde{\mathbf{R}}_{\alpha\eta} + \tilde{\mathbf{R}}_{\alpha\eta} \tilde{\mathbf{b}}_{\eta s}^{(\alpha)}) \times \tilde{\mathbf{b}}_{\nu t}^{(\alpha)} \tilde{\mathbf{R}}_{\alpha\nu} \, dQ_\alpha : \frac{\partial}{\partial \mathbf{r}} \mathbf{u} \tag{15.2}$$

$$(c)\quad \int \Sigma_\eta \zeta_{\alpha\eta} \tilde{\mathbf{R}}_{\alpha\eta} \tilde{\mathbf{R}}_{\alpha\eta} \sum_{st} \frac{\partial}{\partial Q_{\alpha s}} \left(\tilde{G}_{st}^{(\alpha)} \psi_\alpha m_{\alpha\nu}^{-1/2} \mathbf{b}_{\nu t}^{(\alpha)} \cdot \mathbf{F}_{\nu\mu}^{(\alpha)} \right) dQ_\alpha$$

$$= -\sum_{\substack{\eta\nu\mu \\ st}} \left(\frac{\zeta_{\alpha\eta}}{\zeta_{\alpha\nu}} \right)^{1/2} \int \tilde{G}_{st}^{(\alpha)} \psi_\alpha (\tilde{\mathbf{b}}_{\eta s}^{(\alpha)} \tilde{\mathbf{R}}_{\alpha\eta} + \tilde{\mathbf{R}}_{\alpha\eta} \tilde{\mathbf{b}}_{\eta s}^{(\alpha)}) \tilde{\mathbf{b}}_{\nu t}^{(\alpha)} \cdot \mathbf{F}_{\nu\mu}^{(\alpha)} \, dQ_\alpha \tag{15.3}$$

$$(d)\quad -kT \int \Sigma_\eta \zeta_{\alpha\eta} \tilde{\mathbf{R}}_{\alpha\eta} \tilde{\mathbf{R}}_{\alpha\eta} \Sigma_{st} \frac{\partial}{\partial Q_{\alpha s}} \left[\tilde{G}_{st}^{(\alpha)} g_\alpha^{1/2} \frac{\partial}{\partial Q_{\alpha t}} \left(g_\alpha^{-1/2} \psi_\alpha \right) \right] dQ_\alpha$$

$$= kT \sum_{\eta st} \zeta_{\alpha\nu}^{1/2} \int g_\alpha^{1/2} \tilde{G}_{st}^{(\alpha)} \left(\tilde{\mathbf{b}}_{\eta s}^{(\alpha)} \tilde{\mathbf{R}}_{\alpha\eta} + \tilde{\mathbf{R}}_{\alpha\eta} \tilde{\mathbf{b}}_{\eta s}^{(\alpha)} \right) \times \frac{\partial}{\partial Q_{\alpha t}} \left(g_\alpha^{-1/2} \psi_\alpha \right) dQ_\alpha \tag{15.4}$$

Next these four terms are summed on α and divided by 2. When this has been done, (*b*) is equal to the negative of term (3) (symmetrized) in (12.9); similarly (*c*) corresponds to term (5), and (*d*) corresponds to term (4). Therefore when the previous four results are combined with the symmetric part of 12.9, keeping in mind that the antisymmetric part of p is zero, we get

$$\mathsf{p} = \mathsf{p}_0 + \left(\sum_{\alpha \neq 0} n_\alpha \right) kT \mathsf{U} + \frac{1}{2} \sum_{\alpha\eta} \zeta_{\alpha\eta} \left[\frac{d}{dt} \int \tilde{\mathbf{R}}_{\alpha\eta} \tilde{\mathbf{R}}_{\alpha\eta} \psi_\alpha \, dQ_\alpha - \int \tilde{\mathbf{R}}_{\alpha\eta} \tilde{\mathbf{R}}_{\alpha\eta} \psi_\alpha \, dQ_\alpha \cdot \frac{\partial}{\partial \mathbf{r}} \mathbf{u} - \left(\frac{\partial}{\partial \mathbf{r}} \mathbf{u} \right)^\dagger \cdot \int \tilde{\mathbf{R}}_{\alpha\eta} \tilde{\mathbf{R}}_{\alpha\eta} \psi_\alpha \, dQ_\alpha \right]$$

FSDHAEI : (15.5)

or

$$\mathsf{p} = p_0 + \left(\sum_{\alpha \neq 0} n_\alpha \right) kT \mathsf{U} + \frac{1}{2} \Sigma_{\alpha\eta} \zeta_{\alpha\eta} \frac{\mathfrak{d}}{\mathfrak{d} t} \int \tilde{\mathbf{R}}_{\alpha\eta} \tilde{\mathbf{R}}_{\alpha\eta} \psi_\alpha \, dQ_\alpha$$

FSDHAEI : (15.6)

in which $\mathfrak{d}/\mathfrak{d}t$ is the Oldroyd convected derivative[18] of continuum mechanics; it represents the co-deformational derivative of a contravariant convected second-order tensor:

$$\frac{\mathfrak{d}}{\mathfrak{d}t}\underline{\Lambda}=\frac{\partial}{\partial t}\underline{\Lambda}+\mathbf{u}\cdot\frac{\partial}{\partial \mathbf{r}}\underline{\Lambda}-\underline{\Lambda}\cdot\frac{\partial}{\partial \mathbf{r}}\mathbf{u}-\left(\frac{\partial}{\partial \mathbf{r}}\mathbf{u}\right)^{\dagger}\cdot\underline{\Lambda} \tag{15.7}$$

In (15.5) the $\mathbf{u}\cdot(\partial/\partial\mathbf{r})\underline{\Lambda}$ term does not appear because of the assumption of homogeneous flow. An equation similar to (15.6) was first derived by Giesekus[18] for a linear bead–spring model with identical beads; however Eq. 15.6 can be applied to any kind of bead–spring–rod model with any kind of connectivity and any degree of complexity. Equation 15.6 is often easier to use than Eq. 12.9. Equation 15.6 for the momentum-flux tensor and Eq. 11.13 for the distribution function provide the basic kinetic theory equations used recently by Hassager[7,31] in calculations of rheological properties of dilute solutions of flexible macromolecules.

In Section XVI we use Eq. 15.6 to obtain expressions for the zero-shear-rate viscosity and normal stress coefficient. Then in Section XVII we use Eq. 12.9 to get an expression for the infinite-frequency dynamic viscosity.

XVI. THE SECOND-ORDER FLUID CONSTANTS FOR ARBITRARY BEAD–SPRING–ROD MACROMOLECULAR MODELS

From continuum mechanics of incompressible isotropic fluids, it is known that the momentum-flux tensor p can be expanded in a "retarded motion expansion" that is valid for flows slowly varying in time[32]:

$$\mathsf{p}=\mathsf{p}_0+b_0\mathsf{U}-b_1\underline{\gamma}_{(1)}+b_2\underline{\gamma}_{(2)}-b_{11}\underline{\gamma}_{(1)}\cdot\underline{\gamma}_{(1)}+\cdots \tag{I:(16.1)}$$

in which the b's are constants and the $\underline{\gamma}_{(n)}$ are the nth rate of strain tensors of Oldroyd[18,33] defined by

$$\underline{\gamma}_{(1)}\equiv\dot{\underline{\gamma}}=\left(\frac{\partial}{\partial\mathbf{r}}\right)\mathbf{u}+\left(\left(\frac{\partial}{\partial\mathbf{r}}\right)\mathbf{u}\right)^{\dagger} \tag{16.2}$$

$$\underline{\gamma}_{(n)}=\left(\frac{\mathfrak{d}}{\mathfrak{d}t}\right)\underline{\gamma}_{(n-1)} \tag{16.3}$$

in which the $\mathfrak{d}/\mathfrak{d}t$-operator is that defined in (15.7). The three b's are related to the zero-shear-rate limit of the viscometric functions η, Ψ_1, Ψ_2

measured in a steady-state shearing flow $u_1(x_2)=\kappa x_2$:

$$p_{21}-p_{0,21}=-\eta(0)\kappa$$
$$p_{11}-p_{22}=-\Psi_1(0)\kappa^2 \qquad (16.4)$$
$$p_{22}-p_{33}=-\Psi_2(0)\kappa^2$$

and $b_1=\eta(0)$, $b_2=\frac{1}{2}\Psi_1(0)$, $b_{11}=\Psi_2(0)$. The constants b_1 and b_2 are also related to quantities G' and G'', measured in linear viscoelastic measurements in small-amplitude oscillatory shearing motion, and the solvent viscosity η_0:

$$b_1=\lim_{\omega\to 0}\left(\frac{G''}{\omega}-\eta_0\right) \qquad (16.5)$$

$$b_2=\lim_{\omega\to 0}\frac{G'}{\omega^2} \qquad (16.6)$$

where $G'(\omega)\equiv\omega\eta''$ and $G''(\omega)\equiv\omega\eta'$ are the frequency-dependent "storage modulus" and "loss modulus," respectively.[16] Note also that

$$\frac{nkTb_2}{b_1^2}=\frac{J_e^0 RT}{M[\eta]^2 c}\equiv j_{eR}^0 \qquad (16.7)$$

is a dimensionless quantity known experimentally to lie between 0.2 and 0.4 for many solutions; here c is concentration in grams per cubic centimeter and J_e^0 is the "steady-state compliance"[16,34,35] used by polymer chemists. We now want to obtain expressions for the constants b_1 and b_2 in terms of the macromolecular model structure.

First, we show that Eq. 15.6, the generalization of the Giesekus equation, gives $b_{11}=0$ for all models. This is done by comparing Eq. 15.6 with (16.1) for steady-state homogeneous flows, and noting that $\int\Sigma_{\alpha\nu}\zeta_{\alpha\nu}\tilde{\mathbf{R}}_{\alpha\nu}\tilde{\mathbf{R}}_{\alpha\nu}\psi_\alpha\, dQ_\alpha$ must have the form $c_0\mathbf{U}+c_1\dot{\underline{\gamma}}+\cdots$ for small velocity gradients where c_0 and c_1 are independent of $(\partial/\partial\mathbf{r})\mathbf{u}$. Then from (15.6)

$$\mathsf{p}-\mathsf{p}_0=\frac{1}{2}\left[-c_0\,\dot{\underline{\gamma}}+c_1\left[\frac{1}{2}\left(\underline{\omega}\cdot\dot{\underline{\gamma}}-\dot{\underline{\gamma}}\cdot\underline{\omega}\right)-\dot{\underline{\gamma}}^2\right]\right]+\cdots \qquad \text{FSDHAEI}: (16.8)$$

Steady State

and from (16.1)

$$\mathsf{p}-\mathsf{p}_0=-b_1\,\dot{\underline{\gamma}}+b_2\left[\frac{1}{2}\left(\underline{\omega}\cdot\dot{\underline{\gamma}}-\dot{\underline{\gamma}}\cdot\underline{\omega}\right)-\dot{\underline{\gamma}}^2\right]-b_{11}\,\dot{\underline{\gamma}}^2+\cdots \qquad \text{HI}: (16.9)$$

Steady State

Here $\underline{\omega}$ is the vorticity tensor given by $\underline{\omega}=(\partial/\partial\mathbf{r})\mathbf{u}-((\partial/\partial\mathbf{r})\mathbf{u})^{\dagger}$. Hence $b_{11}=0$ and $\psi_2(0)=0$; that is, the fluid obeys the "Weissenberg hypothesis." This was pointed out earlier by Giesekus.[17]

Next we show that it is possible to get b_1 and b_2 by considering steady-state, homogeneous, potential flows only.[7] For this special category of flows with $\mathbf{v}=[\underline{\kappa}\cdot\mathbf{r}]$ and $\underline{\kappa}=\underline{\kappa}^{\dagger}$, from (15.6)

$$\mathsf{p}-\mathsf{p}_0=-\frac{1}{2}\underline{\kappa}\cdot\int\Sigma_{\alpha\nu}\zeta_{\alpha\nu}\tilde{\mathbf{R}}_{\alpha\nu}\tilde{\mathbf{R}}_{\alpha\nu}\psi_\alpha\,dQ_\alpha$$

$$-\frac{1}{2}\int\Sigma_{\alpha\nu}\zeta_{\alpha\nu}\tilde{\mathbf{R}}_{\alpha\nu}\tilde{\mathbf{R}}_{\alpha\nu}\psi_\alpha\,dQ_\alpha\cdot\underline{\kappa} \qquad \text{FSDHAEI: (16.10)}$$

Steady-State Irrotational

and from (16.1)

$$\mathsf{p}-\mathsf{p}_0=-2b_1\underline{\kappa}-4b_2\underline{\kappa}\cdot\underline{\kappa} \qquad \text{HI: (16.11)}$$

Steady-State Irrotational

We consider a single solute species and hence omit the index α in the remainder of this section. We expand Eq. 16.10 using the potential flow solution for ψ given in (12.1) and take the first two terms; then matching the result with (16.11) gives b_1 and b_2. Once b_1 and b_2 are known, then Eq. 16.1 can be used for solving a somewhat wider class of problems.

According to (11.15), for steady-state, homogeneous, potential flow,

$$\psi=n\frac{g^{1/2}\exp\left[(1/2kT)(\underline{\kappa}:\Sigma_\nu\zeta_\nu\tilde{\mathbf{R}}_\nu\tilde{\mathbf{R}}_\nu)-(\phi/kT)\right]}{\int g^{1/2}\exp\left[(1/2kT)(\underline{\kappa}:\Sigma_\nu\zeta_\nu\tilde{\mathbf{R}}_\nu\tilde{\mathbf{R}}_\nu)-(\phi/kT)\right]d\alpha\,d\beta\,d\gamma\,dQ_4\cdots}$$

FSDHAEI: (16.12)

Steady-State Irrotational

in which α, β, γ are the Euler angles and Q_4, $Q_5,\ldots$ are the remaining internal coordinates that describe stretching and bending motions. From (3.14) and (3.21)

$$g=\det[g_{st}]=\sin^2\beta\cdot\det[\Gamma_{uv}]=\Gamma\sin^2\beta \qquad (16.13)$$

where Γ depends on Q_4, $Q_5,\ldots$ but not on α, β, γ. The part of the exponential involving $\underline{\kappa}$ can be expanded in a Taylor series and two terms are retained. In this way, we find, using Eqs. 3.34 and 3.35:

$$\int \sum_{\nu} \zeta_\nu \tilde{\mathbf{R}}_\nu \tilde{\mathbf{R}}_\nu \psi \, dQ = \int \sum_{\nu mn} \zeta_\nu \tilde{R}_{\nu m} \tilde{R}_{\nu n} \check{\boldsymbol{\delta}}_m \check{\boldsymbol{\delta}}_n \psi \, dQ = \frac{n(\mathsf{I}_{n1} + \mathsf{I}_{n2})}{I_{d1} + I_{d2}} \tag{16.14}$$

where

$$\begin{aligned} I_{d1} &= \int \sin\beta \, d\alpha \, d\beta \, d\gamma \int \Gamma^{1/2} \exp\left(\frac{-\phi}{kT}\right) dQ_4 \cdots \\ &= 8\pi^2 \int \Gamma^{1/2} \exp\left(\frac{-\phi}{kT}\right) dQ_4 \cdots \end{aligned} \tag{16.15}$$

$$\begin{aligned} I_{d2} &= \left(\frac{1}{2kT}\right) \underline{\kappa} : \Sigma_{\nu mn} \zeta_\nu \int \check{\boldsymbol{\delta}}_m \check{\boldsymbol{\delta}}_n \sin\beta \, d\alpha \, d\beta \, d\gamma \int \tilde{R}_{\nu m} \tilde{R}_{\nu n} \Gamma^{1/2} \\ &\quad \times \exp\left(\frac{-\phi}{kT}\right) dQ_4 \cdots \\ &= 0 \qquad (\text{since } \underline{\kappa} : \mathsf{U} = \operatorname{tr} \underline{\kappa} = 0) \end{aligned} \tag{16.16}$$

$$\begin{aligned} \mathsf{I}_{n1} &= \sum_{\nu mn} \zeta_\nu \int \check{\boldsymbol{\delta}}_m \check{\boldsymbol{\delta}}_n \sin\beta \, d\alpha \, d\beta \, d\gamma \int \tilde{R}_{\nu m} \tilde{R}_{\nu n} \Gamma^{1/2} \exp\left(\frac{-\phi}{kT}\right) dQ_4 \cdots \\ &= \left(\frac{8\pi^2}{3}\right) \mathsf{U} \Sigma_{\nu m} \zeta_\nu \int \tilde{R}_{\nu m}^2 \, \Gamma^{1/2} \exp\left(\frac{-\phi}{kT}\right) dQ_4 \cdots \end{aligned} \tag{16.17}$$

$$\begin{aligned} \mathsf{I}_{n2} &= \left(\frac{1}{2kT}\right) \underline{\kappa} : \Sigma_{\nu\mu mnpq} \zeta_\nu \zeta_\mu \int \check{\boldsymbol{\delta}}_m \check{\boldsymbol{\delta}}_n \check{\boldsymbol{\delta}}_p \check{\boldsymbol{\delta}}_q \sin\beta \, d\alpha \, d\beta \, d\gamma \\ &\quad \times \int \tilde{R}_{\nu m} \tilde{R}_{\nu n} \tilde{R}_{\mu p} \tilde{R}_{\mu q} \Gamma^{1/2} \exp\left(\frac{-\phi}{kT}\right) dQ_4 \cdots \\ &= \left(\frac{8\pi^2}{30}\right) \left(\frac{1}{kT}\right) \underline{\kappa} \Sigma_{\nu\mu mn} \zeta_\nu \zeta_\mu \int \left(\tilde{R}_{\nu m} \tilde{R}_{\nu n} \tilde{R}_{\mu m} \tilde{R}_{\mu n} - \tilde{R}_{\nu m}^2 \, \tilde{R}_{\mu n}^2\right) \\ &\quad \times \Gamma^{1/2} \exp\left(\frac{-\phi}{kT}\right) dQ_4 \cdots \end{aligned} \tag{16.18}$$

When all these results are substituted into (16.10) and the results compared with (16.11), we get for the constants b_1 and b_2 the following rather simple expressions:

$$b_1 = \frac{n}{6}\{\Sigma_p K_{pp}\} \qquad \text{FSDHAEI}:(16.19)$$

$$b_2 = \frac{n}{40kT}\left\{\Sigma_{pq} K_{pq}^2 - \frac{1}{3}(\Sigma_p K_{pp})^2\right\}$$

$$= \frac{n}{40kT}\left\{\Sigma_{pq}\left(K_{pq} - \frac{1}{3}\delta_{pq}\Sigma_r K_{rr}\right)^2\right\} \qquad \text{FSDHAEI}:(16.20)$$

where n is the number density of the solute molecules, and the $K_{pq} = \Sigma_\nu \zeta_\nu \tilde{R}_{\nu p} \tilde{R}_{\nu q}$ are the quantities defined in (3.33). The braces { } indicate weighted averages over the internal configuration space:

$$\{X\} = \frac{\int \Gamma^{1/2} \exp(-\phi/kT) X \, dQ_4 \cdots}{\int \Gamma^{1/2} \exp(-\phi/kT) \, dQ_4 \cdots} \qquad (16.21)$$

For rigid bodies these averages do not have to be performed. The Γ_{uv} needed to calculate $\Gamma = \det[\Gamma_{uv}]$ are defined by (3.23) and (3.17). It is clear from (16.20) that b_2 is always nonnegative.

Some sample values of b_1 and b_2 are given in Table V. Several comments should be made regarding this table:

(1) In general b_2 is much more sensitive to structure than b_1; usually the normal stresses are higher for the more flexible structures.

(2) Note that a rigid square, a rigid rhombus, a freely jointed rhombus, and an elastic rhombus (all with side length L) have the same viscosity, whereas the normal stresses vary widely.

(3) A rod with three beads and length $2L$ has $b_1 = (1/3)n\zeta L^2$, whereas the freely jointed system with the same overall length has $b_1 = (2/9)n\zeta L^2$; this shows the influence of flexibility on viscosity.

(4) Note that the rigid tridumbbell has a zero normal stress coefficient; this is characteristic of spherically symmetric systems.

(5) The last three models show how the viscosity and normal stress coefficients increase as the structure becomes more lopsided.

(6) Values of the dimensionless group $nkTb_2/b_1^2$ for the models listed in the table vary from 0 to 1. Note that this group does not contain the model parameters ζ or L.

TABLE V.
Values of b_1 and b_2 for Several Models

Model	Meaning of L	$\frac{b_1}{n\zeta L^2}$	$\frac{b_2 kT}{n\zeta^2 L^4}$	$\frac{nkTb_2}{b_1^2}$
Rigid dumbbell[6] with two beads connected by a rod	Length of rod	$\frac{1}{12}$	$\frac{1}{240}$	0.60
Lopsided rigid dumbbell[37] with two dissimilar beads connected by a rod; $(1/\zeta)=(1/\zeta_1)+(1/\zeta_2)$	Length of rod	$\frac{1}{12}$	$\frac{1}{240}$	0.60
Elastic dumbbell[6] with two beads connected by a linear spring with spring constant H_0	$\left(\frac{kT}{H_0}\right)^{1/2}$	$\frac{1}{4}$	$\frac{1}{16}$	1
Kramers' pearl necklace[2,7] with three beads and two rods, freely jointed	Length of each rods	$\frac{2}{9}$	$\frac{1}{1080}\frac{80\pi+3\sqrt{3}}{2\pi+3\sqrt{3}}$	0.42
Three beads and two "Fraenkel springs"[6] in limit that springs are infinitely stiff[7]	Length of infinitely stiff springs	$\frac{2}{9}$	$\frac{17}{810}$	0.43

TABLE V. (*Continued*)

Model	Meaning of L	$\frac{b_1}{n\zeta L^2}$	$\frac{b_2 kT}{n\zeta^2 L^4}$	$\frac{nkTb_2}{b_1^2}$
Three-bead, two-rod model with potential $\phi=\frac{1}{2}K(\chi-\pi)^2$ where χ is included angle[7]	Length of each rod	$\frac{1}{3}\left(1-\frac{kT}{3K}+\cdots\right)$	$\frac{1}{15}\left(1-\frac{7}{6}\frac{kT}{K}+\cdots\right)$	$0.60\left(1-\frac{kT}{2K}+\cdots\right)$
Rouse model[14] of N beads joined linearly by $N-1$ linear springs each with spring constant H_0	$\left(\frac{kT}{H_0}\right)^{1/2}$	$\frac{(N+1)(N-1)}{12}$	$\frac{(N+1)(N-1)(2N^2+7)}{720}$	$\frac{(2N^2+7)}{5(N+1)(N-1)}$ $=0.4\left(1+\frac{9}{2N^2}+\cdots\right)$
Two beads joined by a string[38]	Length of string	$\frac{3}{5}$	$\frac{9}{35}$	0.71
Rigid plane polygon[36] of N beads connected by N rods of length b	Radius of circumscribed circle $L=\frac{b}{2\sin(\pi/N)}$	$\frac{N}{6}$	$\frac{N^2}{240}$	0.15
Four beads joined with rods to form a rigid plane rhombus with included angle θ	Side of rhombus	$\frac{1}{3}$	$\frac{1}{15}\left(1-\frac{3}{4}\sin^2\theta\right)$	$0.60\left(1-\frac{3}{4}\sin^2\theta\right)$

Four beads joined with rods to form a freely jointed plane rhombus	Side of rhombus	$\frac{1}{3}$	$\frac{1}{30}$	0.30
Four beads joined with rods to form a freely jointed rhombus; one opposite pair of beads is joined by a Hookean spring with spring constant H_0 and equilibrium position $L/\sqrt{2}$	Side of rhombus	$\frac{1}{3}$	$\frac{1}{60}\left[1+\left(24I_5+60\sqrt{2}\,I_4+96I_3+24\sqrt{2}\,I_2\right)\left(2I_1+\sqrt{2}\,I_0\right)^{-1}\right]$ where $I_n=\int_{-1/\sqrt{2}}^{1-1/\sqrt{2}} u^n \exp(-a^2u^2)\,du$ and $a^2=\frac{2H_0L^2}{kT}$	$0.30\left[1-\frac{2\sqrt{2}}{105}a^2+\cdots\right]$
Rigid tridumbbell ["Doppelkreuzhantel"[39]], formed by joining three identical rigid dumbbells at their midpoints in a configuration	Length of one dumbbell rod	$\frac{1}{4}$	0	0
	Length of rod between beads	$\frac{7}{3}$	$\frac{16}{15}$	0.20
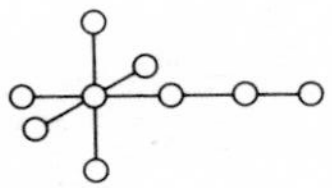	Length of rod between beads	$\frac{73}{27}$	$\frac{2116}{1215}$	0.24
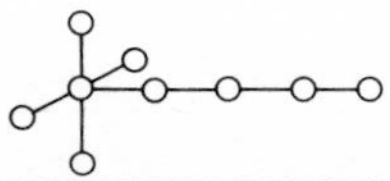	Length of rod between bead	$\frac{103}{27}$	$\frac{5776}{1215}$	0.34

(7) The result we get for b_2 for the rigid plane polygon does not agree with that given by Paul and Mazo[36] (in the "free-draining" limit). The reason for this disagreement is not known.

(8) Note that there is a difference in the value of b_2 for the three-bead–two-rod "pearl necklace" model and that for the corresponding three-bead–two-spring model, when in the latter the springs are allowed to stiffen into rods. Although the difference is small (only about 1.5%) the calculation does show that "freezing out" degrees of freedom is not always identical to the use of constraints.[7]

(9) The three-bead–two-rod model with a bending potential is interesting since it shows that the viscosity and normal stress coefficient both decrease as one goes from a stiff three-bead rod to a "bendable" rod.

Examples

In order to illustrate the use of (16.19) and (16.20), we compute several of the entries in Table V; we discuss one rigid model and one flexible model.

The "Doppelkreuzhantel" (Rigid Tridumbbell)

This model, shown in Fig. 5, consists of six beads whose coordinates are given by

$$\mathbf{R}_{\pm\nu} = \pm\frac{1}{2}L_\nu\check{\boldsymbol{\delta}}_\nu \tag{16.22}$$

where the L_ν are the lengths of the three individual dumbbells, and the $\check{\boldsymbol{\delta}}_\nu$ are the orthogonal unit vectors embedded in the rigid model. The components of the $\mathbf{R}_\nu$ are

$$R_{\pm\nu,n} = \pm\frac{1}{2}L_n\delta_{\nu n} \tag{16.23}$$

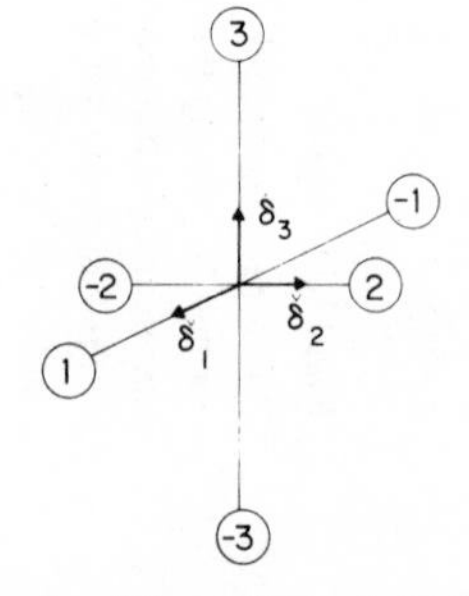

Fig. 5. "Doppelkreuzhantel."

The K-matrix can then be obtained from (3.32):

$$K_{mn}=\frac{1}{2}\zeta L_n^2\delta_{mn} \tag{16.24}$$

so that

$$b_1=\frac{n\zeta}{12}\sum_{\nu=1}^{3}L_\nu^2 \tag{16.25}$$

$$b_2=\frac{n\zeta^2}{160kT}\left[\sum_{\nu=1}^{3}L_\nu^4-\frac{1}{3}\left(\sum_{\nu=1}^{3}L_\nu^2\right)^2\right] \tag{16.26}$$

and when all the L_ν are the same, $b_2=0$.

The Elastic Rhombus

This model was shown in Fig. 3, and the bead locations are given in (3.60) in terms of the internal coordinate σ. The K-matrix has elements:

$$K_{11}=2\zeta L^2\cos^2\sigma \qquad K_{22}=2\zeta L^2\sin^2\sigma$$
$$\text{All other } K_{mn} \text{ are zero} \tag{16.27}$$

and the determinant of the Γ_{uv} from (3.63) is

$$\det[\Gamma_{uv}]=8m^2L^8\sin^2 2\sigma \tag{16.28}$$

The potential energy of the system is

$$\phi=\frac{1}{2}H_0(2L\cos\sigma-\sqrt{2}\,L)^2 \tag{16.29}$$

Then b_1 is found using Eq. 16.21 for the { }-operation:

$$b_1=\frac{n\zeta}{6}\{2L^2\}=\frac{n\zeta L^2}{3} \tag{16.30}$$

which is the same as b_1 for the "limp" (i.e., flexible but with no spring) rhombus or the rigid rhombus. Finally, b_2 is calculated thus:

$$b_2=\frac{n\zeta^2}{40kT}\left[\int_0^{\pi/2}\sin 2\sigma\quad\exp\left(-\frac{H_0(2L\cos\sigma-\sqrt{2}\,L)^2}{2kT}\right)\right.$$

$$\times\left[4L^4(\cos^4\sigma+\sin^4\sigma)-\frac{1}{3}(2L^2)^2\right]d\sigma\Bigg]$$

$$\times\left[\int_0^{\pi/2}\sin 2\sigma\quad\exp\left(-\frac{H_0(2L\cos\sigma-\sqrt{2}\,L)^2}{2kT}\right)d\sigma\right]^{-1}$$

$$=\frac{1}{30}\frac{n\zeta^2L^4}{kT}\left[1-\frac{4\sqrt{2}}{105}\frac{H_0L^2}{kT}+\cdots\right] \tag{16.31}$$

The factor in front is just b_2 for a limp rhombus, and the quantity in brackets is the "correction factor" for a weak Hookean spring.

XVII. THE HIGH-FREQUENCY DYNAMIC VISCOSITY OF ARBITRARY BEAD–SPRING–ROD MACROMOLECULAR MODELS

Consider a small amplitude oscillatory homogeneous flow of an incompressible fluid given by the velocity gradient,

$$\frac{\partial}{\partial \mathbf{r}}\mathbf{u}=\underline{\kappa}_0^\dagger\,\mathscr{R}e\,\{\exp(i\omega t)\} \qquad \text{HI}:(17.1)$$

Here $\underline{\kappa}_0$ is a constant, traceless, second-order tensor, and $\mathscr{R}e$ indicates "the real part of." The stresses that are needed to maintain the motion are also oscillatory and are to first order in the velocity gradient,

$$\mathsf{p}=b_0\mathsf{U}-\left(\underline{\kappa}_0+\underline{\kappa}_0^\dagger\right)\mathscr{R}e\,\{\eta^*\exp(i\omega t)\} \qquad \text{HI}:(17.2)$$

where the "complex viscosity" η^* is given by a real and an imaginary part,

$$\eta^*=\eta'-i\eta'' \tag{17.3}$$

We now want to obtain an expression for η' in the limit as ω tends to infinity.

First, we expand the distribution function for small velocity gradients,

$$\psi = \psi_{\text{eq.}}\Big(1 + \underbrace{\mathcal{Re}\ \{\phi_1 \exp(i\omega t)\}}_{(b)} + O\big(\underline{\kappa}_0^2\big)\Big) \quad \text{(17.4)}$$

(the "1" is term (a) and the $\mathcal{Re}\{\ldots\}$ term is term (b))

Here ϕ_1 is of first order in the velocity gradient. When this expression is introduced in (11.14) and terms of first order in the velocity gradient are equated, we see that ϕ_1 must satisfy

$$i\omega\psi_{\text{eq.}}\phi_1 + \Sigma_{st}\frac{\partial}{\partial Q_s}\left\{\tilde{G}_{st}\left[-kT\psi_{\text{eq.}}\left(\frac{\partial}{\partial Q_t}\phi_1\right) + \psi_{\text{eq.}}\Sigma_\nu \zeta_\nu^{1/2}\tilde{\mathbf{b}}_{\nu t}\tilde{\mathbf{R}}_\nu : \underline{\kappa}_0^\dagger\right]\right\} = 0 \quad \text{FDHAEI:(17.5)}$$

Hence for large frequencies ϕ_1 may be expanded in inverse powers of ω as follows:

$$\phi_1 = i\psi_{\text{eq.}}^{-1}\ \Sigma_{\nu st}\zeta_\nu^{1/2}\frac{\partial}{\partial Q_s}\left(\tilde{G}_{st}\psi_{\text{eq.}}\tilde{\mathbf{b}}_{\nu t}\tilde{\mathbf{R}}_\nu : \underline{\kappa}_0^\dagger\right)\omega^{-1} + O(\omega^{-2}) \quad \text{FDHAEI:(17.6)}$$

In particular we see that ϕ_1 is of order $(1/\omega)$ in the limit as ω tends to infinity. We now introduce Eqs. 17.1, 17.4, and 17.6 into the expression for the momentum-flux tensor given in (12.9). In so doing we retain only terms of first order in the velocity gradient. Furthermore, we consider explicitly the limit in which ω tends to infinity and note the order of magnitude of the various contributions to the stress tensor:

(1) Terms (2) and (3) of (12.9) combined with term (a) in (17.4) contribute terms linear in $\underline{\kappa}_0$.

(2) Terms (4) and (5) of (12.9) combined with term (a) in (17.4) give only isotropic contributions to the stress tensor. For incompressible fluids these terms have no importance.

(3) Terms (4) and (5) of (12.9) combined with term (b) in (17.4) contribute terms linear in $\underline{\kappa}_0$. These terms, however, are of order $(1/\omega)$ in the limit as ω tends to infinity and may hence be neglected.

From the foregoing order-of-magnitude analysis we see that in the limit as ω tends to infinity we have through terms linear in $\underline{\kappa}_0$

$$\mathsf{p}=\mathsf{p}_0+A\mathsf{U}-\left[\Sigma_\nu\zeta_\nu\int\tilde{\mathbf{R}}_\nu\tilde{\mathbf{R}}_\nu\psi_{\text{eq.}}dQ\cdot\underline{\kappa}_0\right.$$

$$\left.+\sum_{\nu\mu st}(\zeta_\nu\zeta_\mu)^{1/2}\int\tilde{G}_{st}\tilde{\mathbf{R}}_\nu\tilde{\mathbf{b}}_{\nu s}\tilde{\mathbf{b}}_{\mu t}\tilde{\mathbf{R}}_\mu\psi_{\text{eq.}}dQ:\underline{\kappa}_0\right]$$

$$\times\,\mathfrak{Re}\,\{\exp(i\omega t)\} \qquad \text{FSDHAEI}:(17.7)$$

Here A is a scalar constant of no importance. The equilibrium distribution function $\psi_{\text{eq.}}$ is given by (16.12) with $\underline{\kappa}$ set equal to zero. We see thus that the high-frequency limit involves only the equilibrium distribution function, as noted earlier by Fixman and Kovac.[40]

We may now express $\tilde{\mathbf{R}}_\nu$ and $\mathbf{b}_{\nu s}$ in terms of the unit vectors $\check{\boldsymbol{\delta}}_m$ and internal coordinates by use of the development in Section III. When this is done the integrations over the Euler angles may be performed by use of 3.34 and 3.35 in a manner similar to that employed in the previous section. After some manipulations the result may be written in the form given in (17.2) with the following results for η' and η'':

$$\lim_{\omega\to\infty}(\eta'-\eta_0)=\frac{n}{6}\{\Sigma_p K_{pp}\}-\frac{n}{60}\sum_{\substack{\nu\mu st\\kl}}(\zeta_\nu\zeta_\mu)^{1/2}$$

$$\times\left\{\Gamma_{st}^{-1}\left[3\tilde{R}_{\nu l}\tilde{b}^{\nu sk}\tilde{b}^{\mu tl}\tilde{R}_{\mu k}+3\tilde{R}_{\nu k}\tilde{b}^{\nu sl}\tilde{b}^{\mu tl}\tilde{R}_{\mu k}-2\tilde{R}_{\nu k}\tilde{b}^{\nu sk}\tilde{b}^{\mu tl}\tilde{R}_{\mu l}\right]\right\}$$

$$\text{FSDHAEI}:(17.8)$$

$$\lim_{\omega\to\infty}\eta''=0 \qquad \text{FSDHAEI}:(17.9)$$

In (17.8) the braces denote the averages over the internal configuration space defined in (16.21). The result for the high-frequency value of the dynamic viscosity η' is somewhat less compact than the results for b_1 and b_2 of the previous section.

It is interesting to note that in this section, it is more convenient to use the momentum-flux tensor of (12.9) rather than the Giesekus expression (Eq. 15.6), which was found to be more convenient in the previous section.

Example

The General Rigid Body

We consider an arbitrary collection of beads with no internal degrees of freedom. For this model, then, the averages indicated by the braces in

(17.8) do not have to be performed. We choose to install the unit vectors $\check{\boldsymbol{\delta}}_m$ along the principal axes of the moment-of-inertia tensor. The matrix with components K_{mn} is then diagonal, and we define

$$[K_{mn}] = [\Sigma_\nu \zeta_\nu \tilde{R}_{\nu m} \tilde{R}_{\nu n}]$$

$$= \begin{bmatrix} K_1 & 0 & 0 \\ 0 & K_2 & 0 \\ 0 & 0 & K_3 \end{bmatrix} \tag{17.10}$$

With this definition the $\tilde{\Gamma}_{uv}$ are

$$[\tilde{\Gamma}_{uv}] = \begin{bmatrix} K_2 + K_3 & 0 & 0 \\ 0 & K_1 + K_3 & 0 \\ 0 & 0 & K_1 + K_2 \end{bmatrix} \tag{17.11}$$

Hence

$$[\tilde{\Gamma}_{uv}^{-1}] = \begin{bmatrix} (K_2 + K_3)^{-1} & 0 & 0 \\ 0 & (K_1 + K_3)^{-1} & 0 \\ 0 & 0 & (K_1 + K_2)^{-1} \end{bmatrix} \tag{17.12}$$

Furthermore, from applications of (3.18) and (3.33), we evaluate the following matrices:

$$\left[\sum_{\nu\mu kl} (\zeta_\nu \zeta_\mu)^{1/2} \tilde{R}_{\nu l} \tilde{b}^{\nu uk} \tilde{b}^{\mu v l} \tilde{R}_{\mu k} \right] = - \begin{bmatrix} 2K_2K_3 & 0 & 0 \\ 0 & 2K_1K_3 & 0 \\ 0 & 0 & 2K_1K_2 \end{bmatrix} \tag{17.13}$$

$$\left[\sum_{\nu\mu kl} (\zeta_\nu \zeta_\mu)^{1/2} \tilde{R}_{\nu k} \tilde{b}^{\nu ul} \tilde{b}^{\mu v l} \tilde{R}_{\mu k} \right] = \begin{bmatrix} K_2^2 + K_3^2 & 0 & 0 \\ 0 & K_1^2 + K_3^2 & 0 \\ 0 & 0 & K_1^2 + K_2^2 \end{bmatrix} \tag{17.14}$$

$$\left[\sum_{\nu\mu kl} (\zeta_\nu \zeta_\mu)^{1/2} \tilde{R}_{\nu k} \tilde{b}^{\nu uk} \tilde{b}^{\mu v l} \tilde{R}_{\mu l} \right] = 0 \tag{17.15}$$

From 17.8, 17.12, 17.13, 17.14, and 17.15, we finally obtain:

$$\lim_{\omega\to\infty} (\eta' - \eta_0) = \frac{n}{6}\Sigma_p K_p - \frac{n}{40}\Sigma_{pq}\frac{(K_p - K_q)^2}{K_p + K_q}$$

$$= \frac{n}{6}(K_1 + K_2 + K_3) - \frac{n}{20}\left[\frac{(K_1 - K_2)^2}{K_1 + K_2} + \frac{(K_1 - K_3)^2}{K_1 + K_3} + \frac{(K_2 - K_3)^2}{K_2 + K_3}\right] \tag{17.16}$$

This result agrees with the high-frequency value of the dynamic viscosity given over the whole frequency range by Hassager[31] for the special case where all masses and friction coefficients are equal, and furthermore two principal moments of inertia are equal.

XVIII. CONCLUSIONS

A kinetic theory for flowing macromolecular solutions has been developed by beginning with a complete phase-space formalism. The principal accomplishments of this development are:

(1) The Kirkwood diffusion equation for the configuration-space distribution function has been rederived by successive integrations of the Liouville equation. The Kirkwood equation has been generalized to include bead–rod–spring systems of any complexity, including beads with unequal masses and friction coefficients, arbitrary connectivity, and internal constraints, as well as polymer–polymer interaction terms. In the derivation Brownian motion terms arise from the integrations over momenta. The principal result is in (11.14), and it is there indicated exactly what assumptions lead to the generalized Kirkwood equation from a considerably more general result in (11.13).

(2) The Kramers' configurational distribution function for steady-state potential flows of dilute solutions has been rederived and appropriately generalized to bead–rod–spring models of arbitrary generality. The principal result, which is a special solution to (11.14), is quoted in (11.15).

(3) The expressions for the momentum-flux tensor previously given by Kramers, Kirkwood, Kotaka, Giesekus, and Fixman have been rederived and generalized to arbitrarily complex bead–rod–spring models. Equation 13.4 contains the Kramers–Kirkwood–Kotaka expression for dilute solutions of macromolecules, and in addition it contains the macromolecule–macromolecule interaction terms proposed by Fixman. It is

shown exactly what assumptions are needed to obtain these earlier results from a much more general expression given in (13.3). In (15.6) a generalization of the Giesekus momentum-flux tensor expression is given; this formula appears to be the most useful for computational purposes. An alternative expression for the momentum-flux tensor is to be found in (12.8); this contains only the same two assumptions as used in the derivation of (13.3); a simplified version of this equation, Eq. 12.9, contains the same assumptions as the generalized Giesekus equation, and this expression is shown to be symmetrical.

(4) The assumptions made at various points in the theoretical development have been carefully catalogued. The principal assumptions which are introduced in order to get the results of earlier configuration-space theories are:

E: Equilibration in momentum space
F: Friction coefficients introduced
S: Short-range forces
A: Acceleration terms omitted
D: Dilute solution
H: Homogeneous flow
I: Incompressibility of fluid

The most serious of these assumptions is probably Assumption F introduced in (10.6). It is possible to modify Eq. 10.6 in such a way as to include hydrodynamic interaction,[41] either by using an Oseen-type approximation or by using a variational approximation such as that devised by Rotne and Prager.[42,43,44] Another modification of Assumption F is the inclusion of the time-dependent generalization of Stokes' law.[45] The use of shielding coefficients has also been explored.[46]

(5) Explicit expressions have been obtained for the coefficients of the first- and second-order terms in the retarded motion expansion, thus providing a direct link between macromolecular structure and rheological parameters. It is found that, as would be expected, the normal stresses are considerably more sensitive to the structural details than are the shear stresses. It is suggested that comparison of experimental and computed steady-state compliance could be useful in elucidating structure or testing the theory.

(6) An explicit expression has also been obtained for the high-frequency limit of the dynamic viscosity for any bead–spring–rod macromolecular model. The results are also given for the special case of arbitrarily complex rigid macromolecules.

ACKNOWLEDGMENTS

One of the authors (CFC) thanks the National Science Foundation for support under Grant GP-28213A1; the other authors (RBB and OH) thank the National Science Foundation for support under Grant GK-24749 and also the Vilas Trust Fund of the University of Wisconsin for financial assistance. We also thank Dr. Said I. Abdel-Khalik of the Chemical Engineering Department and Rheology Research Center for checking some of the derivations. In addition we express our appreciation to Professor R. M. Mazo, Professor E. Paul, Professor W. H. Stockmayer, Professor Hyuk Yu, Professor M. W. Johnson, Jr., Professor A. S. Lodge, Professor J. D. Ferry, Professor R. C. Armstrong, Mr. R. L. Hansen, and Professor H. A. Scheraga for helpful correspondence or conversations.

References

1. J. O. Hirschfelder, C. F. Curtiss, and R. B. Bird, *Molecular Theory of Gases and Liquids*, 2nd ptg., Wiley, New York, 1964, Chapter 11.
2. H. A. Kramers, *Physica*, **11**, 1 (1944); English translation published in *J. Chem. Phys.*, **14**, 415 (1946).
3. J. J. Hermans, *Physica*, **10**, 777 (1943).
4. W. Kuhn, *Kolloid Z.*, **68**, 2 (1934) and **76**, 258 (1936).
5. R. B. Bird, M. W. Johnson, Jr., and C. F. Curtiss, *J. Chem. Phys.*, **51**, 3023 (1969).
6. R. B. Bird, H. R. Warner, Jr., and D. C. Evans, *Fortsch. Hochpolym.-Forsch.*, **8**, 1 (1971).
7. O. Hassager, Ph.D. Thesis, Univ. Wisconsin (1973); *J. Chem. Phys.*, **60**, 2111 (1974).
8. J. G. Kirkwood and J. Riseman, *J. Chem. Phys.*, **16**, 565 (1948); for additional papers by Kirkwood and his collaborators see: J. G. Kirkwood, *Macromolecules*, edited by P. L. Auer, Gordon and Breach, New York, 1967.
9. T. Kotaka, *J. Chem. Phys.*, **30**, 1566 (1959).
10. H. Giesekus, *Kolloid Z.*, **147–149**, 29 (1956); erratum in fn. 18 of *Rheol. Acta*, **1**, 404 (1961).
11. S. Prager, *Trans. Soc. Rheol.*, **1**, 53 (1957).
12. B. H. Zimm, *J. Chem. Phys.*, **24**, 269 (1956).
13. M. C. Williams, *J. Chem. Phys.*, **42**, 2988 (1965); *AIChE Journal*, **21**, 1 (1975).
14. P. E. Rouse, Jr., *J. Chem. Phys.*, **21**, 1272 (1953).
15. A. S. Lodge and Y. Wu, *Rheol. Acta*, **10**, 539 (1971).
16. J. D. Ferry, *Viscoelastic Properties of Polymers*, 2nd ed., Wiley, New York, 1970.
17. H. Giesekus, *Rheol. Acta*, **5**, 29 (1966).
18. J. G. Oldroyd, *Proc. Roy. Soc. (London) Ser. A*, **200**, 523 (1950).
19. O. Hassager and R. B. Bird, *J. Chem. Phys.*, **56**, 2498 (1972).
20. M. Fixman, *J. Chem. Phys.*, **42**, 3831 (1965).
21. C. W. Pyun and M. Fixman, *J. Chem. Phys.*, **42**, 3838 (1965).
22. M. C. Williams, *AIChE J.*, **12**, 1064 (1966); **13**, 534 (1967); **13**, 955 (1967); **14**, 360 (1968); see also H. R. Warner, Jr. [Ph.D. Thesis, Univ. Wisconsin (1971), p. 3] for comments on errata in Williams' work.
23. C. F. Curtiss, unpublished lecture at Gordon Research Conference, 14 July 1970.
24. J. H. Irving and J. G. Kirkwood, *J. Chem. Phys.*, **18**, 817 (1950).
24. A. J. McConnell, *Applications of Tensor Analysis*, Dover, New York, 1957.
26. F. B. Hildebrand, *Methods of Applied Mathematics*, Prentice-Hall, Englewood Cliffs, N.J., 1952, p. 366, Problem 54.
27. S. Imam-Rahajoe, Ph.D. Thesis, Univ. Wisconsin (1967).
28. J. Riseman and J. G. Kirkwood, in *Rheology: Theory and Applications*, edited by F. R. Eirich, Academic, New York, 1956, Chapter 13.

29. J. G. Kirkwood, *J. Chem. Phys.*, **14**, 180 (1946).
30. H. C. Booij and P. H. van Wiechen, *J. Chem. Phys.*, **52**, 5056 (1970).
31. O. Hassager, *J. Chem. Phys.*, **60**, 4001 (1974).
32. R. S. Rivlin and J. L. Ericksen, *J. Rat. Mech. Anal.*, **4**, 323 (1955).
33. R. C. Armstrong and R. B. Bird, *J. Chem. Phys.*, **58**, 2715 (1973). S. I. Abdel-Khalik, O. Hassager, and R. B. Bird, *ibid.*, **61**, 4312 (1974).
34. K. Osaki, Y. Mitsuda, R. M. Johnson, J. L. Schrag, and J. D. Ferry, *Macromolecules*, **5**, 17 (1972).
35. Y. Mitsuda, J. L. Schrag, and J. D. Ferry, *J. Appl. Polym. Sci.*, **18**, 193 (1974).
36. E. Paul and R. M. Mazo, *J. Chem. Phys.*, **51**, 1102 (1969).
37. S. I. Abdel-Khalik and R. B. Bird, *Appl. Sci. Res.* **30**, 268 (1975).
38. H. R. Warner, Jr., *Ind. Eng. Chem. Fund.*, **11**, 379 (1972).
39. H. Giesekus, *Rheol. Acta*, **2**, 101 (1962).
40. M. Fixman and J. Kovac, *J. Chem. Phys.*, **61**, 4939, 4950 (1974); J. Kovac and M. Fixman, *ibid.*, **63**, 935 (1975); M. Fixman, *Proc. Nat. Acad. Sci.*, USA, **71**, 3050 (1974).
41. C. F. Curtiss and R. L. Hansen, (to be published).
42. H. Yamakawa, *Modern Theory of Polymer Solutions*, Harper and Row, New York, 1971, p. 395.
43. M. Fixman and W. H. Stockmayer, *Polymer Conformation and Dynamics in Solution, Ann. Rev. Phys. Chem.*, **21**, 407 (1970), see p. 414.
44. J. Rotne and S. Prager, *J. Chem. Phys.*, **50**, 4831 (1969).
45. S. C. Szu and J. J. Hermans, *J. Polymer Sci. Polym. Phys. Ed.*, **12**, 1743 (1974).
46. S. I. Abdel-Khalik and R. B. Bird, *Biopolymers*, **14**, 1915 (1975).

LIST OF SYMBOLS

Symbol	Meaning	Equation
$A(\chi)$	Arbitrary time-independent dynamical variable	(5.4)
A_{ab}	Elements of "Rouse matrix"	(3.53)
$A_s^{(\alpha)}, B_s^{(\alpha)}, C_s^{(\alpha)}, D_s^{(\alpha)}$	Groupings in Section XI	
$a, b, c, \ldots$	Index for numbering links in chain	(3.47)
$B_{\nu a}$	Matrix elements (Rouse model)	(3.48)
$b_1, b_2, b_{11}, \ldots$	Coefficients in retarded motion expansion	(16.1)
$\mathbf{b}_{\nu s}^{(\alpha i)}, \tilde{\mathbf{b}}_{\nu s}^{(\alpha i)}$	Basis vectors	(2.4), (2.5)
$b_{\nu sm}^{(\alpha)}, \tilde{b}_{\nu sm}^{(\alpha)}$	Components of $\mathbf{b}_{\nu s}^{(\alpha)}, \tilde{\mathbf{b}}_{\nu s}^{(\alpha)}$	(3.7), (3.8)
$b_{(\alpha)}^{\nu sm}, b_{(\alpha)}^{\nu sm}$	Matrix elements	(3.15), (3.16)
C_α	Constants	(2.15), (10.8)
C_{ab}	Matrix elements (Rouse model)	(3.51)
$c, C, ¢$	$\cos\alpha, \cos\beta, \cos\gamma$	(3.2)
c_a	Eigenvalues of the matrix $[C_{ab}]$	(3.52)
d_α	Number of degrees of freedom of species α	(2.1)
$E_t^{(\alpha)}$	Grouping of terms that occur in equation for ψ_α	(11.9), (12.6)
$\mathbf{F}_\nu^{(\alpha i)}$	Force on mass point ν of molecule αi	(4.10)

LIST OF SYMBOLS (*Continued*)

Symbol	Meaning	Equation
$\mathbf{F}^{(\alpha i,\beta j)}$	Force between two molecules	(4.11)
$\mathbf{F}_{\nu\mu}^{(\alpha i,\beta j)}$	Force between two mass points	(4.9)
$\bar{\mathbf{F}}_{\nu\mu}^{(\alpha,\beta)}$	Average force	(9.5)
$\mathfrak{F}$	Average effective force density	(5.19)
$\mathfrak{F}^{(\mathrm{I})}$	Contribution to $\mathfrak{F}$	(7.6)
$\mathfrak{F}^{(\mathrm{II})}$	Contribution to $\mathfrak{F}$	(7.7)
$\mathfrak{F}_{\nu}^{(\alpha i)}$	Effective force	(5.14)
$\mathfrak{F}_{\alpha i\nu}^{(\mathrm{I})}$	Intramolecular force contribution to $m_{\alpha\nu}\dot{\mathbf{V}}_{\alpha i\nu}$	(4.20)
$\tilde{\mathfrak{F}}_{\alpha i\beta j\nu}^{(\mathrm{II})}$	Intermolecular force contributions to $m_{\alpha\nu}\dot{\mathbf{V}}_{\alpha i\nu}$	(4.20)
$\mathfrak{F}_{\alpha i\beta j\nu}^{(\mathrm{II})}$	Contribution to $\mathfrak{F}_{\nu}^{(\alpha i)}$	(5.15)
$f(\chi,t)$	Complete phase-space distribution function	(5.1)
f_{α}	Singlet distribution function in phase-space	(6.1)
$f_{\alpha\beta}$	Pair distribution function in phase space	(6.3)
$\tilde{f}_{\alpha\beta}$	Pair distribution function in phase space	(6.4)
G', G''	Storage and loss moduli	(16.5), (16.6)
G_{α}	Determinant of the $G_{st}^{(\alpha)}$	(10.4)
$G_{st}^{(\alpha i)}, \tilde{G}_{st}^{(\alpha i)}$	Contravarient metric tensor components	(2.19), (2.20)
$\mathcal{G}_{st}^{(\alpha i)}$	Matrix elements	(11.4)
$g_{\alpha i}$	Determinant of the $g_{st}^{(\alpha i)}$	(2.23)
$\tilde{g}_{\alpha}$	Determinant of the $\tilde{g}_{st}^{(\alpha)}$	(3.26)
$g_{st}^{(\alpha i)}, \tilde{g}_{st}^{(\alpha i)}$	Covarient metric tensor components	(2.13), (2.14)
$\mathcal{G}_{st}^{(\alpha i)}$	Matrix elements	(11.3)
H	Hamiltonian	(4.6)
H_{α}	Part of Hamiltonian associated with α	(10.2)
I_{d1}, I_{d2}	Integrals	(16.16)
$\mathsf{I}_{n1}, \mathsf{I}_{n2}$	Integrals	(16.17), (16.18)
i,j	Index for numbering molecules of a given species	(2.3)
$J_{smn}^{(\alpha)}$	Matrix elements	(3.28)
$K_{nm}^{(\alpha)}$	Matrix Elements	(3.33)
$K_{n}^{(\alpha)}$	Eigenvalues of $K_{nm}^{(\alpha)}$ matrix	(17.10)
k	Boltzmann's constant	(10.1)
L	Rod length (elastic rhombus model)	(3.60)
L	Bead separation (dumbbell)	(3.36)

LIST OF SYMBOLS (*Continued*)

Symbol	Meaning	Equation
$\mathfrak{L}$	Liouville operator	(5.2)
m, n, p, q, r	Indices representing 1, 2, 3 only	(2.3)
m_α	Mass of species α	(2.1)
$m_{\alpha\nu}$	Mass of mass point ν of species α	(2.1)
N_α	Number of mass points in species α	(2.1)
n_α	Number density of α	(6.2)
n_0	Number density of solvent molecules	(10.6)
$P_{\alpha is}$	Generalized momenta for molecule αi	(2.1)
$\dot{P}_{\alpha is}$	Time derivative of $P_{\alpha is}$	(4.18)
p	Momentum-flux tensor	(1.2)
p_0	Solvent contribution to p	(16.1)
$\mathsf{p}^{(k)}$	Kinetic contribution to the momentum-flux tensor	(5.18)
$\mathsf{p}_{\mathrm{tr}}^{(k)}$	Translational part of $\mathsf{p}^{(k)}$	(12.1)
$\mathsf{p}_{\mathrm{int}}^{(k)}$	Internal part of $\mathsf{p}^{(k)}$	(12.1)
$\mathsf{p}^{(\mathrm{I})}$	Intramolecular force contribution to p	(5.20)
$\mathsf{p}^{(\mathrm{II})}$	Intermolecular contribution to p	(5.20)
$\mathsf{p}_0^{(\mathrm{II})}$	Solvent contribution to $\mathsf{p}^{(\mathrm{II})}$	(12.4)
$\mathbf{p}_{\alpha i}$	Momentum of center of mass of molecule αi	(2.1)
$\dot{\mathbf{p}}_{\alpha i}$	Time derivative of $\mathbf{p}_{\alpha i}$	(4.17)
$\mathbf{Q}_a$	Bond-orientation vector (Rouse model)	(3.47)
$Q_{\alpha is}$	Generalized coordinates for molecule αi	(2.1)
$\dot{Q}_{\alpha is}$	Time derivative of $Q_{\alpha is}$	(4.1)
$\mathbf{R}_{\alpha\beta}$	Distance between centers of mass of α and β	(6.6)
$\mathbf{R}_{\alpha i\nu}$	Position of mass point ν relative to center of mass of molecule αi	(2.1)
$\tilde{\mathbf{R}}_{\alpha i\nu}$	Position of mass point ν relative to center of resistance of molecule αi	(2.1)
$R_{\alpha\nu m}$	Components of $\mathbf{R}_{\alpha\nu}$	(3.3)
$\tilde{R}_{\alpha\nu m}$	Components of $\tilde{\mathbf{R}}_{\alpha\nu}$	(3.4)
$\mathbf{r}_{\alpha\beta}$	Position of center of mass of system made up of α and β	(6.5)
$\mathbf{r}_{\alpha i}$	Position of center of mass of molecule αi	(2.1)
$\dot{\mathbf{r}}_{\alpha i}$	Time derivative of $\mathbf{r}_{\alpha i}$	
$\tilde{\mathbf{r}}_{\alpha i}$	Position of center of resistance of molecule αi	(2.2)

LIST OF SYMBOLS (*Continued*)

Symbol	Meaning	Equation
$\mathbf{r}_{\alpha i\nu}$	Position of mass point ν in molecule αi	(2.1)
$\mathbf{r}_{\alpha i\nu,\beta j\mu}$	Relative position vector	(4.8)
S_{mn}	Elements of rotation matrix	(3.2)
$s, S, \$$	$\sin\alpha$, $\sin\beta$, $\sin\gamma$	(3.3)
s, t, u, v, w	Indices referring to generalized coordinates	(2.1)
T	Absolute temperature	(10.1)
t	Time	(1.1)
$T_{\alpha i}$	Kinetic energy of molecule αi	(4.3)
U	Unit tensor	(10.5)
$\mathbf{u}$	Fluid velocity	(1.1)
$\mathbf{u}_\alpha$	Velocity of species α	(1.2)
$\mathbf{u}_0$	Solvent velocity	(10.6)
$\mathbf{u}_{\alpha\nu}$	Average velocity of mass points ν of α	(10.7)
$\mathbf{V}_{\alpha i\nu}$	Velocity of mass point ν of molecule αi	(4.1)
$\dot{\mathbf{V}}_{\alpha i\nu}$	Time derivative of $\mathbf{V}_{\alpha i\nu}$	(4.19)
$v_{\alpha\beta}$	Characteristic volume	(9.4)
Z_α	Normalization factor	(10.4)
α, β, γ	Euler angles	(3.2)
$\alpha, \beta, \ldots$	Index indicating chemical species	(1.1)
$\Gamma^{(\alpha)}_{uv}, \tilde{\Gamma}^{(\alpha)}_{uv}$	Matrix elements	(3.23)
$\underline{\gamma}$	Rate-of-deformation tensor	(16.2)
$\underline{\gamma}_{(n)}$	Kinematic tensors	(16.1)
$\delta(x)$	Dirac delta function	(5.6)
$\delta(\mathbf{A})$	Abbreviation for $\delta(A_x)\delta(A_y)\delta(A_z)$	(5.6)
$\boldsymbol{\delta}_n$	Space-fixed unit vectors	(3.1)
$\check{\boldsymbol{\delta}}_n$	Unit vectors embedded in molecule	(3.1)
$\check{\boldsymbol{\delta}}_R, \check{\boldsymbol{\delta}}_\theta, \check{\boldsymbol{\delta}}_\phi$	Unit vectors for dumbbell	(3.36)
δ_{mn}	Kronecker delta	(3.2)
ϵ_{mnp}	Permutation symbol	(3.11)
ζ_α	Friction coefficient of species α	(2.3)
$\zeta_{\alpha\nu}$	Friction coefficient of mass point ν of species α	(2.3), (10.6)
η	Viscosity function	(16.4)
η', η''	Components of the complex viscosity η^*	(16.5), (16.6), (17.3)
θ, ϕ	Spherical coordinates	(3.36)
K	$\cos\sigma$ (in elastic rhombus model)	(3.60)
κ	Velocity gradient in steady shear flow	(16.4)

LIST OF SYMBOLS (*Continued*)

Symbol	Meaning	Equation
$\underline{\kappa}$	Traceless tensor in expressions for homogeneous velocity distribution	(16.10)
$\underline{\kappa}_0$	Velocity-gradient amplitude	(17.1)
Λ_{ts}	Matrix elements	(3.11)
$\boldsymbol{\lambda}_n$	Unit vectors	(3.34)
μ	Reduced mass (for dumbbell)	(3.36)
$\nu, \mu, \eta, \ldots$	Indices used for numbering mass points in molecule	(2.1)
ξ	Integration variable	(7.10)
ρ	Fluid density	(1.2)
ρ_α	Mass concentration of species α	(1.1)
Σ	$\sin\sigma$ (in elastic rhombus model)	(3.60)
σ	Angle (elastic rhombus model)	(3.60)
Φ	Total potential energy	(4.6)
ϕ_α	Intramolecular potential energy for α	(10.2)
$\phi_{\alpha i\nu, \beta j\mu}$	Potential energy of interaction	(4.7)
χ	Set of phase-space coordinates	(5.1)
ψ_1	Primary normal stress coefficient	(16.4)
ψ_2	Secondary normal stress coefficient	(16.4)
ψ_α	Configuration-space distribution function	(6.7)
$\psi_{\alpha\beta}$	Configuration-space pair-distribution function	(6.9)
$\tilde{\psi}_{\alpha\beta}$	Configuration-space pair-distribution function	(6.10)
$\underline{\omega}$	Vorticity tensor	(16.8)
$\partial/\partial\mathbf{r}$	Vector operator with components $\partial/\partial x$, $\partial/\partial y$, $\partial/\partial z$	(1.1)
$\mathfrak{d}/\mathfrak{d}t$	Oldroyd convected derivative	(15.6)
$\dot{}$ (above symbol)	Time derivative	(4.15)
$\tilde{}$ (above symbol)	Quantities associated with center of resistance [except $\tilde{\mathfrak{F}}^{(\mathrm{II})}_{\alpha i\beta j\nu}$, $\tilde{\psi}_{\alpha\beta}$, $\tilde{f}_{\alpha\beta}$]	
$_0$ (subscript)	Solvent	(10.6)
† (superscript)	Transpose of a tensor	
$\langle\ \rangle$	Ensemble average	(5.4)
$\langle\ \rangle^{(\alpha)}$	Average value in one molecule phase space	(6.11)
$[\ ]^{(\alpha)}$	Average value in one molecule	(6.12)
$\{\ \}$	Weighted average	(16.21)

KINETIC THEORY OF GRAVITATIONAL SYSTEMS

M. J. HAGGERTY

Center for Statistical Mechanics and Thermodynamics, University of Texas, Austin, Texas

and

G. SEVERNE

Fakulteit van de Wetenschappen, Vrije Universiteit Brussel, Brussels, Belgium

CONTENTS

I. INTRODUCTION

From the point of view of statistical mechanics, gravitational systems present a particularly intriguing challenge. The analysis of their properties abounds in anomalies and paradoxes. Aside from their intrinsic interest, they form a privileged object of study in view of the variety of counter-examples provided to the behavior norms in statistical mechanics.

At the outset the basic length and time scales reveal quite unusual characteristics. The dimensions of a star cluster or of an elliptical galaxy are small in the relevant hierarchy of length scales. Thus while being the largest of all systems available for study, gravitational systems must often be treated as small systems, in a well-defined sense. If molecular systems evolve in general too quickly for the convenience of the human observer, the opposite is true for gravitational systems. Observations only give photographic snapshots of distinct systems. The characteristics of their evolution must be determined from detail observed at one instant.

The motivations for a statistical mechanical analysis are thus quite different for molecular and for gravitational systems. For the former the task is to determine the microscopic nature of the system from macroscopic time-averaged experimental parameters. For the latter the forces are known, and it is the nature of the evolution that must be elucidated. For both classes of systems, computer simulations have proved to be an invaluable tool for bridging the gaps between theory and observation.

In the historical context, the study of gravitational systems has at different periods provided an essential impetus in the development of statistical mechanics. The work of Jeans[1] and later that of Chandrasekhar[2] have been dominant influences in kinetic theory; in particular, plasma theory in its early development was strongly indebted to stellar dynamics. The situation has since been inverted. Much recent work on the gravitational problem derives from plasma analogies, and gravitational kinetic theory is at present a relatively underdeveloped field. The practical needs and interests of stellar dynamicists in constructing, for example, models for the evolution of star clusters have been satisfied by Chandrasekhar's analysis of collisional processes. Further theoretical refinements have to some extent fallen into an interdisciplinary vacuum.

The basic data provided by observation exhibit striking differences with respect to molecular systems, but also strong analogies. The differences bear foremost upon the spatial distribution, the analogies upon the velocity distribution of the stars. Under the influence of the purely attractive gravitational interactions, stars tend to aggregate in pairs, in small groups, in clusters, in galaxies, and the galaxies themselves form clusters of galaxies. Instead of filling space homogeneously as would the molecules of a gas, stars will form a hierarchical imbrication of clusters. Within a given cluster, the velocity distribution of the subunits constituting the cluster does not appear to differ essentially from the classical Maxwellian. The Schwarzschild distribution, which is often observed, is an anisotropic Maxwellian with velocity dispersions depending upon direction, as a consequence of the conservation of angular momentum.

The self-consistent nature of the mean gravitational field implies that the spatial structure of a nonrotating gravitational system is to a large extent determined by its velocity distribution. First suggested on theoretical grounds,[3] the following evolution picture for a star cluster has now clearly emerged from computer simulations of the N-body gravitational problem. Released from an arbitrary initial state compatible with the stability requirements, the cluster will undergo a spectacular mixing process in which the individual stars can carry out several oscillations taking them across the entire system. This *violent relaxation* phase[4] is accomplished in a few "crossing" times τ_C, and corresponds to a collisionless phase mixing process. It is followed by a smooth *kinetic phase*: under the influence of stellar encounters, typically described by a Fokker–Planck collision term, the velocity distribution tends towards a Maxwellian on the relaxation time scale τ_R. The spatial structure of the cluster that emerges from the violent relaxation phase continues to evolve; a variety of effects, which are contained in the fluid-dynamical description given from the moments of the kinetic equation,[5] come into play. It is this *secular phase* that is the main concern of the astrophysicists. The density profile tends toward the formation of a dense core. Encounters become less and less effective in the outer envelope, where the velocity distribution becomes strongly anisotropic, the elongated stellar orbits giving a greater radial than tangential velocity dispersion. This evolution on the secular time scale τ_S is paralleled by the progressive evaporation of the cluster, its more energetic members escaping to be caught up in the galactic field.

This oversimplified description is nevertheless sufficient to convey the great complexity of the global evolution picture. Insofar as the crossing time τ_C can be identified with the effective duration $\tau_{\text{coll.}}$ of an encounter, the separation of the time scales

$$\tau_{\text{coll.}} = \tau_C \ll \tau_R \ll \tau_S \tag{1.1}$$

associated respectively with the different evolution phases is a favorable factor for the construction of a kinetic theory. If the kinetic evolution tends to establish a Maxwellian velocity distribution, the further development of the cluster shows, however, how precarious the concept of equilibrium is; this holds true even for an isolated system where escaping stars that have acquired a positive total energy are relatively rare.[6] The process of core formation, in particular, is a reminder of the inherent instability of purely attractive interactions, and can prefigure a gravitational collapse.[7]

If, rather than single clusters, one considers hierarchical structures on a larger scale, new classes of phenomena can appear. Such structures can be quite loose and no longer well confined by the mean gravitational field. Their behavior can be approached through the limiting model case of an infinite system. Only for such a model does the long range of the Newtonian pair potential come into full play. [For a finite system the confining self-gravitational field gives rise to an effective cut-off for the interaction range at the system diameter (see Section III.D).] It is now not possible to define a finite characteristic duration of the encounters. The system is intrinsically non-Markoffian, and as such presents a variety of unusual properties.[8] In particular, even in the kinetic phase, there is no longer the possibility of attaining Maxwellian equilibrium.

Referring to a recent and excellent review[9] for a broader perspective on the problem of equilibrium in gravitational systems, we here simply draw the conclusion that, contrary to common practice in the study of molecular systems, it will not be possible in any degree to base the kinetic analysis upon equilibrium results. A useful guiding line remains the analogy with the electron gas of plasma theory. The formal analogy is complete, up to the difference in the sign of the interactions. One of the fascinating aspects of the statistical mechanics of gravitational systems consists precisely in following up the consequences of a simple sign difference at the level of the Hamiltonian. The plasma analogy must be treated with circumspection, but it does constitute a valuable element of perspective and a bridge between molecular systems with short range forces and gravitational systems.

As in plasmas the stability analysis of the collisionless system is fundamental to the elaboration of a coherent kinetic theory and is thus presented here in certain detail. The kinetic analysis takes on quite different forms depending upon whether one considers finite or infinite systems. For finite systems the main difficulty will be the proper treatment of the confining, self-gravitational field. For infinite systems the focus of interest lies upon the non-Markoffian behavior induced by the long range of the interactions. The wealth of interplay between the two descriptions is suggested in the conclusion.

In a publication of this nature, it should be stressed that none of the kinetic-theoretical developments that go beyond the classical analysis of Chandrasekhar[2] have yet gained any measure of general acceptance.[10] Due to the impracticality of presenting different formalisms in a text of reasonable length, the material presented is centered upon the personal work of the authors. A short historical survey is inserted to provide a proper perspective.

Throughout we adhere to the simple model of a nonrelativistic Newtonian system constituted of identical point masses. Unless otherwise specified, we restrict attention to nonrotating systems isolated in space.

II. COLLISIONLESS SYSTEMS

The simplest possible description of a system within the framework of statistical mechanics consists in ignoring all effects due to collisions. For long-range forces, such a reduction focuses attention on the role of the self-consistent field, which in many circumstances is the dominating effect. While in plasmas the field usually vanishes in mean value (local electroneutrality), its fluctuations give rise to the density waves that determine important aspects of the plasma behavior. For gravitational systems the mean field is nonvanishing; in fact it drains the quasitotality of the interaction energy. The analysis of its fluctuations will bring out the fundamental gravitational instability that largely accounts for the global structure of gravitational systems as observed.

In the absence of collisions, a system is fully characterized by its one-body distribution function $f(\mathbf{x}_1, \mathbf{v}_1, t)$. The time evolution of f is given simply by the continuity equation in the one-body phase space. For the associated distribution $f(\mathbf{x}, \mathbf{v}, t)$ in physical space, and for a nonrotating system one thus has

$$\frac{\partial f}{\partial t} + \mathbf{v} \cdot \frac{\partial f}{\partial \mathbf{x}} + \mathbf{F} \cdot \frac{\partial f}{\partial \mathbf{v}} = 0 \tag{2.1}$$

where the acceleration field $\mathbf{F}(\mathbf{x}, t)$ is self-consistently determined by the Poisson equation

$$\operatorname{div} \mathbf{F}(\mathbf{x}, t) = -4\pi G m \int d\mathbf{v} f(\mathbf{x}, \mathbf{v}, t) \tag{2.2}$$

The collisionless Boltzmann equation, (2.1), as this gravitational analog of the plasma Vlassov equation is known, has been studied from a number of quite different points of view. Here we shall retain only certain aspects that are directly relevant to kinetic theory.

A first observation to be made is that, in opposition to plasmas, gravitational systems are necessarily inhomogeneous. Indeed in a spatially homogeneous system, the acceleration field must be uniform. Its divergence then vanishes. For nonrotating systems, the Poisson equation (2.2) then predicts vanishing mass density $m\int d\mathbf{v}f$, showing that homogeneous nonrotating systems cannot exist. It is of interest to recall that for plasmas the corresponding argument imposes a vanishing charge density, that is, electroneutrality. The intrinsic inhomogeneity of gravitational systems is thus a direct expression of the absence of masses of opposite signs, that is, of the purely attractive nature of the forces.

There exists a vast body of literature dealing with the stationary solutions of the collisionless Boltzmann equation. Indeed, their study constitutes the central problem in the theory of the structure of stellar systems. Here we shall only quote one elementary but basic result, known as Jeans' theorem: The stationary solutions of (2.1) are arbitrary functions of the one-valued integrals of the motion.[11] Thus in the absence of any symmetry, and for a Gaussian velocity dependence, one will take the Boltzmann distribution

$$f_B^{(0)}(\mathbf{x},\mathbf{v}) \propto \exp\frac{-\left[v^2+2U(\mathbf{x})\right]}{2\sigma^2}, \qquad \mathbf{F}^{(0)}=\frac{-\partial U}{\partial \mathbf{x}} \tag{2.3}$$

as stationary solution to (2.1), while for spherical symmetry the appropriate form is

$$f_S^{(0)}(\mathbf{x},\mathbf{v}) \propto \exp-\tfrac{1}{2}\left[a_0(v^2+2U)+a_0'I^2\right], \qquad \mathbf{I}=\mathbf{x}\times\mathbf{v} \tag{2.4}$$

The dependence upon the modulus I of the angular momentum introduces an anisotropy in velocity space even in the absence of rotation. Both expressions (2.3) and (2.4) are still purely formal, since it remains to determine a potential function $U(\mathbf{x})$ such that the Poisson equation (2.2) is satisfied.

The collision processes do not respect spherical symmetry, which, as seen in Section I, tends to establish itself in the secular stage of evolution. The (quasi-)stationary solutions of the kinetic equations will thus derive from the simple Boltzmann distribution (2.3) and not from (2.4). It should be mentioned in this connection that neither (2.3) nor (2.4), taken strictly, are realistic for an isolated system in that they predict infinite mass. (A Gaussian velocity distribution with uniform dispersion corresponds to the continuum model of the isothermal gas sphere.[12])

From the point of view of the dynamical behavior, the collisionless description of a gravitational system gives the "violent relaxation" characteristic of the short time evolution.[4] This is a nondissipative process of chaotic phase mixing that lasts only a few crossing times for a stable system. The resulting density distribution and the corresponding field $\mathbf{F}(\mathbf{x},t)$ will then engage upon a slow, smooth evolution (apart from possible fluctuations). The slow time variation of the field in the smooth relaxation phase is of essential importance in the kinetic analysis.

For a description of the violent relaxation phase, one has at one's disposal, besides the N-body computer simulations, only numerical analyses for the one-dimensional case.[13] The outcome of the violent relaxation was analyzed by Lynden-Bell[4] using a coarse-grained phase space distribution $\bar{f}(\mathbf{x},\mathbf{v},t)$. Maximizing the entropy $\langle \log \bar{f} \rangle$, he obtained the following equilibrium distribution:

$$\bar{f}^{(0)} = \eta \frac{\exp[-\beta(\epsilon-\mu)]}{1+\exp[-\beta(\epsilon-\mu)]}, \qquad \epsilon = \frac{1}{2}v^2 + U(\mathbf{x}) \tag{2.5}$$

Here β and μ are Lagrange multipliers, while η is the initial value $\bar{f}(t=0)$ of the phase density. The point of interest here is the analogy between (2.5) and the Fermi–Dirac distribution of quantum statistics. The resemblance comes from the existence of an exclusion principle in phase space expressing the conservation of the fine-grained density. Degeneracy with respect to the Boltzmann distribution does not, however, turn out to be of physical significance.

A third broad aspect of collisionless theory is the stability analysis, which we now turn to in some detail.

A. Simplified Stability Analysis

To discuss the linear stability of gravitational systems, we consider small perturbations δf, $\delta\mathbf{F}$ to the stationary state $f^{(0)}$, $\mathbf{F}^{(0)}$, the behavior of which is determined by the linearized form of the collisionless Boltzmann equation:

$$\frac{\partial}{\partial t}\delta f + \mathbf{v}\cdot\frac{\partial}{\partial \mathbf{x}}\delta f + \delta\mathbf{F}\cdot\frac{\partial f^{(0)}}{\partial \mathbf{v}} + \mathbf{F}^{(0)}\cdot\frac{\partial}{\partial \mathbf{v}}\delta f = 0 \tag{2.6}$$

$$\operatorname{div}\delta\mathbf{F}(\mathbf{x},t) = -4\pi G m \int d\mathbf{v}\,\delta f(\mathbf{x},\mathbf{v},t) \tag{2.7}$$

A straightforward transposition of plasma theory is not possible due to the inhomogeneity of the stationary state: a Fourier transformation would leave us with an integral equation.

The difficulty can be turned by considering a rotating system (see Section II. C for further discussion). In an alternative approach we here introduce a restriction to perturbations of space variation rapid with respect to that of the stationary state $f^{(0)}(\mathbf{x},\mathbf{v})$, $\mathbf{F}^{(0)}(\mathbf{x})$. This makes it possible to take Fourier–Laplace transforms according to

$$\phi_{\mathbf{k}}(z)=\int_0^{\infty} dt\exp(izt)\int d\mathbf{x}\exp(-i\mathbf{k}\cdot\mathbf{x})\phi(\mathbf{x},t) \tag{2.8}$$

ignoring the position dependence of the stationary functions and to write for the transform of (2.6):

$$\left(-iz+i\mathbf{k}\cdot\mathbf{v}+\mathbf{F}^{(0)}\cdot\frac{\partial}{\partial\mathbf{v}}\right)\delta f_{\mathbf{k}}(z,\mathbf{v})-h_{\mathbf{k}}(\mathbf{v})+\delta\mathbf{F}_{\mathbf{k}}(z)\cdot\frac{\partial f^{(0)}}{\partial\mathbf{v}}=0 \tag{2.9}$$

with $h_{\mathbf{k}}(\mathbf{v})$ the initial value of the perturbation

$$h_{\mathbf{k}}(\mathbf{v})\equiv\delta f_{\mathbf{k}}(\mathbf{v},t=0) \tag{2.10}$$

The restriction to rapidly varying perturbations can be formulated as a *scaling condition*, limiting the validity of (2.9) to wave vectors **k** such that

$$\frac{kL_0}{2\pi}\gg 1 \tag{2.11}$$

where L_0 gives the scale on which the stationary state varies. In order of magnitude L_0 is comparable with the dimensions of the system.

We shall first consider a simplified model in which we omit the term containing the mean field in (2.9). This leaves the sign of the interaction (in the Poisson equation) as the only difference with respect to the plasma, and the standard analysis[14] is directly applicable. Writing

$$\mathbf{v}=u\mathbf{1}_{\mathbf{k}}+\mathbf{v}_{\perp}\qquad \mathbf{v}_{\perp}\cdot\mathbf{k}=0 \tag{2.12a}$$

$$\bar{\phi}(u)=\int d\mathbf{v}_{\perp}\phi(\mathbf{v}) \tag{2.12b}$$

and noting that $\delta\mathbf{F}_{\mathbf{k}}$ is parallel to **k**, one integrates (2.9) over the velocity components $\mathbf{v}_{\perp}$ perpendicular to **k**:

$$i(ku-z)\delta\bar{f}_{\mathbf{k}}(z,u)=\bar{h}_{\mathbf{k}}(u)-\delta F_{\mathbf{k}}\frac{\partial\bar{f}^{(0)}}{\partial u} \tag{2.13}$$

Substitution in the Fourier–Laplace transform of the Poisson equation (2.7) yields

$$\delta F_{\mathbf{k}}(z) = \mathfrak{D}^{-1}(z,k) H_{\mathbf{k}}(z) \tag{2.14}$$

$$\mathfrak{D}(z,k) = 1 + \frac{\omega_C^2}{k^2} \int_{-\infty}^{\infty} du \frac{\partial \bar{\varphi}^{(0)}/\partial u}{u - z/k} \tag{2.15}$$

$$H_{\mathbf{k}}(z) = \frac{\omega_C^2}{n^{(0)} k^2} \int_{-\infty}^{\infty} du \frac{\bar{h}_{\mathbf{k}}(u)}{u - z/k} \tag{2.16}$$

with the characteristic frequency ω_C defined by

$$\omega_C^2(\mathbf{x}) = 4\pi G m n^{(0)}(\mathbf{x}) \tag{2.17}$$

$n^{(0)}(\mathbf{x})$ being the stationary value of the number density

$$n(\mathbf{x},t) = \int d\mathbf{v} f(\mathbf{x},\mathbf{v},t) \tag{2.18}$$

To facilitate the discussion we have in (2.15) assumed the factorization

$$f^{(0)}(\mathbf{x},\mathbf{v}) = n^{(0)}(\mathbf{x}) \varphi^{(0)}(\mathbf{v},\mathbf{x}) \tag{2.19}$$

where the stationary velocity distribution $\varphi^{(0)}(\mathbf{v},\mathbf{x})$ can in general be position dependent.

The stability of the system is determined by the position of the zeros of the dispersion function $\mathfrak{D}(z,k)$. This quantity differs from the plasma dielectric function by the sign of the second term. The sign difference is, of course, sufficient to alter radically the stability properties of the system. It will be recalled that the condition for stability is[14]

$$\mathfrak{D}(z = ku_0, k) = 1 + \frac{\omega_C^2}{k^2} \int_{-\infty}^{\infty} du \frac{\partial \bar{\varphi}^{(0)}/\partial u}{u - u_0} > 0 \tag{2.20}$$

u_0 being an extremum of the velocity distribution $\bar{\varphi}^{(0)}$. We can generally consider that $\bar{\varphi}^{(0)}(u)$ has a single maximum; the integral appearing in (2.20) is then negative definite (numerator and denominator of the integrand are always of opposite sign); and instability obtains for all perturbations such that

$$k^2 < k_C^2(\mathbf{x}) \equiv \omega_C^2 \left| \int_{-\infty}^{\infty} du \frac{\partial \bar{\varphi}^{(0)}/\partial u}{u - u_0} \right| \tag{2.21}$$

This condition expresses the *gravitational instability*: The system is unstable to perturbations of wavelengths in excess of a critical value $2\pi/k_C$. The stability condition is here local: The minimum wavelength k_C depends on position through ω_C and also $\varphi^{(0)}$. Taken in average value, k_C sets a linear dimension L_J of order $2\pi/k_C$ as the limit to the maximum size a stable system can obtain.

This basic result remains unmodified, as we shall see, when one retains the mean field term omitted from (2.9). The validity of the proof is in some degree restricted by the scaling condition (2.11): The latter does not allow us to make any statement as to the behavior of systems of dimensions $L_0 \lesssim 2\pi/k_C$ with respect to the longer wavelength perturbations that can arise.

The dependence of the critical wavenumber k_C upon the precise nature of the velocity distribution is predictably weak.[15] For the case of a Maxwellian

$$\varphi^{(0)}(\mathbf{v}) = (2\pi\sigma^2)^{-3/2} \exp\frac{-v^2}{2\sigma^2} \tag{2.22}$$

the dispersion function, (2.15), can be expressed with the aid of the derivative of the plasma dispersion function $Z(\zeta)$:

$$\mathfrak{D}(z,k) = 1 + \left(\frac{\omega_C^2}{2\sigma^2 k^2}\right) Z'\left(\frac{z}{\sigma k\sqrt{2}}\right) \tag{2.23}$$

$$Z(\zeta) = 2i\exp(-\zeta^2)\int_{-\infty}^{i\zeta} dt \exp -t^2 = \frac{1}{\pi^{1/2}}\int_{-\infty}^{\infty} du \frac{\exp -u^2}{u-\zeta} \tag{2.24}$$

One then finds that the critical wavevector is

$$k_J^2 \equiv -\left(\frac{\omega_C^2}{2\sigma^2}\right) \operatorname{Re} Z'(\operatorname{Im} z = 0) = \frac{\omega_C^2}{\sigma^2} \tag{2.25}$$

giving the stability criterion

$$k^2 > k_J^2(\mathbf{x}) = \frac{4\pi G m n^{(0)}(\mathbf{x})}{\sigma^2} \tag{2.26}$$

The Jeans wavenumber k_J thus defined is a local quantity through the dependence $n(\mathbf{x})$ of the density on position.

B. The Effect of the Mean Field

The omission of the mean field in the preceding analysis provides for a simplified model and for a direct comparison with plasma theory, but is not a consistent approximation. We shall here summarize the main results obtained when the field is retained.[15]

The integration of (2.9) over the perpendicular velocity components now gives, instead of the algebraic relation (2.13) for $\delta f_{\mathbf{k}}(z,u)$, the differential equation

$$sE\frac{\partial}{\partial u}\delta\bar{f}_{\mathbf{k}}+i(ku-z)\delta\bar{f}_{\mathbf{k}}=h_{\mathbf{k}}-\delta F_{\mathbf{k}}\frac{\partial\bar{f}^{(0)}}{\partial u} \tag{2.27}$$

where sE is the component of the mean field parallel to $\mathbf{k}$:

$$E=|\mathbf{F}^{(0)}\cdot\mathbf{1}_{\mathbf{k}}| \qquad s=\operatorname{sgn}(\mathbf{F}^{(0)}\cdot\mathbf{k}) \tag{2.28}$$

The differential equation (2.27) can immediately be solved yielding the dispersion function:

$$\mathfrak{D}(z,k)=1+\frac{\omega_C^2}{n^{(0)}E^2}\int_{-s\infty}^{0}dw\int_{-\infty}^{\infty}du\,w\bar{f}^{(0)}(u+w)\times\exp\left\{i\left[w^2+2w\left(u-\frac{z}{k}\right)\right]\frac{k}{2sE}\right\} \tag{2.29}$$

This becomes, for a Maxwellian velocity distribution (2.22),

$$\mathfrak{D}=1+(\omega_C^2/2\sigma^2k^2\chi^2)Z'\left(\frac{z}{\sigma k\chi\sqrt{2}}\right) \tag{2.30}$$

$$\chi=\left(1+\frac{isE}{\sigma^2k}\right)^{1/2} \tag{2.31}$$

The dispersion relation $\mathfrak{D}=0$ is now accessible only to numerical analysis. It is found[15] that the mean field enhances the stability of the system. This can be readily understood in terms of the following picture of the instability mechanism. A sufficiently large scale density fluctuation breaks the balance between the pressure and the gravitational forces. The attractive center constituted by a positive fluctuation will tend to grow by accreting more and more matter from neighboring regions of diminishing density (leading at the limit to the fragmentation of a large system). The mean field constitutes an antagonistic force tending to maintain the existing structure, and its explicit retention in the calculation thus yields an enhancement of the stability.

This effect, which is position dependent, reduces the value of the critical wavevector $k_J(\mathbf{x})$, as determined for zero field, by up to a factor of 1.4 in the outer regions of the system (see Fig. 1). The essential point, however, is that the stability criterion for zero field is qualitatively confirmed: A system of dimensions $L_0 > 2\pi k_J^{-1}$, with k_J an average value defined from (2.26), is unstable to plane wave perturbations of $k < k_J$.

Stable gravitational systems can support density waves that, for the nonrotating systems and the plane wave perturbations we consider here, are, however, very strongly damped (see Fig. 1). (Our discussion does not apply to spiral waves in rotating systems.) The mean field gives rise to an interesting symmetry-breaking effect. The damping mechanism in the collisionless theory is, as for plasmas, that of Landau damping: It results

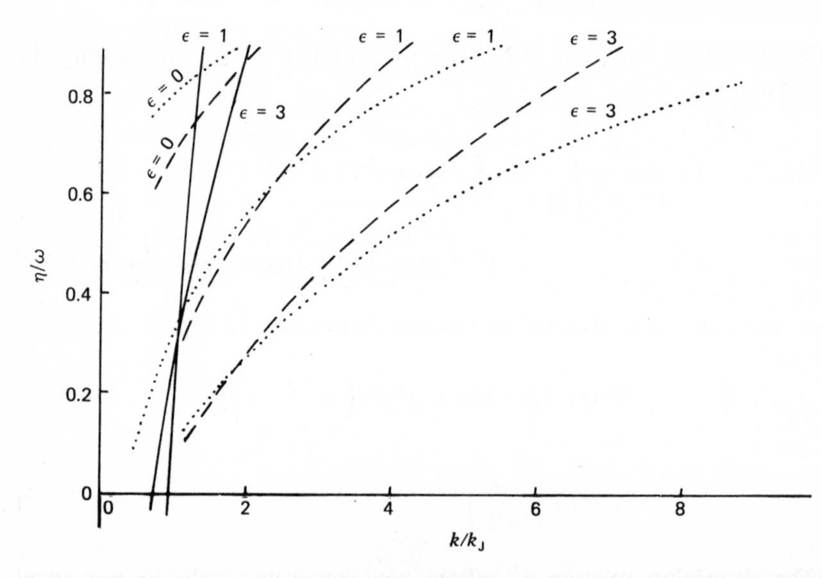

Fig. 1. Relative damping for inwardly directed waves ($\epsilon > 0$) and azimuthal waves ($\epsilon = 0$). (Outward waves have $\eta/\omega > 1$.) The case of zero field corresponds to azimuthal propagation. Full, dashed, and dotted lines correspond to the zero-, first-, and second-order modes, ordered according to increasing ω for given k. The dimensionless field parameter ϵ is defined as $\epsilon = E/\sigma^2 k_J$. The curves are obtained by numerical solution of the dispersion relation (2.30) for the Maxwellian, with $z = \omega - i\eta$. The stability is determined by the zero-order mode, k_C being reduced to 0.70 k_J for $\epsilon = 3$. Although from $\epsilon = 0$ to $\epsilon = 3$ the reduction of the damping is appreciable, the relative damping remains, in the most favorable case (first-order mode, $\epsilon = 3$), in excess of 0.25 for $k > 2k_J$. The restriction to $\epsilon_{\max} = 3$ corresponds to the outer regions $r \simeq R$ of the model density distribution $n(r) = (3N/\pi R^3)(1 + 4r^2/R^2)^{-5/2}$, where R is an effective radius containing 0.72 of the total mass for this model.

from the interaction of the density wave with the "resonant" particles moving with the wave at velocities close to its phase velocity. The faster particles give up energy to the wave, while the more numerous slower particles absorb energy, the net result being a damping of the wave. To see how this mechanism is modified by the mean field, which is radial and directed toward the center of the system, consider a density wave propagating inwards along the field. The resonant particles are in this case being accelerated by the field. Hence, superimposed upon the wave–particle interaction, there will be a net transfer of energy from the mean field to the wave, and correlatively a reduction of the damping. For an outwardly propagating wave on the contrary, the resonant particles are being slowed down by the field, so that the wave effectively feeds energy into the field and sees its damping increased. There is no effect upon azimuthal waves perpendicular to the field. [This is also immediately apparent from (2.27), which involves only the mean field component $\pm E$ along the direction of $\mathbf{k}$.]

The preferential propagation of inwardly directed perturbation is a quite large effect (see Fig. 1). The relative damping, however, remains strong, in excess of 0.25, if to respect stability we require $k > 2k_C$. This is too large a value for effective propagation, the e-folding time being less than one half-period. The role of density waves in a stable nonrotating system must, therefore, be considered as negligible. This conclusion is of importance for the derivation of the kinetic equations for the system.

An interesting computational aspect of the field-dependent problem consists in the raising of the degeneracy that certain velocity distributions introduce. Thus for a Lorentzian and zero field, the dispersion function is simply

$$\mathfrak{D} = 1 + \omega_C^2 (z + i\sigma k)^{-2} \tag{2.32}$$

so that only two modes appear, both purely damped:

$$\operatorname{Re} z = 0 \qquad \operatorname{Im} z = (\pm \omega_C - k\sigma) \tag{2.33}$$

When one takes the field into account, one finds an infinity of modes deriving from the dispersion function

$$\mathfrak{D} = 1 + \frac{\omega_C^2}{2isEk} Z'(\zeta), \qquad \zeta = \frac{1 - is}{2(Ek)^{1/2}} (z + ik\sigma) \tag{2.34}$$

The appearence of the plasma dispersion function Z irrespective of the nature of the velocity distribution, seems to follow necessarily from the presence of the imaginary exponential in (2.29); this itself is the direct consequence of the fact that due to the field the dispersion analysis derives from a differential equation (2.27) instead of from a purely algebraic relation.

C. Other Aspects of the Stability Analysis

The approach we have outlined above has two basic limitations. The scaling condition (2.11) implies that for stable systems one is led to extrapolate the theory beyond its domain of validity. Second, all results necessarily pertain to a local description of the system. The study of the global stability of inhomogeneous equilibria has been undertaken by different authors[16] in terms of a general energy principle. This approach is, however, quite unrelated to the problems of kinetic theory, and we must refer the reader to the original literature.

The dispersion analysis of the collisionless Boltzmann equation was originally carried out by Lynden-Bell for a rotating system.[17] Aside from their intrinsic interest, rotating systems present the major computational advantage of allowing for homogeneous equilibria. Indeed, consider the stationary collisionless Boltzmann equation in the rotating referential:

$$\mathbf{v}\cdot\frac{\partial f^{st}}{\partial \mathbf{x}}+(\mathbf{F}^{st}+\omega^2\mathbf{x}-2\boldsymbol{\omega}\times\mathbf{v})\cdot\frac{\partial f^{st}}{\partial \mathbf{v}}=0 \tag{2.35}$$

The total acceleration field will be independent of position if at each point the mean gravitational field and the centrifugal acceleration cancel:

$$\mathbf{F}^{st}=-\omega^2\mathbf{x} \tag{2.36}$$

Inserted into the Poisson equation, Eq. 2.36 leads to the nonvanishing uniform density $\Omega^2/2\pi Gm$. Using $v_{\parallel}=\mathbf{v}\cdot\boldsymbol{\omega}/\omega$, $v_{\perp}$, ϕ, as cylindrical polar coordinates in velocity space, one can verify that any function $f^{st}(v_{\parallel},v_{\perp})$ independent of ϕ and of position determines a homogeneous equilibrium. Note that Eq. 2.36 implies that one considers either a flat, two-dimensional system, or a uniform infinite cylinder of axis parallel to $\boldsymbol{\omega}$.

The linearized collisionless Boltzmann equation now reads

$$\left[\frac{\partial}{\partial t}+\mathbf{v}\cdot\left(\frac{\partial}{\partial \mathbf{x}}\right)-2(\boldsymbol{\omega}\times\mathbf{v})\cdot\left(\frac{\partial}{\partial \mathbf{v}}\right)\right]\delta f+\delta\mathbf{F}\cdot\left(\frac{\partial f^{st}}{\partial \mathbf{v}}\right)=0 \tag{2.37}$$

and is the exact transposition of the equation for a homogeneous plasma in a constant uniform magnetic field. For the infinite cylinder and $\mathbf{k}\cdot\boldsymbol{\omega}\neq 0$, one recovers[17] the criterion (2.26)(but written for a uniform system) as the necessary and sufficient condition for stability. The modes $\mathbf{k}\cdot\boldsymbol{\omega}=0$ turn out to be stable for all $\mathbf{k}$ provided that the condition

$$\omega^2>\pi Gmn \tag{2.38}$$

holds.[18] For a flat system the rotation also completely suppresses the instability provided that

$$\omega > \frac{1.18\, Gmn_s}{\sigma_r} \tag{2.39}$$

with n_s the surface density and σ_r^2 the radial velocity dispersion.[19] All the preceding results refer to stability against plane wave perturbation; the analysis of spiral modes, leading to the density wave theories of spiral galaxies,[20] constitutes a quite different problem.

The original demonstration of the gravitational instability by Jeans[1] derives from a fluid-dynamical description of the system. The normal mode analysis of the linearized fluid-dynamical equation yields, for an infinite homogeneous medium[21] of density n_0, the simple dispersion relation

$$\omega^2 = \sigma^2 k^2 - 4\pi Gmn_0 \tag{2.40}$$

This gives as critical wavenumber for stability

$$k_J^2 = \frac{4\pi Gmn_0}{\sigma^2} \tag{2.41}$$

in agreement with the result (2.26). The results of the stability analysis for rotating systems are also recovered.[21] The Landau damping mechanism does not appear at this level of description.

Finally, the virial theorem gives a nonperturbative approach to the problem of stability. From the well-known relation between kinetic and potential energies

$$2E_K = -E_V \tag{2.42}$$

one obtains for the continuum model of an isothermal sphere of mass Nm

$$3Nm\sigma^2 = \left(\frac{3}{5R}\right) GN^2m^2 \tag{2.43}$$

and, hence, for the radius of the stable system

$$R^2 = \frac{15\sigma^2}{4\pi Gmn_0} \tag{2.44}$$

In the following we shall use the characteristic Jeans' length:

$$L_J = \sigma(Gmn_0)^{-1/2} \tag{2.45}$$

to give an order of magnitude estimation of the dimensions of a stable system. (Note that for convenience we have taken $L_J = 2\pi^{1/2}/k_J$). If the stability analysis gives L_J as a maximum dimension, the virial theorem, however, indicates that a smaller system will expand to attain dimensions L_J. The role of L_J is remarkably borne out by observation for a wide variety of systems. Since rotation exerts a stabilizing influence in the plane, for rotating systems L_J characterizes the minimum dimension along the rotation axis. The nature of the system considered appears through m, the mass of the constituent elements. Thus, for gas clouds, star clusters or galaxies, and clusters of galaxies, the relevant entities of mass m are respectively molecules, stars, and the galaxies themselves. The largest systems can be formed by an imbricated hierarchy of clusters within clusters.

The fact that a stable system of given density n_0 and velocity dispersion σ^2 is approximately of linear dimensions $L_J(n_0, \sigma)$ will play an essential role in the elaboration of a kinetic theory.

III. KINETIC THEORY FOR A FINITE INHOMOGENEOUS SYSTEM

By the simple fact that one considers a finite and isolated system, the mathematical difficulties associated with the "infinite" range of the Newtonian potential are suppressed: The dimensions L_J of the system determine the maximum separation between two stars, and the potential $V(r)$ is effectively cut off at $r = L_J$. The concept of a definite boundary is, of course, an idealization but the approximation involved in neglecting pair interactions with the small numbers of stars that orbit out to distances in excess of L_J is quite insignificant. Moreover, it is legitimate here to ignore the time variation of the system dimensions, since this takes place on the slow, secular time scale.

The finite system is confined by its mean self-gravitational field which plays a dominant role. If the time variation of the mean field can in good approximation be considered as slow, no simplifying assumptions can realistically be made concerning its spatial variation, that is on the scale L_J. The essential difficulty of the statistical mechanical analysis will prove to center upon the conceptually simple problem of describing binary encounters in a smooth but strong field of arbitrary spatial variation.

The problem will be turned by arguing that the major contribution to collisional processes must come from encounters between stars that are not too distant, in a well-defined sense. This is a plausible physical assumption since, contrary to the situation in an unbounded system, the number of distant stars is limited. With this "proximate encounter" approximation, it is possible to obtain a suppression of the field effects in the collision integral, and to derive a kinetic equation generalizing the classical Fokker–Planck equation of stellar dynamics.[2,10]

The usual thermodynamic limit of a large system ($N \to \infty, \Omega \to \infty, N/\Omega = \text{constant}$, Ω being the volume) cannot be applied to a finite system,[22] in particular in view of the stability condition (2.21). It is the largeness of N and the passage to the asymptotic time limit that here suffice to determine the irreversible behavior. We do not attempt to make a rigorous discussion of this problem; the condition of large N is used explicitly to justify the weak-coupling approximation, and underlies the entire description in terms of a mean field and of smooth distribution functions.

The theoretical development is formulated for a nonrotating, isolated system of particles interacting through purely gravitational forces, and is thus most directly applicable to a globular cluster of stars. We shall, however, see that in part it can also be extended to rotating galaxies and to clusters of galaxies; we therefore shall continue to use the general denomination "gravitational system."

A. Characteristic Parameters and Scales

From the point of view of kinetic theory, gravitational systems do possess one basic element of simplicity: The strong coupling effects appear to be entirely expressed by the existence of stable, bound structures. For a given system, which is not too small, and with the effect of the mean field subtracted out, a weak-coupling condition is generously satisfied. The basic dimensionless parameter for gravitational systems is the Jeans number (see Eq. 2.45):

$$N_J = n_0 L_J^3 = n_0^{-1/2}(mG)^{-3/2}\sigma^3 \gg 1 \tag{3.1}$$

For a stable nonrotating system $N_J \sim N$ gives the number N of stars in the system of volume $\Omega \sim L_J^3$ and mean density $n_0 = N/\Omega$. The Jeans number is thus necessarily large in any system for which a statistical analysis is meaningful. Following Miller[9] we characterize the strength of the coupling by the ratio:

$$\frac{|E_V^{\text{corr}}|}{E_K^{\text{rel}}} = \frac{Gm}{\bar{r}\langle v_{12}^2\rangle} \cong N_J^{-2/3} \ll 1 \tag{3.2}$$

E_V^{corr} is the mean potential energy of near neighbors, while E_K^{rel} is the mean kinetic energy of their relative motion. Since the near-neighbor separation $\bar{r} \cong n_0^{-1/3}$, we see that, on the average, the coupling is of order $N_J^{-2/3}$ and can thus safely be considered as weak. (This does not exclude the occurrence of bound states; in the kinetic analysis, binaries, for example, are to be treated as single entities of a different mass species.)

The weakness of the coupling does not imply that the total potential energy per star is small. On the contrary, the virial theorem (2.42) shows that

$$\frac{E_V^{\text{tot}}}{E_V^{\text{corr}}} = \frac{2E_K}{|E_V^{\text{corr}}|} = 2N_J^{2/3} \gg 1 \tag{3.3}$$

The quasitotality of the potential energy lies in the mean field. Neglecting higher order correlations, one can write for the energy of the mean field E_V^{m}

$$|E_V^{\text{m}}| \cong |E_V^{\text{tot}}| - |E_V^{\text{corr}}| \gg |E_V^{\text{corr}}| \tag{3.4}$$

The Jeans length L_J, which determines the dimensions of the system in order of magnitude, defines the fundamental length scale. The associated time scale is the "crossing" time

$$\tau_C = L_J \langle v^2 \rangle^{-1/2} \cong (Gmn_0)^{-1/2} \tag{3.5}$$

and corresponds to the frequency ω_C of Section II.

A second characteristic length that will play an important role is the impact parameter L_{min} corresponding to a 90° deflection in a binary encounter:

$$L_{\text{min}} = \frac{G(m_1 + m_2)}{(\mathbf{v}_1 - \mathbf{v}_2)^2}, \qquad \langle L_{\text{min}} \rangle \cong \frac{L_J}{N_J} \tag{3.6}$$

Indeed the weak binary gravitational interactions becomes strong at short distances, and L_{min} constitutes a natural limit to the weak-coupling theory. In particular, it is well known that for inverse square forces the weak-coupling (Born) approximation to the cross-section coincides with the exact analysis if one introduces a short-distance cut-off at L_{min}.

The relaxation processes induced by two-body encounters will a priori develop on a kinetic time scale τ_R' inversely proportional to the concentration n_0. From (3.1) and (3.5) one would therefore expect the relaxation time τ_R' and the associated length scale L_R', the mean free path, to be given by

$$\tau_R' = \tau_C N_J \qquad L_R' = L_J N_J \tag{3.7}$$

An estimate of this nature is an anticipation of the results of a kinetic theory—one of its main objectives is precisely the determination of a characteristic relaxation time—and here it proves to be too large by a factor log N_J. The *a priori* expressions do bring out a typical aspect of the gravitational kinetic problem. The mean free path exceeds the linear dimensions of the system by a factor N_J, that is, by several orders of magnitude. This is to be accounted for by the curvature of the unperturbed trajectories: A mean free path then corresponds to a large number of revolutions within the finite system. The curvature itself is one of the aspects of the dominant role of the mean field.

However, it is with respect to the fluid-dynamical scale that gravitational systems contrast most strongly with their molecular counterparts. The scale of variation of the inhomogeneity of any system is at most comparable with its linear dimensions, so that

$$L_H \lesssim L_J \tag{3.8}$$

The fluid-dynamical length scale in thus essentially short: In the plasma analogy hydrodynamics would develop within a Debye sphere.

The fluid-dynamical time scale τ_H, on the other hand is a long time scale: It determines the rate of evolution in the secular phase as discussed in Section I; that is, it coincides with τ_S (and not with the crossing time τ_C corresponding to L_J). It will be specified according to the usual ordering

$$\tau_H \cong \tau_S \gg \tau_R \gg \tau_C \tag{3.9}$$

Indeed, in an aged system the fluid-dynamical quantities necessarily change more slowly than the associated velocity-dependent quantities of which they are the moments, and that vary on the scale τ_R. Thus the density $n(\mathbf{x}, t)$ must be a smoother function of time than the one-body distribution $f(\mathbf{x}, \mathbf{v}, t)$. For the particular conditions of Larson's numerical analysis of the fluid-dynamical equations[5], we note that τ_S is about two orders of magnitude greater than the relaxation time at the center of the cluster.

B. The Statistical Mechanical Model

The most convenient point of departure for a statistical-mechanical theory is here the BBGKY hierarchy. Expressed in terms of the one-body distribution function $f(i) = f(\mathbf{x}_i, \mathbf{v}_i, t)$, and of the pair and triplet correlations $g(i,j)$, $h(i,j,k)$, as defined by the usual cluster decomposition

$$f(1,2)=f(1)f(2)+g(1,2) \tag{3.10a}$$

$$\begin{aligned} f(1,2,3)=&f(1)f(2)f(3)+f(1)g(2,3)+f(2)g(3,1)\\ &+f(3)g(1,2)+h(1,2,3) \end{aligned} \tag{3.10b}$$

the first two equations of the hierarchy read:

$$[\partial_t+\mathbf{v}_1\cdot\nabla_1+\mathbf{F}_1\cdot\boldsymbol{\partial}_1]f(1)=\int d\mathbf{v}_2 d\mathbf{x}_2 \Theta_{12}\, g(1,2) \tag{3.11a}$$

$$\begin{aligned} \left[\partial_t+\sum_{i=1}^{2}(\mathbf{v}_i\cdot\nabla_i+\mathbf{F}_i\cdot\boldsymbol{\partial}_i)\right]g(1,2)=&\Theta_{12}[f(1)f(2)+g(1,2)]\\ &+\int d\mathbf{v}_3 d\mathbf{x}_3[\Theta_{13}f(1)g(2,3)+\Theta_{23}f(2)g(1,3)+(\Theta_{13}+\Theta_{23})h(1,2,3)] \end{aligned} \tag{3.11b}$$

We are considering an isolated nonrotating system of identical masses m and use the following notations:

$$\partial_t=\frac{\partial}{\partial t} \qquad \nabla_i=\frac{\partial}{\partial \mathbf{x}_i}, \qquad \boldsymbol{\partial}_i=\frac{\partial}{\partial \mathbf{v}_i}$$

$$\Theta_{ij}=\nabla_i V^{ij}\cdot\boldsymbol{\partial}_{ij} \qquad \boldsymbol{\partial}_{ij}=\boldsymbol{\partial}_i-\boldsymbol{\partial}_j \tag{3.12}$$

with V^{ij} the Newtonian potential

$$V^{ij}=-Gm|\mathbf{x}_i-\mathbf{x}_j|^{-1} \tag{3.13}$$

As in Section II $\mathbf{F}_i$ is the self-consistent acceleration field acting upon particle i; its explicit expression being

$$\mathbf{F}_i=-\boldsymbol{\Delta}_i\int d\mathbf{v}_j d\mathbf{x}_j V^{ij}f(j)=\mathbf{F}(\mathbf{x}_i,t) \tag{3.14}$$

To obtain a systematic approximation scheme for the hierarchy, we carry out a classical order-of-magnitude estimation according to the double criterion

$$G\to 0, \qquad \int d\mathbf{x}_j V^{ij}\cdots \equiv G\int d\mathbf{x}_j\cdots = O(1) \tag{3.15}$$

quite analogous to that of plasma theory.[14] The weak coupling property (3.2) allows us to treat the gravitational constant $G \sim N_J^{-2/3}$ as a small coupling parameter; the dominant role of the mean field (3.14) requires us to consider that, due to the long range of the interactions, a volume integration $\int d\mathbf{x}_j V^{ij}$ will compensate the smallness of G. The criterion (3.15) is to be completed by taking

$$f(i) = O(1), \qquad g(i,j) = O(G), \qquad h(i,j,k) = O(G^2) \tag{3.16}$$

The smallness of the pair correlation function follows most immediately from that of the correlational energy: Indeed, using Eqs. (3.10a) and (3.14), the relation (3.4) is expressed by

$$E_V^{\text{corr}} \equiv m \int d\mathbf{x}_1 d\mathbf{v}_1 \int d\mathbf{x}_2 d\mathbf{v}_2 \tfrac{1}{2} V^{12} g(1,2)$$
$$\ll m \int d\mathbf{x}_1 d\mathbf{v}_1 f(1) \int d\mathbf{x}_2 d\mathbf{v}_2 \tfrac{1}{2} V^{12} f(2) \equiv E_V^{\text{m}} \tag{3.17}$$

showing that $g = O(G)$. Alternatively, the first term on the right-hand side of (3.11b) allows us to identify g as $O(G)$. Similarly, in the equation for the triplet correlation $h(1,2,3)$, there appears a term such as $\Theta_{23}\, g(1,2)\, f(3)$, showing that h is necessarily $O(G^2)$.

Applying the criteria of (3.15) and (3.16) to the BBGKY hierarchy, we see that at zero order in uncompensated G, the hierarchy reduces to the left-hand side of (3.11a). The collisionless Boltzmann equation studied in the preceding section thus effectively constitutes the dominant approximation to the complete dynamical description of the system.

The one-body distribution $f(1)$ couples to the pair correlation $g(1,2)$ at first order in uncompensated G, that is, at order $N_J^{-2/3}$. In (3.11b) for $g(1,2)$, two terms can be neglected as being of order G^2: the coupling to the triplet correlation $h(1,2,3)$, and the term $\Theta_{12}\, g(1,2)$, which gives rise to the higher order Born approximations in the two-body scattering problem. This leaves us with

$$\left[\partial_t + \sum_{i=1}^{2} (\mathbf{v}_i \cdot \nabla_i + \mathbf{F}_i \cdot \partial_i)\right] g(1,2) = \Theta_{12} f(1) f(2)$$
$$+ \int d\mathbf{v}_3 d\mathbf{x}_3 \left[\Theta_{13} f(1)\, g(2,3) + \Theta_{23} f(2)\, g(1,3)\right] \tag{3.18}$$

The description provided by the closed set of coupled equations (3.11a) and (3.18) is familiar from plasma theory. In the latter context, the effective interactions and the correlations are of short range due to Debye screening. On this short-length scale it is usually a good approximation to consider the system as homogeneous. With this assumption, which also entails the vanishing of the field $\mathbf{F}_i$, the integrodifferential equation (3.18) can be solved explicitly[14]; the effect of the collective interactions expressed by the integral on the right-hand side of (3.18) is to introduce the dielectric function in the Balescu–Lenard–Guernsey collision term for plasmas, and thereby, the essential difference with respect to a simple weak-coupling model.

The situation is very different for gravitational systems. As observed, the hydrodynamical scale is here a short-length scale, so that the inhomogeneity may never be neglected. Moreover the mean field is always a dominant effect, so that the plasma approximation is inappropriate. We are, therefore, led to consider the simplified description obtained by omitting the collective effects in (3.18). Insofar as these are associated to the interaction of stars with vibration modes, they will not be important in a stable system. As discussed in Section II.B, the density waves in a nonrotating system are too strongly damped to be of significance in the collisional processes. However the collective behavior also introduces polarization effects. These are not negligible, but as has been shown by Gilbert[23] (and is further discussed in Section V), they can in good approximation be expressed by means of a renormalization of the stellar mass.

The self-consistent determination of the field $\mathbf{F}(\mathbf{x}_i, t)$ that appears in both hierarchy equations leaves us with one last but formidable difficulty. An essential simplification follows from the global evolution picture discussed in the introduction (Section I). After the violent relaxation phase lasting only a few crossing times, the density distribution, and thus also the field, settle down into a slow and smooth evolution. Our present concern is the behavior of the velocity dependent distributions in this smooth kinetic phase. Here, as in the stability analysis of Section II, we can separate out in the self-consistent field a dominant smooth contribution from a fluctuating term:

$$\mathbf{F}_i \equiv \mathbf{F}(\mathbf{x}_i, t) = \overline{\mathbf{F}}(\mathbf{x}_i, t) + \delta\mathbf{F}(\mathbf{x}_i, t) \rightarrow \overline{\mathbf{F}}(\mathbf{x}_i, t) \qquad (3.19)$$

It is consistent with our omission of collective effects in the hierarchy to ignore the fluctuations $\delta\mathbf{F}$, physically of little importance in stable systems.

The smooth mean field component $\overline{\mathbf{F}}(\mathbf{x}_i, t)$ is a typical fluid-dynamic quantity, being directly related to the density by the Poisson equation, and as such it develops on the secular time scale τ_S. By (3.9) we can thus consider $\overline{\mathbf{F}}(t)$ as quasistationary both with respect to the one-body distribution $f(t)$ and to the pair correlation $g(t)$. At this approximation, one can ignore the self-consistent determination (3.14) of $\overline{\mathbf{F}}$ in terms of f and consider $\overline{\mathbf{F}}$ as a *stationary field of given spatial variation*. This is a radical simplification for the analysis of $f(t)$. The operator describing the free motion in the mean field

$$\mathcal{L}_i \equiv \mathbf{v}_i \cdot \nabla_i + \overline{\mathbf{F}}(\mathbf{x}_i) \cdot \partial_i \tag{3.20}$$

can now be treated as a linear, time-independent operator in both hierarchy equations, (3.11a) and (3.18). Since in the latter the collective effects are to be omitted, the hierarchy finally reduces to the form

$$(\partial_t + \mathcal{L}_1) f(1) = \int d\mathbf{v}_2 \, d\mathbf{x}_2 \Theta_{12} g(1,2) \tag{3.21a}$$

$$(\partial_t + \mathcal{L}_1 + \mathcal{L}_2) g(1,2) = \Theta_{12} f(1) f(2) \tag{3.21b}$$

corresponding to the weak-coupling description of the two-body problem in an external field. The further approximations that will be necessary in order to cope with the arbitrary spatial dependence of the field will more appropriately be introduced and discussed in the next section.

C. Proximate Encounter Approximation

By substituting the solution of the pair equation (3.21b) into the singlet equation (3.21a), we get

$$(\partial_t + \mathcal{L}_1) f(\mathbf{x}_1, \mathbf{v}_1, t) = D(\mathbf{x}_1, \mathbf{v}_1, t)$$
$$+ \int_0^t d\tau \int d\mathbf{v}_2 \, d\mathbf{x}_2 \Theta_{12} \exp[-(\mathcal{L}_1 + \mathcal{L}_2)\tau] \Theta_{12} f(\mathbf{x}_1, \mathbf{v}_1, t-\tau) f(\mathbf{x}_2, \mathbf{v}_2, t-\tau) \tag{3.22}$$

with the so-called "destruction" term D expressing the influence of the initial correlations:

$$D(\mathbf{x}_1, \mathbf{v}_1, t) = \int d\mathbf{v}_2 \, d\mathbf{x}_2 \Theta_{12} \exp[-(\mathcal{L}_1 + \mathcal{L}_2)t] \, g(\mathbf{x}_1, \mathbf{v}_1; \mathbf{x}_2, \mathbf{v}_2; 0) \tag{3.23}$$

The evolution equation (3.22) is a typical non-Markoffian relation, in the sense that the time evolution at an instant t is determined by the history of the system over its entire past.

The formal treatment of the free motion operators $\mathcal{L}_i$ makes it possible to postpone the unavoidable approximations that will have to be made. However, no assumptions are to be introduced concerning the free uncorrelated motion. We therefore pass to an interaction representation for f:

$$f(\mathbf{x}_i, \mathbf{v}_i, t) = \exp(-\mathcal{L}_i t) \tilde{f}(\mathbf{x}_i, \mathbf{v}_i, t) \tag{3.24}$$

In terms of the $\tilde{f}$, Eq. 3.22 can be written

$$\exp(-\mathfrak{L}_1 t)\partial_t \tilde{f}(\mathbf{x}_1,\mathbf{v}_1,t) = D(\mathbf{x}_1,\mathbf{v}_1,t) + \int_0^t d\tau \int d\mathbf{v}_2 d\mathbf{x}_2 \Psi_{12}$$
$$\times \exp[-(\mathfrak{L}_1+\mathfrak{L}_2)t]\tilde{f}(\mathbf{x}_1,\mathbf{v}_1,t-\tau)\tilde{f}(\mathbf{x}_2,\mathbf{v}_2,t-\tau) \quad (3.25)$$

where we have separated out a generalized collision operator

$$\Psi_{12} \equiv \Theta_{12}\exp[-(\mathfrak{L}_1+\mathfrak{L}_2)\tau]\Theta_{12}\exp[+(\mathfrak{L}_1+\mathfrak{L}_2)\tau] \quad (3.26)$$

We next make some transformations in view of approximating the correlated motion as expressed by Ψ_{12}. We note that Leibnitz' rule for the differentiation of a product can be written

$$\frac{d^n}{dy^n}u(y)v(y) = \int dy'\delta(y-y')\left(\frac{d}{dy'}+\frac{d}{dy}\right)^n u(y')v(y) \quad (3.27)$$

This allows us to reformulate (3.26) as

$$\Psi_{12} = \Theta_{12}\int d(1,2)'\exp[-(\mathfrak{L}_1'+\mathfrak{L}_2'+\mathfrak{L}_1+\mathfrak{L}_2)\tau]\frac{\partial V(r')}{\partial \mathbf{r}'}$$
$$\cdot\, \boldsymbol{\partial}_{12}\exp[(\mathfrak{L}_1+\mathfrak{L}_2)\tau] \quad (3.28)$$

where use has been made of (3.12) and of the shorthand notations

$$d(1,2)' \equiv d\mathbf{x}_1' d\mathbf{v}_1' d\mathbf{x}_2' d\mathbf{v}_2' \delta(\mathbf{x}_1'-\mathbf{x}_1)\delta(\mathbf{v}_1'-\mathbf{v}_1)\delta(\mathbf{x}_2'-\mathbf{x}_2)\delta(\mathbf{v}_2'-\mathbf{v}_2) \quad (3.29a)$$

$$\mathbf{r} = \mathbf{x}_1 - \mathbf{x}_2, \qquad \mathbf{r}' = \mathbf{x}_1' - \mathbf{x}_2' \quad (3.29b)$$

The primed operators $\mathfrak{L}_i'$ differ from the $\mathfrak{L}_i$ only by the substitution of primed for unprimed variables.

The simplification of the collision operator Ψ_{12} depends crucially upon the properties of the commutators

$$[\boldsymbol{\partial}_{12}, \mathfrak{L}_1+\mathfrak{L}_2] = \boldsymbol{\nabla}_1 - \boldsymbol{\nabla}_2 \equiv \boldsymbol{\nabla}_{12} \quad (3.30)$$

$$[\mathbf{F}_i\cdot\boldsymbol{\partial}_i, \mathbf{v}_i\cdot\boldsymbol{\nabla}_i] = \mathbf{F}_i\cdot\boldsymbol{\nabla}_i - (\mathbf{v}_i\cdot\boldsymbol{\nabla}_i)(\mathbf{F}_i\cdot\boldsymbol{\partial}_i) \quad (3.31)$$

It is only in the case when these commutators themselves commute with the operators involved that one can go beyond the purely formal expression (3.28). This requires that in some sense the mean field can be treated as uniform in (3.28). Since the mean field varies on the scale L_J, this would imply that the encounters described by Ψ_{12} can effectively be limited to stellar separations $r_0 \ll L_J$, r_0 being a characteristic distance over which the field variation can be neglected.

In principle such a restriction on the encounters is excluded by the long range of the interactions, which extend over the entire volume of the system, of dimensions L_J. However, it is reasonable to expect that the major contribution to the collision term will be for pair separations r that are not too large. The specification

$$r < r_0 \ll L_J \tag{3.32}$$

will be taken as characterizing *proximate encounters* (p.e.). To take the mean field as uniform in the collision operator Ψ_{12} is thus in fact to make a *proximate encounter approximation*. The error thereby introduced should affect only the small contribution coming from distant encounters. It will be observed that not only are the individual effects of distant stars smaller, but also a greater degree of statistical cancellation can be expected.

The p.e. approximation would not be good for an infinite homogeneous model, where it is of course quite superfluous since there is no mean field to complicate the description. In such a model, the weakness of the distant interactions is compensated by their number to give a divergent total cross-section. Anticipating the discussion of Section III.F, we note that for a homogeneous model the p.e. approximation would bring down the long-distance cut-off from L_J to r_0, reducing the relaxation time by an appreciable factor $(\log L_J/r_0)$. For the finite systems we are here considering it would seem difficult to assess the p.e. approximation otherwise than by the numerical analysis of a detailed model.

To apply the p.e. approximation in the expression (3.28) of the collision operator, we ignore the position dependence of the mean field, writing $\overline{\mathbf{F}}_1 = \overline{\mathbf{F}}_2 = \overline{\mathbf{F}}$. One then obtains the following simplifications:

$$\begin{aligned}
\Psi_{12} &= \Theta_{12} \int d(1,2)' \exp\left[-\left(\mathbf{v}_1' \cdot \nabla_1' + \mathbf{v}_2' \cdot \nabla_2' + \overline{\mathbf{F}} \cdot \partial_1' + \overline{\mathbf{F}} \cdot \partial_2' \right) \tau \right] \\
&\quad \times \frac{\partial V(r')}{\partial \mathbf{r}'} \cdot (\partial_{12} + \tau \nabla_{12}) \\
&= \Theta_{12} \int d(1,2)' \{ \exp[-(\mathbf{v}_1' \cdot \nabla_1' + \mathbf{v}_2' \cdot \nabla_2') \tau] \} \\
&\quad \times \{ \exp[-\mathbf{F} \cdot (\partial_1' + \partial_2') \tau] \} \left\{ \exp\left[\frac{1}{2} \mathbf{F} \cdot (\nabla_1' + \nabla_2') \tau^2 \right] \right\} \frac{\partial V(r')}{\partial \mathbf{r}'} \cdot (\partial_{12} + \tau \nabla_{12}) \\
&= \Theta_{12} \int d(1,2)' \exp\left[-\tau (\mathbf{v}_1' - \mathbf{v}_2') \cdot \frac{\partial}{\partial \mathbf{r}'} \right] \frac{\partial V(r')}{\partial \mathbf{r}'} \cdot (\partial_{12} + \tau \nabla_{12})
\end{aligned} \tag{3.33}$$

In the first step we have used (3.30) to commute the exponential through the $\boldsymbol{\partial}_{12}$ operator; for the second step we have applied the Baker–Hausdorff theorem using Eq. 3.31 with $\nabla\bar{\mathbf{F}}=0$. Recalling Eq. 3.29*a*, we can now write the evolution equation (3.25) as

$$\exp(-\mathcal{L}_1 t)\partial_t\tilde{f}(\mathbf{x}_1,\mathbf{v}_1,t)=D(\mathbf{x}_1,\mathbf{v}_1,t)+\int d\mathbf{v}_2 d\mathbf{x}_2\int d\mathbf{r}\delta(\mathbf{r}-\mathbf{x}_1+\mathbf{x}_2)$$

$$\times\int_0^t d\tau(8\pi^3)^{-1}\tilde{\Psi}(\mathbf{r},\tau)\exp[-(\mathcal{L}_1+\mathcal{L}_2)t]\tilde{f}(\mathbf{x}_1,\mathbf{v}_1,t-\tau)\tilde{f}(\mathbf{x}_2,\mathbf{v}_2,t-\tau) \tag{3.34}$$

with

$$\tilde{\Psi}(\mathbf{r},\tau)=8\pi^3\Theta_{12}(\mathbf{r})\left[\exp\left(-\tau\mathbf{v}_{12}\cdot\frac{\partial}{\partial\mathbf{r}}\right)\right]\left[\Theta_{12}(\mathbf{r})+\tau\left(\frac{\partial V}{\partial\mathbf{r}}\right)\cdot\boldsymbol{\nabla}_{12}\right] \tag{3.35}$$

$$\mathbf{v}_{12}=\mathbf{v}_1-\mathbf{v}_2,\qquad \boldsymbol{\nabla}_{12}=\boldsymbol{\nabla}_1-\boldsymbol{\nabla}_2 \tag{3.36}$$

As was to be expected, the generalized collision operator in the p.e. approximation of (3.34) is independent of the mean field: Indeed, a uniform acceleration field will not modify the relative motion expressed by $\tilde{\Psi}(\mathbf{r},\tau)$. We note that the expression (3.34) can in fact directly be obtained from the exact formal collision operator (3.26) by simply annulling the field. In view of the self-consistent nature of the mean field, one can raise the objection that it is not consistent with the p.e. approximation to retain the delocalization effects that alone subsist in (3.34). This question is deferred to the general discussion of Section III.F. Finally, it should be stressed that the p.e. approximation concerns only the treatment of the correlations and that no approximation is made upon the free individual particle motion.

D. The Passage to the Kinetic Description

The consideration of a finite system trivially suppresses the difficulties associated to the long range of the gravitational forces: The maximum separation is restricted to the linear dimensions of the system. Quite generally one can replace a cut r^{-1} potential by a screened potential:

$$V(r)=\begin{cases} r^{-1}, & r\leqslant L_J \\ 0, & r>L_J\end{cases}\to V(r)=\frac{\exp(-\kappa r)}{r},\qquad \kappa=\frac{2\pi}{L_J} \tag{3.37}$$

The approximation hereby introduced is certainly acceptable in the light of the stronger p.e. approximation we have just made; given the relative indeterminacy of the dimensions of the system, one can insert the convenient factor 2π in the definition of the screening constant κ.

We can now pass to the usual Fourier–Laplace representation of the collision operator.[24] The Fourier transformation, which is introduced to give a meaning to the free motion propagator $\exp(-\tau\mathbf{v}_{12}\cdot\partial/\partial\mathbf{r})$ in (3.35), does not bear upon the distribution functions, but only on the interaction potential (3.37); we pass to a Fourier integral transform V_q by

$$V(r)=\int d\mathbf{q}\,V_q\exp(i\mathbf{q}\cdot\mathbf{r}) \qquad V_q=-Gm\left[2\pi^2(q^2+\kappa^2)\right]^{-1} \tag{3.38}$$

For the collision operator we introduce a Laplace transformation according to

$$\Phi(t)=\int\frac{dz}{-2\pi}\exp(-izt)\phi(z), \qquad \phi(z)=\int_0^{\infty}dt\exp(izt)\Phi(t) \tag{3.39}$$

In the absence of other indications, the contour for the z integration is a line antiparallel to and just above the real axis. With these transformations, the non-Markoffian evolution equation (3.34) can be written

$$\begin{aligned}\exp(-\mathfrak{L}_1 t)\partial_t\tilde{f}(\mathbf{x}_1,\mathbf{v}_1,t)=D(\mathbf{x}_1,\mathbf{v}_1,t)+\int d\mathbf{v}_2 d\mathbf{x}_2\int_0^t d\tau\int\frac{dz}{-2\pi}\exp(-iz\tau)\\ \times(8\pi^3)^{-1}\int d\mathbf{q}\exp\left[i\mathbf{q}\cdot(\mathbf{x}_1-\mathbf{x}_2)\right]\tilde{\psi}_{\mathbf{q}}(z)\\ \times\exp\left[-(\mathfrak{L}_1+\mathfrak{L}_2)t\right]\tilde{f}(\mathbf{x}_1,\mathbf{v}_1,t-\tau)\tilde{f}(\mathbf{x}_2,\mathbf{v}_2,t-\tau)\end{aligned} \tag{3.40}$$

$$\tilde{\psi}_{\mathbf{q}}(z)\equiv 8\pi^3\int d\mathbf{l}\,\theta^{12}_{\mathbf{l}-\mathbf{q}}\frac{-i}{\mathbf{l}\cdot\mathbf{v}_{12}-z}\left[\theta^{12}_{\mathbf{l}}-\frac{i}{\mathbf{l}\cdot\mathbf{v}_{12}-z}V_l\mathbf{l}\cdot\nabla_{12}\right] \tag{3.41}$$

$$\theta^{12}_{\mathbf{l}}\equiv V_l\mathbf{l}\cdot\boldsymbol{\partial}_{12} \tag{3.42}$$

The expression (3.41) for the collision operator is essentially the standard form for inhomogeneous systems[26] (the commutator contribution, which is the second term, is however not usually included in the definition of $\tilde{\psi}_{\mathbf{q}}$).

The non-Markoffian equations (3.34) or (3.40) still provide a reversible description of the evolution of the system. With the screened potential (3.37), it is easily shown [cf. Appendix, (A.30) and (A.36)] that one satisfies the conditions

$$D(\tau)\rightarrow 0, \qquad \tilde{\Psi}_{\mathbf{q}}(\tau)\rightarrow 0, \qquad \tau \gg \tau_C \tag{3.43}$$

which are required for the long-time irreversible kinetic approximation to be valid. The time scale for the memory of the destruction term and collision operator is determined by the screening constant $\kappa \sim L_J^{-1}$, that is, by the crossing-time scale.

Let us note that the mean field, which has been eliminated in $\tilde{\psi}_{\mathbf{q}}$ through the p.e. approximation, should reduce, rather than enhance, the memory effects. The physical picture underlying the restriction of the memory effects in a finite system to τ_C is quite clear. Indeed the crossing-time scale τ_C gives a measure of the orbital periods, and in a many-body system one does not expect the pair interactions to remain coherent over more than one revolution.[27]

Given the properties in (3.43), we can now apply the customary *Markoffianization procedure*[24] to pass from (3.40) to the kinetic equation. One omits the destruction term $D(t)$ and extends the upper bound on the time integral to $t\rightarrow\infty$. Since the use of an interaction representation in (3.40) ensures that the free motion is treated exactly, the distribution functions in the collision term can be synchronized: $\tilde{f}(t-\tau)\rightarrow\tilde{f}(t)$. This Markoffianization procedure is known[24] to neglect corrections in τ_C/τ_R, with τ_R the relaxation time determined by the kinetic equation. Such corrections are here $O(N_J^{-1})$ and thus quite negligible. One obtains the following after carrying out the z and τ integrations and reverting to ordinary representation by (3.24):

$$\begin{aligned}&\left(\partial_t+\mathbf{v}_1\cdot\nabla_1+\overline{\mathbf{F}}_1\cdot\partial_1\right)f(\mathbf{x}_1,\mathbf{v}_1,t)\\&\quad=\int d\mathbf{v}_2 d\mathbf{x}_2(8\pi^3)^{-1}\int d\mathbf{q}\exp[i\mathbf{q}\cdot(\mathbf{x}_1-\mathbf{x}_2)]\tilde{\psi}_{\mathbf{q}}(i0)f(\mathbf{x}_1,\mathbf{v}_1,t)f(\mathbf{x}_2,\mathbf{v}_2,t)\end{aligned} \tag{3.44}$$

The collision operator $\tilde{\psi}_{\mathbf{q}}(z)$ has been evaluated in the Appendix, (A.27) and (A.28). In its asymptotic expression for $z=+i0$ we explicit only the logarithmic term

$$\tilde{\psi}_{\mathbf{q}}(i0)=\hat{\psi}_{\mathbf{q}}(i0)+2\pi G^2m^2\partial_{12}\cdot\frac{\mathsf{I}v_{12}^2-\mathbf{v}_{12}\mathbf{v}_{12}}{v_{12}^3}\left(\log\frac{iKv_{12}-\mathbf{q}\cdot\mathbf{v}_{12}}{i\kappa v_{12}-\mathbf{q}\cdot\mathbf{v}_{12}}\right)\cdot\partial_{12} \tag{3.45}$$

with I the unit tensor. The divergence in the $\mathbf{l}$ integration (3.41) at large wavenumber, which is characteristic for the weak-coupling approximation, has been cut off at $l=K$. As discussed in Section III.A, K is determined by the minimum impact parameter $L_{\min}$ of (3.6):

$$K = \frac{2\pi}{L_{\min}} = \frac{\pi v_{12}^2}{Gm} \tag{3.46}$$

Since in average value one has

$$\left\langle \frac{\mathbf{q}\cdot\mathbf{v}_{12}}{v_{12}} \right\rangle \sim L_J^{-1} \sim \kappa \ll \langle K \rangle \tag{3.47}$$

and since by (3.6)

$$\log\left(\frac{\langle K\rangle}{\kappa}\right) = \log\left(\frac{L_J}{\langle L_{\min}\rangle}\right) \cong \log N_J \gg 1 \tag{3.48}$$

one sees that one can write

$$\log \frac{iKv_{12} - \mathbf{q}\cdot\mathbf{v}_{12}}{i\kappa v_{12} - \mathbf{q}\cdot\mathbf{v}_{12}} = \log\left(\frac{K}{\kappa}\right) + O(1) \tag{3.49}$$

where the notation $O(1)$ is introduced to indicate that we have only explicited the dominant contribution proportional to log N_J. At this approximation the asymptotic collision operator is simply

$$\tilde{\psi}_{\mathbf{q}}(i0) = \psi_W \log \frac{K}{\kappa} + O(1) \tag{3.50}$$

$$\psi_W \equiv 2\pi G^2 m^2 v_{12}^{-3} \boldsymbol{\partial}_{12} \cdot \left(\mathbf{l} v_{12}^2 - \mathbf{v}_{12}\mathbf{v}_{12}\right) \cdot \boldsymbol{\partial}_{12} \tag{3.51}$$

The dominant term explicited in (3.50) coincides with the Landau collision operator of plasma theory. It will here prove convenient to separate out the characteristic logarithm, and thereby to introduce the "weak-coupling collision operator" ψ_W of (3.51).

At dominant order in log N_J, the collision operator no longer depends on $\mathbf{q}$, the wavevector integration in (3.44) gives rise simply to a localization condition $8\pi^3\delta(\mathbf{x}_1 - \mathbf{x}_2)$, and we recover the classical weakly-coupled Boltzmann equation

$$\left(\partial_t + \mathbf{v}_1 \cdot \boldsymbol{\nabla}_1 + \overline{\mathbf{F}}_1 \cdot \boldsymbol{\partial}_1\right) f(\mathbf{x}_1, \mathbf{v}_1, t)$$

$$= \log\left(\frac{K}{\kappa}\right) \int dv_2 \psi_W f(\mathbf{x}_1, \mathbf{v}_1, t) f(\mathbf{x}_1, \mathbf{v}_2, t) + O(1) \tag{3.52}$$

with $O(1)$ denoting the nonexplicited subdominant terms. With the dominant collision term, we recover, as will be discussed in detail in Section III.F, the standard description of encounters in stellar dynamics. Among the well-known properties of (3.52), we recall the existence of an H-theorem, driving the system to a stationary, barometric equilibrium distribution.

Before discussing the relevance of these concepts to real gravitational systems, it is of interest to consider the nature of the stationary solution to the complete kinetic equation, subdominant terms included.

E. The Generalized Barometric Distribution

This section constitutes a parenthesis that, although raising an interesting theoretical point, is unimportant to the physics of the problem. Essentially, we show that the equilibrium one-body distribution for a system in an external field is given, in similar manner as the correlations at ordinary equilibrium, by an expansion in the coupling parameter.

Let us first recall that in our model the mean field appears as a given external field so that the Boltzmann distribution

$$f^{(0)}(i) \equiv f^{(0)}(\mathbf{x}_i, \mathbf{v}_i) = n^{(0)}(\mathbf{x}_i)(2\pi\sigma^2)^{-3/2} \exp\frac{-v_i^2}{2\sigma^2} \tag{3.53}$$

$$n^{(0)}(\mathbf{x}_i) = Q_1^{-1} \exp\frac{-U(\mathbf{x}_i)}{\sigma^2} \tag{3.54}$$

will appropriately be referred to as the (simple) barometric distribution. $U(\mathbf{x}_i)$ is the potential of the mean field,

$$\overline{\mathbf{F}}_i \equiv \overline{\mathbf{F}}(\mathbf{x}_i) = -\nabla_i U(\mathbf{x}_i) \tag{3.55}$$

while Q_1 assures the normalization of both $f^{(0)}$ and $n^{(0)}$ to N, the total number of stars. We note that separately

$$\mathcal{L}_i f^{(0)}(i) = 0 \qquad \psi_W f^{(0)}(1) f^{(0)}(2) = 0 \tag{3.56}$$

so that $f^{(0)}$ is a solution to the simple weak-coupling description of (3.52).

The collision term of the generalized equation (3.44) is at barometric equilibrium:

$$B^{(0)}(\mathbf{x}_1, \mathbf{v}_1) \equiv \int d\mathbf{v}_2 \, d\mathbf{x}_2 (8\pi^3)^{-1} \int d\mathbf{q} \exp[i\mathbf{q}\cdot(\mathbf{x}_1 - \mathbf{x}_2)] \tilde{\psi}_{\mathbf{q}}(i0) f^{(0)}(1) f^{(0)}(2) \tag{3.57}$$

So using (3.41)

$$B^{(0)}(\mathbf{x}_1,\mathbf{v}_1)=\sigma^{-2}\int d\mathbf{v}_2 d\mathbf{x}_2\int d\mathbf{q}\exp[i\mathbf{q}\cdot(\mathbf{x}_1-\mathbf{x}_2)]\int d\mathbf{l}\,\theta^{12}_{\mathbf{l}-\mathbf{q}}\ iV_l \times\left[1+i(\mathbf{l}\cdot\mathbf{v}_{12}-i0)^2\mathbf{l}\cdot(\overline{\mathbf{F}}_1-\overline{\mathbf{F}}_2)\right]f^{(0)}(1)f^{(0)}(2) \qquad (3.58)$$

The second contribution, originating from the commutator (3.30), now introduces the difference $\overline{\mathbf{F}}_1-\overline{\mathbf{F}}_2$. It would not be consistent with the p.e. approximation (3.32), by which the mean field was taken to be uniform ($\overline{\mathbf{F}}_1\cong\overline{\mathbf{F}}_2$), to retain this term here. The first term simplifies by applying the convolution theorem to give, with $\mathbf{x}_{12}=\mathbf{x}_1-\mathbf{x}_2$,

$$B^{(0)}(\mathbf{x}_1,\mathbf{v}_1)=-\sigma^{-2}\int dv_2 d\mathbf{x}_2\left[\frac{\partial V(x_{12})}{\partial\mathbf{x}_{12}}\right]\cdot\boldsymbol{\partial}_{12}V(x_{12})f^{(0)}(1)f^{(0)}(2)$$

$$=f^{(0)}(1)\sigma^{-4}\mathbf{v}_1\cdot\nabla_1\int d\mathbf{v}_2 d\mathbf{x}_2\frac{1}{2}(V^{12})^2 f^{(0)}(2) \qquad (3.59)$$

The nature of this result suggests that the stationary solution of the kinetic equation will take the form

$$f^{st}(\mathbf{x},\mathbf{v})=f^{(0)}(\mathbf{x},\mathbf{v})\left[1+G^2h^{(2)}(\mathbf{x})\right] \qquad (3.60)$$

where $h^{(2)}(\mathbf{x})$ depends only on position and is of order G^2. Indeed, substitute (3.60) into the kinetic equation (3.44):

$$\left(\partial_t+\mathbf{v}_1\cdot\nabla_1+\overline{\mathbf{F}}_1\cdot\boldsymbol{\partial}_1\right)f^{st}(\mathbf{x}_1,\mathbf{v}_1)=B^{(0)}(\mathbf{x}_1,\mathbf{v}_1)+O(G^4) \qquad (3.61)$$

and identify in powers of uncompensated G (recall that $V^{12}=O(G),\overline{F}=O(1)$). The zero-order equation corresponds to the first relation (3.56); there are no terms of order G; at order G^2 one gets

$$f^{(0)}(1)\mathbf{v}_1\cdot\nabla_1 G^2h^{(2)}(\mathbf{x}_1)=f^{(0)}(1)\sigma^{-4}\mathbf{v}_1\cdot\nabla_1\int d\mathbf{x}_2\frac{1}{2}(V^{12})^2\int d\mathbf{v}_2 f^{(0)}(2) \qquad (3.62)$$

We thus find for the stationary solution

$$f^{st}(\mathbf{x}_1,\mathbf{v}_1)=f^{(0)}(\mathbf{x}_1,\mathbf{v}_1)\left\{1+\frac{1}{2}\sigma^{-4}\left[G^2C_2+\int d\mathbf{x}_2(V^{12})^2 n^{(0)}(\mathbf{x}_2)\right]\right\} \qquad (3.63)$$

with the integration constant C_2 determined by the condition that both f^{st} and $f^{(0)}$ are normalized to N.

This result is of quite general nature: The fact that we are concerned with a gravitational rather than a molecular system appears only through the absence of terms of $O(G)$. For a molecular system such terms would be introduced by the mean field; whereas, here the mean field is $O(1)$ and plays the role of an external field.

We must therefore expect to recover the stationary distribution (3.63) from the N-body canonical ensemble for barometric equilibrium (of temperature $kT = m\sigma^2$ and with Q_N ensuring the normalization in configuration space),

$$f_N^{eq} = (2\pi\sigma^2)^{-3N/2} Q_N^{-1} \exp\left\{ -\sum_{i=1}^{N} \sigma^{-2}\left[\frac{1}{2}v_i^2 + U(\mathbf{x}_i) + \frac{1}{2}\sum_{j\neq i}^{N} V^{ij}\left(|\mathbf{x}_i - \mathbf{x}_j|\right)\right]\right\} \tag{3.64}$$

provided that we correctly identify the external potential $U(\mathbf{x}_i)$ with the potential of the mean field. The one-body distribution deriving from (3.64) is

$$f^{eq}(\mathbf{x}_1, \mathbf{v}_1) = N\int (d\mathbf{x}\, d\mathbf{v})^{N-1} f_N^{eq} \equiv f^{(0)}(\mathbf{x}_1, \mathbf{v}_1) h^{eq}(\mathbf{x}_1) \tag{3.65}$$

with [cf. (3.54)]

$$h^{eq}(\mathbf{x}_1) = NQ_1Q_N^{-1}\int (d\mathbf{x})^{N-1} \exp\left[-\sigma^{-2}\left(\sum_{i=2}^{N} U(\mathbf{x}_i) + \frac{1}{2}\sum_{i\neq j}^{N}\sum^{N} V^{ij}\right)\right] \tag{3.66}$$

The position dependence of f^{eq} is not given simply by the Boltzmann factor; there appear "imperfect gas" corrections h^{eq}. It is only for vanishing interactions that f^{eq} coincides with the simple barometric $f^{(0)}$.

Expanding in (3.66) both the integral and the configuration sum Q_N up to second order in G and using (3.54), we obtain

$$h^{eq}(\mathbf{x}_1) = 1 + \sigma^{-2}\left[GC_1 - \int d\mathbf{x}_2 V^{12} n^{(0)}(\mathbf{x}_2)\right] + \frac{1}{2}\sigma^{-4}\left[G^2C^2 + \int d\mathbf{x}_2 (V^{12})^2 n^{(0)}(\mathbf{x}_2) + \int d\mathbf{x}_2\, d\mathbf{x}_3 (V^{12}V^{13} + 2V^{12}V^{23}) n^{(0)}(\mathbf{x}_2) n^{(0)}(\mathbf{x}_3)\right] \tag{3.67}$$

where the C_i come from the expansion of Q_N and can be determined by the requirement that h^{eq} be normalized to N.

It remains to express the condition that the external field is in fact the mean field:

$$U(\mathbf{x}_i) = \int d\mathbf{x}_j V^{ij} n^{(0)}(\mathbf{x}_j) \tag{3.68}$$

This we can do simply by annulling the potential of the mean field, wherever it appears in the form given by the right-hand side of (3.68), since it is already accounted for by the $U(\mathbf{x}_i)$ dependence of the $n^{(0)}(\mathbf{x}_i)$. Correspondingly, we must also annul the $O(G)$ contribution from Q_N. The expression (3.67) then reduces identically to the quantity multiplying $f^{(0)}$ in (3.63), proving that our generalized barometric distribution derives from the canonical ensemble.

Given the smallness of the coupling parameter $G \sim N_J^{-2/3}$, the difference between the generalized and simple barometric distributions f^{st} and $f^{(0)}$ is not of physical significance. The question at issue is, however, of some theoretical interest and is a check on the internal consistency of the theory.

F. Comparison with the Classical Theory

The standard approach in stellar dynamics to the study of encounters[2,10] proceeds from the exact asymptotic analysis of the two-body scattering process in a homogeneous, field-free system. Although the various velocity moments of interest are directly evaluated without reference to a kinetic equation, the analysis made is equivalent to a description based upon the classical Boltzmann equation. It thus assumes localized encounters (scale L_{coll}) separated by rectilinear trajectories ($L_{\text{coll}} \ll L_R$, the mean free path) in a system that is homogeneous ($L_{\text{coll}} \ll L_H$) and without field. The inadaptation of this description to long-range inverse square forces shows up through the appearance of a long-distance divergence in the scattering cross-section. This is resolved by observing that the linear dimensions L_J of the system set an upper bound to the impact parameter and provide a natural cut-off.

This classical analysis can also be carried through in the weak-coupling approximation.[10] The incorrect treatment of close encounters introduces now also a divergence at short distances. This difficulty is to be resolved simply by introducing the cut-off $L_{\min}$ of (3.6), chosen so as to reproduce the results of the exact analysis of the two-body problem.

It is the weak-coupling variant of the standard analysis that can immediately be recovered from the generalized kinetic equation (3.44). It suffices to pass to a homogeneous description, for which the mean field

$\overline{\mathbf{F}}_i = 0$ and

$$f(\mathbf{x}_i, \mathbf{v}_i, t) = n_0 \varphi(\mathbf{v}_i, t), \qquad n_0 = \frac{N}{\Omega} \quad \text{(number density)} \tag{3.69}$$

Equation 3.44 then reduces to

$$\partial_t \varphi(\mathbf{v}_1, t) = n_0 \int d\mathbf{v}_2 \psi_0(i0) \varphi(\mathbf{v}_1, t) \varphi(\mathbf{v}_2, t) \tag{3.70}$$

where ψ_0 differs from $\tilde{\psi}_0$ by the omission of the commutator term (second term) in (3.41). Referring to the appendix, (A.33), the homogeneous collision operator $\psi_0(i0)$ is

$$\psi_0(i0) = \psi_W \left[\log\left(\frac{K}{\kappa} \right) - \frac{1}{2} \right] \cong \psi_W \log \frac{K}{\kappa} \tag{3.71}$$

so that we again find the weak-coupling collision operator of (3.50).

The assumptions implied in obtaining Eq. 3.70 are precisely those of the classical analysis: The system is taken as homogeneous and finite; the dynamics are described asymptotically (due to the Markoffianization). Effectively, the weak-coupling Boltzmann equation (3.70) coincides with the Fokker–Planck equation (see, for example, Ref. 14, Section 37) classically used in stellar dynamics. We should add that the velocity-dependent cut-off $K = \pi v_{12}^2 / Gm$ (Eq. 3.46) is usually replaced by an average value. This approximation has, however, recently been shown to result in a nonnegligible error.[28]

In comparing what we may thus call the "classical" equation (3.70) and the generalized kinetic equation (3.44), we first note that the two descriptions coincide at dominant order in $\log N_J$. It is only at subdominant order that the effects of the delocalization [$\mathbf{x}_2 \neq \mathbf{x}_1$ in (3.44)] of the encounters appear. Effects from the mean field appear neither in the classical equation, derived for a homogeneous system, nor in the generalized equation (3.44), from which they have been eliminated by the p.e. approximation.

The generalized weak-coupling equation (3.44) does not constitute a systematic description of the system to order G^2 for two main reasons. As previously discussed, it omits the polarization effects introduced by the right-hand side of (3.18), which are known to give an increased effective mass.[23] Second, the p.e. approximation gives a collision term where the inhomogeneity of the system (delocalization effects) is treated exactly but not the mean field. Since mean field and density are in fact self-consistently determined through the Poisson equation, it may well be questioned if it is significant to retain, through the subdominant contributions to $\tilde{\psi}_{\mathbf{q}}$, the effects of the delocalization but not of the field. Both effects are of course of the same order. But considering the manner in which they appear in the collision operator [the field comes in only through the "propagators" $\exp(-\mathcal{L}_i \tau)$, while the delocalization appears also in the interactions; see the $\mathbf{q}$ dependence of θ in (3.41)], it is excluded that they could exactly compensate. It seems, therefore, well worth investigating all effects that can be calculated, bearing in mind, however, that one has only an incomplete description.

Let us now consider the relevance of the theoretical results with respect to the "experimental" situation as provided by computer simulation. The predictions from the direct N-body dynamics are in good agreement—in remarkably good agreement given the relatively small numbers $N=250$ or $N=500$ to which the simulations are restricted—with those of the Monte-Carlo method based upon the classical analysis of collisions.[29] This suggests that the classical equation is much better than its derivations, either restricted by the p.e. approximation or by the assumption of homogeneity, and that polarization effects are not important.

In general the subdominant terms of the generalized kinetic equation will give only corrective effects: preliminary calculations for the relaxation time give typically an increase of 10% to 20%. For certain parameters the dominant collision terms gives a vanishing contribution, so that the only effect comes from the subdominant terms. This is the case for the balance equations for momentum and kinetic energy. It will be of interest to examine the consequence of this upon Larson's fluid-dynamical analysis of the secular behavior.

G. Further Discussion

The systematic approach we have developed here gives a better understanding and an extension of the classical kinetic description. It also provides a relatively firm basis for the discussion of important points of principle, such as the validity of the H-theorem, and of the thermodynamical description, as applied to finite stable gravitational systems.

The appearance of a "good" collision operator, giving rise to the H-property, is conditioned by the effective finite duration of the encounters, as expressed by (3.43). This in turn depends essentially upon two factors: explicitly, upon the effective finite range of the potential (3.37), that is, upon the finite volume of the system; implicitly, upon the largeness of $N \sim N_J$, which conditions the validity of a weak-coupling approximation (3.3), as of the entire description in terms of a mean field and of smooth distribution functions. The hypothetical retention of strong coupling, expressing for instance the formation of binaries (rare events in a large system but with a possible strong cumulative effect), would not be expected to lead to a "good" collision operator.

Given these two basic characteristics, finite effective interaction range and weak coupling, the approximations and omissions of the theory (proximate encounters, collective effects) appear as being of secondary nature. Moreover, in view of the experimental evidence both from observation and from numerical simulation, there is no reason to doubt the validity of the classical kinetic equation at dominant order ($\log N_J \gg 1$), nor, therefore, the validity of the H-property at this order.

In the generalized kinetic equation only the dominant contribution is purely dissipative, as may be seen by considering the invariance properties with respect to time reversal. The subdominant terms will thus be expected to give an oscillatory character to the time evolution. Although no proof can as yet be advanced, the indications are very strong that a weak H-theorem exists: From the kinetic description obtained here, it may be inferred that a finite stable system will display an oscillatory relaxation towards a Maxwellian velocity distribution.

This statement is immediately to be qualified by recalling that the kinetic phase is not an end point, but only an intermediate phase in the global evolution of a system such as a star cluster. Most of the characteristic properties of a cluster, such as the formation of a dense core or the anisotropy of the velocity distribution, develop in the secular phase, to be described by the fluid-dynamic equations.[5] The H-theorem of the kinetic phase is in large measure not relevant to the secular evolution, which can well culminate in Lynden-Bell's gravothermal instability.[7] The appropriate evolutionary criteria are here to be sought at the level of the fluid-dynamic equations, and a clear distinction must be drawn between the existence of a kinetic H-theorem and of a general evolutionary principle applying to the secular phase.

It is important to note that the encounters tend to destabilize the system, assumed to be stable with respect to the collisionless description. The tendency to establish a Maxwellian gives rise to the phenomenon of stellar evaporation[30] whereby a cluster progressively loses its more energetic members to the galactic field. The precise form of the velocity distribution is of importance in determining the structure of the star cluster,[31] which slowly evolves with time.

It is thus clear that the H-theorem expresses only a tendency, which is counteracted by a variety of effects. At the end of the kinetic phase and during the secular phase, the velocity distribution will at best present only small deviations with respect to a quasistatic Maxwellian equilibrium.

The problem presents itself somewhat differently for galaxies. Stellar encounters here give a relaxation time some orders of magnitude in excess of the Hubble age of the universe.[2] The structure of most galaxies is much more complex than that of star clusters, and a variety of other mechanisms can be invoked to account for the relaxed nature of the stellar velocity distribution (see Section V for further discussion). It is of interest to observe that the generalized kinetic equation can also be applied to rotating systems. Except in the case of rapid rotation, the acceleration field due to rotation can in first approximation be grouped together with the mean field; it is thus eliminated in the expression of the collision operator as a result of the p.e. approximation.

Reverting to the relatively well-defined situation for star clusters, we finally observe that the tendency toward a quasistatic equilibrium provides a justification for the use of a thermodynamical description. At the level of the dominant order weak-coupling equation (3.52), this description will be that of an ideal gas in an external field. The subdominant contributions here will give imperfect gas corrections that, being of order G^2, should be quite negligible. The paradoxical ideality of a gravitational system is of course entirely dependent upon the possibility of treating the mean self-gravitational field as an external field.

IV. KINETIC THEORY FOR INFINITE SYSTEMS

In the preceding section we considered isolated N-body systems limited in spatial extent by their own mean gravitational fields. Here we develop a model that will include longer range interactions between units aggregated in subsystems of a very large system, for example, galaxies, or bound pairs of galaxies, that are in local clouds and also clustered on larger scales. Since N_J is relatively small, we relax the requirement that the mean field be dominant. The large system is closer to being statistically homogeneous. Units are considered to travel rather freely between aggregates (clouds), rather than being confined to semiclosed orbits.

Although the size of the system or parts of it would normally change on the short time scale $\tau_C[\tau_C \approx (Gmn)^{-1/2}$, where m is the mass of a unit (or "particle") and n is their number density], we retain the weak-coupling approximation. Marginally stable situations are being considered, where there is formation and disruption of quasistable clouds by "tidal" interactions between units in different clouds.

Collective effects are thus omitted, except insofar as polarization is taken into account when m is specified. We are interested in describing secular evolutions using a picture of distant encounter effects rather than a more macroscopic picture of unstable fluid flows.

The new results to be presented are promising in several ways. They are (a) *simple*: the most interesting formulas are very short, and approximations are easily generated for finding explicit predictions; (b) *bizarre*, in the sense that they are radically different from the results of most weak-coupling theories, and therefore of great academic interest as counterexamples to hypothetical general results; (c) *physically reasonable*, in the sense that where they have a definite sign or order of magnitude it agrees with the one expected from other less sophisticated considerations, and (d) possibly *important* compared to competing and masking effects in non-empty ranges of system parameters. If N, the number of units in the large system, is allowed to remain large, one should not be afraid to extrapolate weak-coupling results to a range of intermediate coupling. Then, definitive tests will be relatively easier to invent since the postulated processes are relatively rapid.

A. An Approach to Intermediate Coupling

"Intermediate coupling" means that all time scales are similar, that several incomplete statistical descriptions of the same physical process will exist, and that

$$N \gg N_J \gtrsim O(1) \tag{4.1}$$

The evolution in such situations has on occasions been extrapolated from weak-coupling results.[32] Strictly representing weak (and therefore slow) processes, they will be applied to describe substantial momentum transfer between units. Let us give two examples:

(1) In expanding clusters of galaxies, the velocity dispersion is small, because of transport by any earlier large peculiar velocities of galaxies to regions where the center-of-mass velocity of each galaxy is similar to the local mean velocity. Thus, by (3.1) N_J is small, and one expects that few-body interactions may be dominant in generating subclusters.[33–35] Pair-correlational energy evolution in weak coupling may extrapolate to the binding energy evolution of the subclusters.

(2) Marginally bound systems with large halos are disrupted by tidal interactions with nearby similar systems (e.g., see Ref. 36). One would like to approach theoretical evolution predictions by studying interactions between particles in different systems, and thus over ranges much greater than the local Jeans length for the particles, calculated from the velocity dispersion and density within a system. It may not even be necessary to extrapolate the weak-coupling approximation; it could be that

$$N \gg N_J \gg 1 \tag{4.2}$$

The smaller mean global density would allow the support of the whole against rapid collapse, due to appreciable relative velocities of the systems with respect to each other—perhaps existing because of a global angular momentum.

The dismissal of the commutation problems associated with the mean field, and which for finite systems led to the imposition of a "proximate encounter" approximation (Section III.C), is now plausible for other reasons. The weak coupling between units is here consistent with the domination of mean field terms by streaming terms in the propagation of pair correlations.

B. Kinetic Description of Homogeneous Systems

In this section we study the velocity distribution for an infinite *homogeneous* nonrotating system of equal masses m. Inhomogeneities are left out until Section IV.E; we begin by ignoring the presence of such stabilizing aggregations, and rather study the unusual effort that the weak-coupling description exerts to describe the growing spatial correlations that would be expected from linear-instability descriptions. The divergence associated to the infinite range interactions is controlled by retaining a non-Markoffian evolution equation. It is found that the total kinetic energy increases monotonically on the kinetic time scale. Correspondingly, the magnitude of the correlational energy increases. The evolution of Boltzmann's H suggests that this behavior is of a dissipative or irreversible nature.

1. *The Equation of Evolution*

Since the system is homogeneous, the one particle distribution function reduces to the velocity distribution $\varphi(\mathbf{v}_1,t)$ (multiplied by a constant number density n_0). The general equation of evolution for the model is (cf. Eq. 3.40)

$$\partial_t \varphi(\mathbf{v}_1,t) = n_0^{-1} D(\mathbf{v}_1,t) + \int d\mathbf{v}_2 \int_0^t d\tau \int \frac{dz}{-2\pi}$$

$$\times \exp(-iz\tau)\psi(z)\varphi(\mathbf{v}_1,t-\tau)\varphi(\mathbf{v}_2,t-\tau) \tag{4.3}$$

$$\psi(z) = -8\pi^3 i n_0 \int d\mathbf{l}\, V_l \mathbf{l}\cdot\boldsymbol{\partial}_{12} (\mathbf{l}\cdot\mathbf{v}_{12} - z)^{-1} V_l \mathbf{l}\cdot\boldsymbol{\partial}_{12} \tag{4.4}$$

where again D is the "destruction term" (3.23) dependent upon the initial correlations $g(t=0)$; $\boldsymbol{\partial}_{12} \equiv \partial/\partial\mathbf{v}_1 - \partial/\partial\mathbf{v}_2$, and V_l is the Fourier transform of the interaction potential:

$$V_l = \lim_{\kappa\to 0} \frac{-Gm}{2\pi^2} \frac{1}{l^2+\kappa^2} \tag{4.5}$$

We temporarily retain the parameter κ so as to obtain a comparison with results in Section III.

We have

$$\psi(z)=G^2m^2n_0\partial_{12}\cdot\mathsf{T}(z)\cdot\partial_{12} \tag{4.6}$$

where

$$\mathsf{T}(z)=\lim_{\kappa\to 0}\frac{2}{i\pi}\int d\mathbf{l}\frac{1}{\mathbf{l}\cdot\mathbf{v}_{12}-z}\frac{\mathbf{ll}}{(l^2+\kappa^2)^2} \tag{4.7}$$

With the approximation

$$\int_{l<K}d\mathbf{l}\cdots\cong\int_{-\infty}^{\infty}d\left(\frac{\mathbf{l}\cdot\mathbf{v}_{12}}{v_{12}}\right)\int_0^K l_\perp dl_\perp\int_0^{2\pi}d\alpha\cdots \tag{4.8}$$

where K^{-1} is the short-range cutoff (3.46) and α is an azimuthal angle, we get

$$\mathsf{T}(z)=T_1(z)(\mathsf{I}v_{12}^2-\mathbf{v}_{12}\mathbf{v}_{12})+T_2(z)\mathbf{v}_{12}\mathbf{v}_{12} \tag{4.9}$$

$$T_1(z)=\lim_{\kappa\to 0}\frac{2\pi}{v_{12}^3}\left(\log\frac{z+iKv_{12}}{z+i\kappa v_{12}}-\frac{1}{2}\frac{iKv_{12}}{z+iKv_{12}}\right) \tag{4.10}$$

$$T_2(z)=\lim_{\kappa\to 0}\frac{2\pi}{v_{12}^3}\left(\frac{iKv_{12}}{z+iKv_{12}}-\frac{i\kappa v_{12}}{z+i\kappa v_{12}}\right) \tag{4.11}$$

(cf. Eqs. A.26 and A.32 in the appendix).

Next the inverse Laplace transformation is to be done. For finite systems it would be appropriate to take the asymptotic limit

$$\psi(z)\to\psi(i0) \tag{4.12}$$

corresponding to the pole in $z=0$ from $\int d\tau\exp(-iz\tau)$ in (4.3). Then the contribution $T_2(i0)$ would vanish. Here, on the other hand, the limit $\kappa\to 0$ gives a finite value

$$T_2(i0)=\frac{2\pi}{v_{12}^3} \tag{4.13}$$

The order of the limiting processes is thus seen to play a critical role, since it is precisely the contribution from (4.13) that gives the most interesting part of the anomalous behavior.

With the time dependence fully explicit, Eq. 4.3 becomes (cf. Eq. A.29)

$$\partial_t \varphi(\mathbf{v}_1,t) = n_0^{-1} D(\mathbf{v}_1,t) + \int d\mathbf{v}_2 \int_0^t d\tau G^2 m^2 n_0 \boldsymbol{\partial}_{12}$$

$$\cdot \left\{ \frac{2\pi}{v_{12}^3}\left[\mathbf{I} v_{12}^2 - \mathbf{v}_{12}\mathbf{v}_{12}\right]\left[\frac{1}{\tau} - \left(\frac{1}{\tau} + \frac{Kv_{12}}{2}\right)\exp(-Kv_{12}\tau)\right]\right.$$

$$\left. + 2\pi v_{12}^{-2}\mathbf{v}_{12}\mathbf{v}_{12} K \exp(-Kv_{12}\tau)\right\} \cdot \boldsymbol{\partial}_{12}\varphi(\mathbf{v}_1, t-\tau)\varphi(\mathbf{v}_2, t-\tau) \qquad (4.14)$$

It is appropriate to neglect the variation of φ on the short time scale $(Kv_{12})^{-1}$:

$$\partial_t \varphi(\mathbf{v}_1,t) = n_0^{-1} D(\mathbf{v}_1,t) + \int d\mathbf{v}_2 \phi \varphi(\mathbf{v}_1,t)\varphi(\mathbf{v}_2,t)$$

$$+ n_0 \int d\mathbf{v}_2 \left[\int_0^t d\tau \frac{1-\exp(-Kv_{12}\tau)}{\tau} \psi_W \varphi(\mathbf{v}_1,t-\tau)\varphi(\mathbf{v}_2,t-\tau)\right.$$

$$\left. - \tfrac{1}{2}\psi_W \varphi(\mathbf{v}_1,t)\varphi(\mathbf{v}_2,t)\right] \qquad (4.15)$$

where ψ_W is the "good" weak-coupling operator (3.51), and ϕ is the part corresponding to T_2:

$$\phi = 2\pi G^2 m^2 n_0 \boldsymbol{\partial}_{12} \cdot \frac{\mathbf{v}_{12}\mathbf{v}_{12}}{v_{12}^3} \cdot \boldsymbol{\partial}_{12} \qquad (4.16)$$

Essentially, we have replaced (4.6) by

$$\psi(z) = \phi + n_0 \psi_W \log \frac{iKv_{12} + z}{z e^{1/2}} \qquad (4.17)$$

The $1/\tau$ contribution in the collision integral of (4.14) or (4.15) determines a long memory tail, and leaves a characteristic "collision" time scale for distant encounters undefined: $\tau_{coll} \to \infty$. Physically, this corresponds to the impossibility of defining complete collisions for long range forces. The properties of (4.15) must be investigated at the non-Markoffian level of description.

2. *Effects of Initial Correlations*

For the destruction term $D(\mathbf{v}_1,t)$, the wavevector integration cannot be carried out without specifying to some degree the nature of the initial correlations $g(\mathbf{x}_1,\mathbf{x}_2,\mathbf{v}_1,\mathbf{v}_2,0)$. If one assumes the same functional dependence as in equilibrium:

$$g(0) = -n_0^2\sigma^{-2}V^{12}\varphi(\mathbf{v}_1,0)\varphi(\mathbf{v}_2,0) \tag{4.18}$$

with σ^2 the velocity dispersion; then, proceeding as for ψ, (cf. also Eq. 3.23), one gets[8]

$$D(\mathbf{v}_1,t) = -4\pi^2 n_0^2\sigma^{-2}\int d\mathbf{v}_2\int dz$$

$$\times \exp(-izt)\int d\mathbf{l}\, V_l \mathbf{l}\cdot\boldsymbol{\partial}_{12}(\mathbf{l}\cdot\mathbf{v}_{12}-z)^{-1}V_l\varphi(\mathbf{v}_1,0)\varphi(\mathbf{v}_2,0) \tag{4.19a}$$

$$= (2\pi i)^{-1}G^2m^2n_0^2\sigma^{-2}\int d\mathbf{v}_2\int dz \exp(-izt)\boldsymbol{\partial}_{12}\cdot\mathbf{S}(z)\varphi(\mathbf{v}_1,0)\varphi(\mathbf{v}_2,0) \tag{4.19b}$$

where (cf. Eq. A.25)

$$\mathbf{S}(z) = \lim_{\kappa\to 0}\frac{2}{i\pi}\int d\mathbf{l}\frac{1}{\mathbf{l}\cdot\mathbf{v}_{12}-z}\,\frac{\mathbf{l}}{(l^2+\kappa^2)^2} \cong \frac{2\pi}{z}\frac{\mathbf{v}_{12}}{v_{12}} \tag{4.20}$$

So D maintains a constant value

$$D(\mathbf{v}_1,t) = 2\pi G^2m^2n_0^2\sigma^{-2}\int d\mathbf{v}_2\boldsymbol{\partial}_{12}\cdot\left(\frac{\mathbf{v}_{12}}{v_{12}}\right)\varphi(\mathbf{v}_1,0)\varphi(\mathbf{v}_2,0) \tag{4.21}$$

The anomalous character of the collision operator finds a counterpart in the equally abnormal behavior of the destruction term, at least for the class of initial correlations (4.18).

On the other hand, if an "initial time" $t_0 \leqslant 0$ can be found when there are no pair correlations, then D does not contribute to the evolution equations.[37]

3. *Deflection of the Approach to Equilibrium*

Let us note one useful mathematical feature of the results (4.15) and (4.17): While the existence of the term ϕ and of the $\log z$ factor are both related to the long range of the potential, the two complications appear in separate terms. Somewhat paradoxically, the essentially non-Markoffian evolution is the part that contains the "good" operator ψ_W. As will be argued in Section IV.C, that is enough to cause the term to drive $\varphi(\mathbf{v}_1,t)$

toward the Maxwellian equilibrium. The term is zero at equilibrium and conserves total kinetic energy.

It is the other term $\int d\mathbf{v}_2 \phi\varphi\varphi$ that causes a deflection away from the Maxwellian. Being smaller than the relaxation term by a factor of order $\log N_J^2$, it becomes important in the late stages of evolution when $\psi_W \varphi\varphi$ has sufficiently decayed. Indeed, for the equilibrium distribution

$$\phi\varphi^{(0)}(\mathbf{v}_1)\varphi^{(0)}(\mathbf{v}_2) = -8\pi G^2 m^2 n_0 \sigma^{-2} v_{12}^{-1}\left(1 - \frac{1}{4}\sigma^{-2} v_{12}^2\right)\varphi^{(0)}(\mathbf{v}_1)\varphi^{(0)}(\mathbf{v}_2) \neq 0 \tag{4.22}$$

and we shall see (Section IV.D) that it causes the total kinetic energy $N\int d\mathbf{v}_1 \frac{1}{2} m v_1^2 \varphi(\mathbf{v}_1)$ to increase, whether or not $\varphi = \varphi^{(0)}$.

The property (4.22) stands in direct relation to the persistence (4.21) of the destruction term. Indeed the evolution equation (4.15) retains a property of the Liouville equation: The canonical distribution (equilibrium φ and g) is stationary:

$$n_0^{-1} D\{g^{(0)}\} + \int d\mathbf{v}_2 \phi\varphi^{(0)}\varphi^{(0)} = 0 \tag{4.23}$$

We still must be prepared to look for reasonable approximations for $\varphi(\mathbf{v}_1, t)$ in order to calculate the novel macroscopic evolution due to ϕ. Thus the $\psi_W \log z$ term cannot be summarily dismissed. In the next section we shall see in what measure the contributions from the branch point at $z = 0$ affects the relaxation process.

C. Linear Model of Non-Markoffian Behavior

A linear model is introduced via the weak-coupling master equation[24] that retains the linearity of the N-body Liouville equation. In this way we avoid linearizing with respect to the difference between $\varphi(\mathbf{v}, t)$ and some particular fixed distribution such as $\varphi^{(0)}(\mathbf{v})$. The master equation for the N-body velocity distribution is

$$\partial_t \varphi_N(t) = D_N(t) + \int_0^t d\tau \int \frac{dz}{-2\pi} \exp(-iz\tau)\psi_N(z)\varphi_N(t-\tau) \tag{4.24}$$

where ψ_N contains a sum over all particle pairs. Equation 4.3 for $\varphi(\mathbf{v}_1, t)$ is recovered by integrating over $\mathbf{v}_2, \mathbf{v}_3, \ldots, \mathbf{v}_N$ and factorizing:

$$\varphi_2(\mathbf{v}_1, \mathbf{v}_2, t) = \varphi(\mathbf{v}_1, t)\varphi(\mathbf{v}_2, t) \tag{4.25}$$

The simplification then consists of assuming that one can correctly approximate the time evolution of $\varphi(\mathbf{v}_1,t)$ by considering the linear equation

$$\partial_t \rho(t) = \int_0^t d\tau \int \frac{dz}{-2\pi} \exp(-iz\tau)\left(\log\frac{z+i\gamma}{z}\right)\bar{\psi}\rho(t-\tau) \tag{4.26}$$

where $\bar{\psi}$ has simple eigenfunctions such that

$$\rho = \sum_{\nu=0}^{\infty} \rho^{(\nu)}|\nu\rangle \tag{4.27}$$

$$\bar{\psi}|\nu\rangle = -\nu|\nu\rangle, \qquad \nu \geqslant 0 \tag{4.28}$$

Here, $\gamma \cong \langle K v_{12}\rangle$, γ^{-1} being the scale of duration of close encounters. If we identify $\rho(t)$ with $\varphi_2(\mathbf{v}_1,\mathbf{v}_2,t)$ and $\bar{\psi}$ with $n_0\psi_W$, then integrating Eq. 4.26 with respect to $\mathbf{v}_2$ would reduce it to (4.3) and (4.17) with D and ϕ suppressed.

The evolution equation (4.26) is now solved by projecting upon the eigenfunction space $|\nu\rangle$ for a particular $\nu \neq 0$. Taking the Laplace transform gives

$$-iz\tilde{\rho}^{(\nu)}(z) - \rho^{(\nu)}(t=0) = -\nu\tilde{\rho}^{(\nu)}(z)\log\left(\frac{z+i\gamma}{z}\right) \tag{4.29}$$

Then applying the inverse transformation, we have

$$\rho^{(\nu)}(t) = \frac{1}{2\pi i}\int dz \frac{\exp(-izt)}{z + i\nu\log(1+i\gamma/z)}\rho^{(\nu)}(0) \tag{4.30}$$

where again the contour is from right to left above the real axis and may be closed by a semicircle in the lower half plane with radius greater than γ.

The passage to the dimensionless frequency $\zeta = z/\nu$,

$$\rho^{(\nu)}(t) = \frac{1}{2\pi i}\int d\zeta \frac{\exp(-i\zeta\nu t)}{\zeta - i\log\zeta + i\log(\zeta + i\beta)}\rho^{(\nu)}(0) \tag{4.31}$$

introduces the characteristic parameter

$$\beta = \frac{\gamma}{\nu} \tag{4.32}$$

The weak-coupling approximation for the model implies that not only β is large but also $\log\beta$:

$$\beta = O\left(N_J^2\right) \gg 1, \qquad \log\beta = O(\log N_J) \gg 1 \tag{4.33}$$

We now sketch out the main steps in the integration of (4.31), referring to the original paper[38] for details. The integrand presents two logarithmic branch points at $\zeta = -i\beta$ and $\zeta = 0$. It also has two poles at $\zeta = \zeta_{\pm}$, given by the solutions of the transcendental equation

$$\frac{\zeta}{i} = \log\left(\frac{\zeta}{i\beta}\right) - \log\left(1 + \frac{\zeta}{i\beta}\right) \tag{4.34}$$

Using Eq. 4.33 one finds that

$$\zeta_{\pm} = -i\left[B - \frac{1}{2}\pi^2 B^{-2} + O\left(B^{-3}\right)\right] \pm \pi\left[1 - B^{-1} + O\left(B^{-2}\right)\right] \tag{4.35}$$

with B defined by

$$B + \log B = \log\beta \tag{4.36}$$

or by iteration

$$B = \log\beta - (\log\log\beta)\left[1 + (\log\beta)^{-1} + O\left(\left[\log\beta\right]^{-2}\right)\right] \tag{4.37}$$

To evaluate (4.31) we integrate counterclockwise around a branch cut on the imaginary axis and around the poles (4.35). The contribution from the cut is

$$\int_0^\beta dx \frac{\exp(-x\nu t)}{\left(x + \log\left[x/(\beta - x)\right]\right)^2 + \pi^2} = \alpha \int_0^\infty dx \frac{\exp(-x\nu t)}{(x - B)^2 + \pi^2} - \alpha'$$

$$1 - 2B^{-1}\log B < \alpha < 1, \qquad 0 < \alpha' < B^{-2} \tag{4.38}$$

which we can closely approximate by the integral on the right-hand side. The latter can be expressed in terms of the exponential integral of complex argument

$$E_1(u) = \int_u^\infty du' \frac{\exp(-u')}{u'} = \frac{\exp(-u)}{u}\left[1 - \frac{1}{u} + O\left(\frac{1}{u^2}\right)\right] \tag{4.39}$$

The contribution from the poles (4.35) is evaluated by straightforward residue integration, and one finally obtains[38]

$$\rho^{(\nu)}(t)=\{2(1-B^{-1})\exp(-B\nu t)\cos\pi(1-B^{-1})\nu t$$
$$-\alpha\pi^{-1}\mathrm{Im}\left[\exp[(-B+i\pi)\nu t]E_1(-B\nu t+i\pi\nu t)\right]+O(B^{-2})\}\rho^{(\nu)}(0) \tag{4.40}$$

The leading term in this expression, which is the contribution from the poles, is an oscillatory relaxation that is so strongly damped (when $B \sim \log N_J^2 \gg \pi$) that it practically reduces to the exponential factor. The second term, originating from the branch cut, asymptotically vanishes only as $1/t$, as can be seen from the expansion (4.39). For long times it will eventually dominate the pole contribution.

Numerical values are given in Table I for various values of $B \sim \log N_J^2$ and of $B\nu t$. One sees that for $N_J > 10^4$ the branch cut contribution becomes dominant only after at least four relaxation times $(B\nu)^{-1}$. It will, however, in general mask the oscillatory nature of the relaxation.

TABLE I.
Relative Importance of the Pole (p) and the Branch-cut (b) Contributions

	$B=4\pi$ ($N_J \approx 180$)		$B=8\pi$ ($N_J \approx 14{,}000$)		$B=12\pi$ ($N_J \approx 10^6$)	
$B\nu t$	(p)	(b)	(p)	(b)	(p)	(b)
1	0.66	0.34	0.70	0.35	0.72	0.33
2	0.22	0.15	0.25	0.14	0.25	0.14
3	0.71	0.075	0.090	0.066	0.094	0.061
4	0.20	0.045	0.031	0.033	0.034	0.028
5	0.005	0.031	0.010	0.019	0.012	0.015

In a global characterization of the evolution determined by (4.26), we see that the oscillatory behavior usually associated with non-Markoffian effects[39] is here completely masked. The evolution begins as an exponential relaxation, on a time scale

$$\tau_R=(B\nu_{\min})^{-1}=\frac{\tau_R'}{\log N_J^2} \tag{4.41}$$

with the time scale τ_R' thus defined determined by the smallest nonzero eigenvalue of $\bar{\psi}$. In its final phase, however, the evolution towards equilibrium is characterized by a long, slowly decaying tail in $1/t$.

For many situations, the long tail plays no role, since it becomes important only when the damping has nearly been completed. Correspondingly, one may ignore the branch-cut contribution in (4.40) as being a "low-amplitude" correction. Since the oscillation is not important, one sees that the dominant behavior of (4.26) is correctly rendered by the equation:

$$\partial_t \rho^{(\nu)}(t) = -B\nu\rho^{(\nu)}(t) = -\nu\left(\log\frac{\gamma}{\nu}\right)\rho^{(\nu)}(t) \tag{4.42}$$

when t is of order τ_R.

This Markoffian approximation to the original essentially non-Markoffian description (4.26) can be reproduced by a simple *ansatz*:

> Insert into (4.26) a memory cut-off $\exp(-\epsilon_0\tau)$ and apply the standard Markoffianization procedure as in Section III.D.

Neglecting where possible γ^{-1} in comparison to ϵ_0^{-1}, and ignoring again the branch-cut contributions, one obtains

$$\partial_t \rho = \left(\log\frac{\gamma}{\epsilon_0}\right)\bar{\psi}\rho(t) \tag{4.43}$$

Equation 4.42 is thus recovered if one identifies ϵ_0^{-1} with a characteristic relaxation time scale τ_R. One can then determine τ_R from the Markoffian relation (4.43) itself (cf. Ref. 40):

$$\tau_R^{-1} \cong \nu \log \gamma\tau_R \tag{4.44}$$

giving by iteration,

$$\tau_R^{-1} = \epsilon_0 = \nu\left[\log\left(\frac{\gamma}{\nu}\right) + O(1)\right] \tag{4.45}$$

in agreement with the direct result (4.41), with ν a characteristic eigenvalue of $\bar{\psi}$.

The foregoing ansatz is suggested for situations where a complete mathematical treatment is intractable. Within our approximations it will prove of value in Section IV.D, where attention is focussed on the more important effects of the Markoffian contribution ϕ.

Finally, it is of interest to compare our simplified result (4.43) with the standard collision term (3.70). To the relaxation time τ_R one can associate a mean free path $L_R = \tau_R\langle v^2\rangle^{1/2}$. Comparing Eq. 4.43 with (3.70), we see that our approximate (and partial) result for an infinite system can be reproduced from the standard analysis by displacing the long-distance cut-off from $O(L_J)$, a typical dimension of the system, to $L_R = \epsilon_0^{-1}\langle v^2\rangle^{1/2}$, the mean free path. This fits in well with the intuitive picture[40] that one can construct a relaxation process in a system where the duration of the "collisions" is arbitrarily long. The renormalization effect expressed by (4.44) can be understood as resulting from an overlapping of incomplete collisions.

D. Macroscopic Evolution

We now return to the non-Markoffian description (4.15), or Eqs. 4.3 and 4.17, for an initially uncorrelated system: $D=0$. It has just been ascertained that the term containing the "good" operator ψ_W, defined by (3.51), will drive the system towards equilibrium, the unusual form of the logarithmic coefficient making a correction that was estimated to be small.

The rate of change of mean kinetic energy per particle,

$$\partial_t E_K = \int d\mathbf{v}_1 \left(\frac{1}{2} m v_1^2\right) \partial_t \varphi(\mathbf{v}_1, t) \tag{4.46}$$

is anyway independent of such estimates. Indeed, with the abbreviations

$$\mathbf{w} = \frac{1}{2}(\mathbf{v}_1 + \mathbf{v}_2) \tag{4.47}$$

$$\varphi\varphi(t) = \varphi\left(\mathbf{w} + \frac{1}{2}\mathbf{v}_{12}, t\right)\varphi\left(\mathbf{w} - \frac{1}{2}\mathbf{v}_{12}, t\right) \tag{4.48}$$

we have from (4.15), with $D=0$,

$$\begin{aligned} \partial_t E_K &= \frac{1}{2} m \int d\mathbf{v}_1 d\mathbf{v}_2 \left(w^2 + \frac{1}{4} v_{12}^2\right) \left[\phi\varphi\varphi(t) + \psi_W \int_0^t d\tau \cdots \right] \\ &= \frac{1}{8} m \int d\mathbf{v}_1 d\mathbf{v}_2 v_{12}^2 \phi\varphi\varphi(t) \end{aligned} \tag{4.49}$$

There is no contribution from $\psi_W \int_0^t d\tau \cdots$; being a "good" collision operator, ψ_W conserves kinetic energy, as can be verified by integrating by parts. As for ϕ (cf. Eq. 4.16), we arrive at the very simple form

$$\partial_t E_K = 4\pi G^2 m^3 n_0 \int d\mathbf{w}\, d\mathbf{v}_{12} v_{12}^{-1} \varphi\varphi(t) > 0 \tag{4.50}$$

which is always positive since $\varphi\varphi(t) \geqslant 0$ everywhere.

To show the generality of such results (when $D=0$), let us quote a similar result[41] for stars with various masses interacting in an infinite, rapidly rotating system. For the kinetic energy along the axis of rotation, neglecting exchange with kinetic energy normal to the axis,

$$n_0\partial_t\tilde{E}_K=\sum_{\nu_1}\sum_{\nu_2}\int_{-\infty}^{\infty}dp_1\int_{-\infty}^{\infty}dp_2 2\pi G^2 m_{\nu_1}m_{\nu_2}(m_{\nu_1}+m_{\nu_2})N_{\nu_1}N_{\nu_2}\Omega^{-2}$$
$$\times\delta\left(\frac{p_1}{m_{\nu_1}}-\frac{p_2}{m_{\nu_2}}\right)\tilde{\varphi}_{\nu_1}(p_1,t)\tilde{\varphi}_{\nu_2}(p_2,t)>0 \qquad (4.51)$$

where m_{ν_i} is the mass of a star of type ν_i, N_{ν_i} is the mean number of such stars in a volume Ω, and $\tilde{\varphi}_{\nu_i}$ is the corresponding distribution of momentum components p_i along the axis of rotation.

Such results suggest the existence of a general relation

$$\partial_t\sigma^3=O(G^2 m\rho)>0 \qquad (4.52)$$

where σ^2 is the mean velocity dispersion, ρ is the mass density of the most effective scatterers (units), and m is an effective mass per scatterer. We shall verify (Eq. 4.107) that even in the presence of inhomogeneities, such results survive, with other terms to be added. The dominant contribution comes from pairs of particles with low relative velocities [note for instance the factor v_{12}^{-1} in (4.50)].

The retention of the destruction term, that is, of nonvanishing initial correlations, can, of course, affect the result (4.50); indeed, for canonical equilibrium it is to be replaced by $\partial_t E_K=0$ (cf. Eq. 4.23).

By calculating the pair correlation and hence the potential energy,[8,37] one can verify the necessary conservation of total energy:

$$\partial_t E_V=-\partial_t E_K<0 \qquad (4.53)$$

The appearance of a production of negative correlational energy is physically plausible, if unusual in weak-coupling theories. When initial correlations are negligible in some sense, the dynamical interactions will on average cause the pairs to fall towards each other, building up kinetic and potential energy. This would be the only contribution from distant pairs if the time available is such that the collision process is very incomplete.

In infinite systems with weak coupling, the rates are slow compared to the Jeans time $L_J\sigma^{-1}$. But extrapolations and extensions could make the results significant (cf. Sections IV.A and IV.E). In particular, for systems that are slowly expanding, the average separation of pairs with small relative velocities is reduced, N_J is small, and all time scales are similar (intermediate coupling). The pair correlational energy could be identified with future binding energies of new clusters. The larger masses of such new units would be a factor favoring the similar creation of larger clusters of the units, despite the decreasing density and initial statistical bias toward larger separations and relative velocities at each stage. For more details concerning such nonlinear processes of hierarchical clustering, we must refer to the literature (Refs. 35 and 33, and papers quoted therein).

Our weak-coupling homogeneous model is also of interest as a simple but cogent example of highly atypical behavior in statistical mechanics. The kinetic energy production develops on the time scale τ_R', exceeding the relaxation time scale by a factor of order $\log N_J^2$. This separation of time scales allows for a two-stage picture of the evolution of $\varphi(\mathbf{v},t)$. [For a detailed model analysis of the combined effect, see Ref. 42.] The operator ψ_W drives the system towards equilibrium on the time scale τ_R (Section IV.C). The antagonistic effect of ϕ begins to compete when φ is already close to equilibrium. There is a continual flow of energy from potential to kinetic form, and some order in velocity space ($\varphi \neq \varphi^{(0)}$) is maintained as a by-product.

The tendency away from equilibrium, nevertheless, involves a truly irreversible process, as appears upon considering the evolution of Boltzmann's H quantity, defined to within a constant C by

$$H(t) = N\int d\mathbf{v}_1 \varphi(\mathbf{v}_1,t)\log\varphi(\mathbf{v}_1,t) + C \tag{4.54}$$

Here we need to apply the Markoffianization ansatz, which in no way affects the ϕ term of special interest here. Proceeding as for (4.43), and symmetrizing,[37] we get

$$\frac{dH}{dt} = N\int d\mathbf{v}_1 \left[\log\varphi(\mathbf{v}_1,t)\right]\partial_t\varphi(\mathbf{v}_1,t) \tag{4.55}$$

$$\frac{dH}{dt} = \frac{1}{2}N\int d\mathbf{w}\,d\mathbf{v}_{12}\left[\log\varphi\varphi(t)\right] \times\left[\phi + n_0\psi_W\log\left(e^{-1/2}Kv_{12}\tau_R\right)\right]\varphi\varphi(t) \tag{4.56}$$

Recalling again that any function of $|\mathbf{v}_{12}|$ can be commuted through the differential operators in ψ_W (Eq. 3.51), we find

$$\frac{dH}{dt} = \left(\frac{dH}{dt}\right)_{\phi} + \left(\frac{dH}{dt}\right)_{\psi_W} \tag{4.57}$$

$$\left(\frac{dH}{dt}\right)_{\psi_W} = -4\pi G^2 m^2 n_0 N\int d\mathbf{w}\,d\mathbf{v}_{12} \times\left[\log\frac{\pi v_{12}^3\tau_R}{Gm\sqrt{e}}\right]\frac{v_{12}^2\mathsf{I} - \mathbf{v}_{12}\mathbf{v}_{12}}{v_{12}^3\varphi\varphi(t)} : \frac{\partial\varphi\varphi(t)}{\partial\mathbf{v}_{12}}\frac{\partial\varphi\varphi(t)}{\partial\mathbf{v}_{12}} \tag{4.58}$$

Apart from the logarithm (cf. Eq. 3.46), the integrand of (4.58) is positive everywhere because of the Schwartz inequality. The logarithm is positive except in a very small interval around $v_{12}=0$, where the integrand remains bounded. It follows that

$$\left(\frac{dH}{dt}\right)_{\psi_W} \leqslant 0 \tag{4.59}$$

the equality holding at the Maxwellian equilibrium.

For $(dH/dt)_\phi$ the substitution of (4.16) immediately yields

$$\left(\frac{dH}{dt}\right)_\phi = -4\pi G^2 m^2 n_0 N \int d\mathbf{w}\, d\mathbf{v}_{12} \frac{(\mathbf{v}_{12}\cdot \partial\varphi\varphi(t)/\partial\mathbf{v}_{12})^2}{v_{12}^3 \varphi\varphi(t)} \tag{4.60}$$

so

$$\left(\frac{dH}{dt}\right)_\phi < 0 \tag{4.61}$$

Thus for the total H, we have proved that

$$\frac{dH}{dt} < 0 \tag{4.62}$$

The inequality here holds even when φ is Maxwellian. The model of a homogeneous, initially uncorrelated system obeys a partial H-theorem; the H quantity decreases systematically but is not bounded below.

The result (4.62) is, of course, directly related to the systematic creation of kinetic energy (4.50). Indeed, if $\varphi(t)=\varphi^{(0)}$ for a particular time, the substitution of $[\log\varphi^{(0)}]$ into the general expression (4.55) yields

$$\frac{dH}{dt} = -\frac{3N}{2E_K}\frac{dE_K}{dt} \quad (\text{when } \varphi=\varphi^{(0)}) \tag{4.63}$$

It can be verified that the results (4.50), (4.57), (4.58), and (4.60) satisfy the relation (4.63) when $\varphi\varphi(t)=\varphi^{(0)}\varphi^{(0)}$. The latter can be interpreted as describing a quasistatic thermodynamical evolution

$$T\frac{dS}{dt} = N\frac{dE_K}{dt} \quad \left(\frac{d\Omega}{dt}=0, \quad \varphi=\varphi^{(0)}\right) \tag{4.64}$$

where $T=m\sigma^2/k_B$, $S=-k_B H$.

In Fig. 2 a schematic plot is made of directions of evolutions in E_K-H space, a representation introduced by Miller.[43] The velocity dispersion increases indefinitely. More precisely, it can be said that there is an

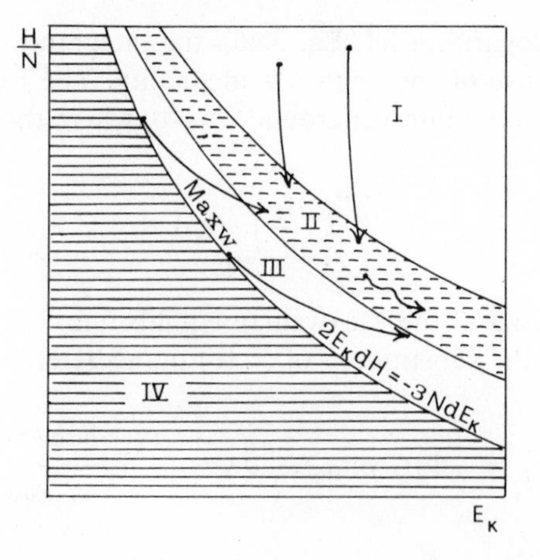

Fig. 2. Schematic illustration of the monotonic increase in mean kinetic energy E_K and of the monotonic decrease in Boltzmann's H. The line between regions III and IV represents Maxwellian velocity distributions $\varphi^{(0)}$. Region IV contains no allowed distributions. The points represent hypothetical distributions at time t, and the arrows show the direction of evolution of H and E_K from these points. Region I is dominated by the effect of decreasing H, especially from the contribution in (4.56) containing the logarithm. Region III is dominated by the kinetic energy production (4.50); the H quantity decreases there also, primarily because of increasing disorder in velocity space due to the heating. The evolution in the transition region II is less rigorously established, and the region has not been shown to have boundaries independent of the other moments of φ. But since it has been rigorously shown that the horizontal evolution is toward the right, therefore $dH/dt<0$, at least in a time-averaged sense, will hold even when branch-cut contributions and superimposed oscillations are not neglected.

approach to an infinitely dispersed velocity distribution, a measure of the deviation being given by the Liapunov function $\exp(H/N)$. We also note that in Fig. 2 the evolution from points on the Maxwellian curve

$$-k_B H^{(0)} = S^{(0)} = \frac{3}{2} N k_B \log E_K^{(0)} + C' \tag{4.65}$$

is tangent to the curve, and one will expect a parallel evolution from points nearby. Such nearby points are normally accessible because of (4.59); but as to points precisely on the curve, it should be borne in mind that $\varphi(t) = \varphi^{(0)}(\sigma = \hat{\sigma}(t))$ does not satisfy the evolution equation for any function $\hat{\sigma}$.

To complete the picture of the system evolution, we remark that the pair correlations also grow, building up the ratio of negative correlational energy to kinetic energy, until the validity of the weak-coupling approximation breaks down. The unbounded variations of E_V, E_K, and H appear here as specifically related to the infinite size (and constant mean density) of the system. They develop on the slow time scale τ_R'; for instance,

$$\frac{dH^{(0)}}{dt} = -(54\pi)^{1/2}G^2m^{7/2}n_0N\left(E_K^{(0)}\right)^{-3/2} \tag{4.66}$$

and from (4.63)

$$E_K(t+\Delta t) \sim \left(\frac{\Delta t}{\tau_R'}\right)^{2/3} E_K(t) \tag{4.67}$$

In conclusion let us again recall that these bizarre effects will be uncompetitive in a linearly perturbed, strictly weakly coupled homogeneous system, with respect to instabilities. However, in extrapolations and extensions to inhomogeneous systems or to intermediate coupling, they provide a simple picture that could be a realistic description of nonlinear processes of formation and disruption of clusters.

E. Kinetic Description for Inhomogeneous Systems

We now extend the description of an infinite system to include inhomogeneities. Their effects could be important for instance in the process of disruption or enhancement of the binding of a particular cluster, as a result of the interactions between stars of the clusters and massive external sources. Such "tidal" perturbations of the clusters can remain important on the kinetic time scale.

Although it would be appropriate here to introduce different mass species, for simplicity we keep to the single mass formalism. The model occupies a position intermediate between the models of a finite inhomogeneous system and of a homogeneous infinite system. Not only does it offer interesting perspectives on both, but it introduces an entirely new feature: an enhanced relaxation mechanism, specifically related to the long range of the gravitational interaction.

1. The Non-Markoffian Equations of Evolution

Since we have in mind loosely bound structures, where units can be exchanged rather freely between the clusters containing them, we shall in the present analysis suppress the mean field $\overline{\mathbf{F}}(\mathbf{x},t)$. No approximation is, however, to be made upon the scale and the amplitude of the inhomogeneities in the system. The mean field could be retained if the related commutation problems are ignored, as was done in Section III with different justification. It can always be restored in the "collisionless" part of the evolution equations for the distribution function $f(\mathbf{x},\mathbf{v},t)$ and its moments.

The analysis of Section III for finite systems provides us with our starting point. It will be sufficient to set $\overline{\mathbf{F}}=0$ and to suppress the restriction $\Omega \cong L_J^3$ on the volume. There is no limitation now on the range of the pair interactions. It is here convenient to make the Fourier transformation:

$$\rho_{\mathbf{k}}(\mathbf{v},t)=(8\pi^3)^{-1}\int_\Omega d\mathbf{x}\exp(-i\mathbf{k}\cdot\mathbf{x})f(\mathbf{x},\mathbf{v},t) \tag{4.68}$$

$$\gamma_{\mathbf{k}_1,\mathbf{k}_2}(\mathbf{v}_1,\mathbf{v}_2,t)=(8\pi^3)^{-2}\int\int_\Omega d\mathbf{x}_1 d\mathbf{x}_2\exp[-i(\mathbf{k}_1\cdot\mathbf{x}_1+\mathbf{k}_2\cdot\mathbf{x}_2)]$$
$$\times g(\mathbf{x}_1,\mathbf{v}_1,\mathbf{x}_2,\mathbf{v}_2,t) \tag{4.69}$$

g is the pair correlation function defined by (3.10a). It is appropriate to separate out the spatial average:

$$n_0\varphi(\mathbf{v},t)=\Omega^{-1}\int_\Omega d\mathbf{x}f(\mathbf{x},\mathbf{v},t)=\left(\frac{8\pi^3}{\Omega}\right)\rho_0(\mathbf{v},t) \tag{4.70}$$

$$f(\mathbf{x},\mathbf{v},t)=n_0\varphi(\mathbf{v},t)+⨍ d\mathbf{k}\exp(i\mathbf{k}\cdot\mathbf{x})\rho_{\mathbf{k}}(\mathbf{v},t) \tag{4.71}$$

$⨍ d\mathbf{k}$ denoting a principal part with respect to the point $\mathbf{k}=0$. The inhomogeneities are not restricted to regions of finite volume, and one must be careful to retain contributions to $\partial_t\varphi(v,t)$ of the form

$$\Omega^{-1}⨍ d\mathbf{q}\cdots\rho_{-\mathbf{q}}\rho_{\mathbf{q}} \tag{4.72}$$

which are usually neglected for molecular systems.[14,24]

Proceeding from (3.40), (3.24), or equivalently from the general description for a weakly coupled inhomogeneous system,[26] we thus write:

$$(\partial_t+i\mathbf{k}\cdot\mathbf{v}_1)\rho_{\mathbf{k}}(\mathbf{v}_1,t)=D_{\mathbf{k}}(\mathbf{v}_1,t)+\int_0^t d\tau\int\frac{dz}{-2\pi}\exp(-iz\tau)$$
$$\times\int d\mathbf{v}_2\Big[\psi(z-\mathbf{k}\cdot\mathbf{v}_1)\rho_{\mathbf{k}}(\mathbf{v}_1,t-\tau)\varphi(\mathbf{v}_2,t-\tau)$$
$$+⨍ d\mathbf{q}\psi_{\mathbf{q}}(z-\mathbf{k}\cdot\mathbf{v}_1+\mathbf{q}\cdot\mathbf{v}_{12})\rho_{\mathbf{k}-\mathbf{q}}(\mathbf{v}_1,t-\tau)\rho_{\mathbf{q}}(\mathbf{v}_2,t-\tau)\Big] \tag{4.73}$$

$$\partial_t\varphi(\mathbf{v}_1,t)=n_0^{-1}D(\mathbf{v}_1,t)+\int_0^t d\tau\int\frac{dz}{-2\pi}\exp(-iz\tau)$$
$$\times\int d\mathbf{v}_2\Big[\psi(z)\varphi(\mathbf{v}_1,t-\tau)\varphi(\mathbf{v}_2,t-\tau)$$
$$+\left(\frac{8\pi^3}{\Omega n_0}\right)⨍ d\mathbf{q}\psi_{\mathbf{q}}(z+\mathbf{q}\cdot\mathbf{v}_{12})\rho_{-\mathbf{q}}(\mathbf{v}_1,t-\tau)\rho_{\mathbf{q}}(\mathbf{v}_2,t-\tau)\Big] \tag{4.74}$$

where

$$D_{\mathbf{k}}(\mathbf{v}_1,t)=4\pi^2\int dz\exp(-izt)\int d\mathbf{v}_2\int d\mathbf{l}\,\theta_{\mathbf{l}}^{12}$$
$$\times(\mathbf{l}\cdot\mathbf{v}_{12}+\mathbf{k}\cdot\mathbf{v}_1-z)^{-1}\gamma_{\mathbf{l}+\mathbf{k},-\mathbf{l}}(\mathbf{v}_1,\mathbf{v}_2,t=0) \tag{4.75}$$

$$\psi_{\mathbf{q}}(z)=-8\pi^3 i\int d\mathbf{l}\,\theta_{\mathbf{l}-\mathbf{q}}^{12}\,(\mathbf{l}\cdot\mathbf{v}_{12}-z)^{-1}\theta_{\mathbf{l}}^{12} \tag{4.76}$$

$$\theta_{\mathbf{l}}^{12}=V_l\mathbf{l}\cdot\boldsymbol{\partial}_{12},\qquad V_l=-Gm(2\pi^2l^2)^{-1} \tag{4.77}$$

The evolution equations (4.73) and (4.74) introduce the collision operator $\psi_{\mathbf{q}}(z)$ and the destruction term $D_{\mathbf{k}}(t)$ that generalize to inhomogeneous situations the quantities $\psi(z)$, $D(t)$ of the homogeneous theory. (Note that $\psi(z)=n_0\psi_{\mathbf{q}\to 0}(z)$ except at $z=0$.) Both equations remain, as in the homogeneous case, intrinsically non-Markoffian. This we now proceed to show, and first consider the destruction term (4.75).

To discuss the expression (4.75), we shall again, as for the homogeneous case (Section IV.B.2), consider initial correlations having the same functional form as in equilibrium:

$$\gamma_{\mathbf{l}+\mathbf{k},-\mathbf{l}}(\mathbf{v}_1,\mathbf{v}_2,0)=-\sigma^{-2}\Big[V_l\rho_{\mathbf{k}}(\mathbf{v}_1,0)n_0\varphi(\mathbf{v}_2,0)$$
$$+\int d\mathbf{q}\,V_{|\mathbf{l}-\mathbf{q}|}\rho_{\mathbf{k}+\mathbf{q}}(\mathbf{v}_1,0)\rho_{-\mathbf{q}}(\mathbf{v}_2,0)\Big] \tag{4.78}$$

Inserted into (4.75), this gives (cf. Eq. 4.20)

$$D_{\mathbf{k}}(\mathbf{v}_1,t)=(2\pi i)^{-1}G^2m^2\sigma^{-2}\int d\mathbf{v}_2\int dz\exp(-izt)\boldsymbol{\partial}_{12}\cdot\Big[\mathbf{S}(z-\mathbf{k}\cdot\mathbf{v}_1)$$
$$\times\rho_{\mathbf{k}}(\mathbf{v}_1,0)n_0\varphi(\mathbf{v}_2,0)+\int d\mathbf{q}\,\mathbf{S}_{\mathbf{q}}(z-\mathbf{k}\cdot\mathbf{v}_1)\rho_{\mathbf{k}+\mathbf{q}}(\mathbf{v}_1,0)\rho_{-\mathbf{q}}(\mathbf{v}_2,0)\Big] \tag{4.79}$$

The first contribution contains the quantity $\mathbf{S}$ encountered in the homogeneous case, and from (4.20) we have

$$\mathbf{S}(z-\mathbf{k}\cdot\mathbf{v}_1)\cong\frac{2\pi}{z-\mathbf{k}\cdot\mathbf{v}_1}\frac{\mathbf{v}_{12}}{v_{12}} \tag{4.80}$$

This gives rise in (4.79) to a factor $\exp(i\mathbf{k}\cdot\mathbf{v}_1t)$ that simply propagates the initial state forward in time. Thus, just as for the purely homogeneous contribution $D(\mathbf{v}_1,t)$ (cf. Section IV.B.2) appearing in (4.74) for the velocity distribution, there is a persistence of the memory of the initial state.

The second and specifically inhomogeneous contribution to (4.79) introduces the integral $\mathbf{S}_{\mathbf{q}}(z)$, which is evaluated in the appendix, (A.38). Although fully explicit forms are not obtained (except in the case $\mathbf{q}$ parallel to $\mathbf{v}_{12}$), it is established that the only singularities on the real axis are logarithmic. By (A.29) this second contribution then vanishes asymptotically at least as $1/t$.

In what follows we shall consider an initially uncorrelated system ($D=D_{\mathbf{k}}=0$), in order to focus attention on the quite surprising properties of the collision integrals.

The strongly non-Markoffian nature of the collision term in (4.73) is most conveniently exhibited by showing in what manner the straightforward Markoffianization procedure (cf. Section III.D)

$$\int_0^t d\tau \cdots \rightarrow \lim_{\epsilon\to 0}\int_0^\infty d\tau \exp(-\epsilon\tau)\cdots \tag{4.81a}$$

$$\rho_{\mathbf{k}}(\mathbf{v}_i, t-\tau)\rightarrow \exp(i\mathbf{k}\cdot\mathbf{v}_i\tau)\rho_{\mathbf{k}}(\mathbf{v}_i, t) \tag{4.81b}$$

fails. With the destruction term omitted, one gets

$$\begin{aligned}(\partial_t + i\mathbf{k}\cdot\mathbf{v}_1)\rho_{\mathbf{k}}(\mathbf{v}_1,t) &= G^2m^2\int d\mathbf{v}_2 \lim_{\epsilon\to 0}\\ &\times\Big\{\boldsymbol{\partial}_{12}\cdot\big[\mathsf{T}(i\epsilon)\cdot\boldsymbol{\partial}_{12}+\mathsf{T}'(i\epsilon)\cdot\mathbf{k}\big]\rho_{\mathbf{k}}(\mathbf{v}_1,t)\varphi(\mathbf{v}_2,t)\\ &+\int d\mathbf{q}\,\boldsymbol{\partial}_{12}\cdot\big[\mathsf{T}_{\mathbf{q}}(i\epsilon)\cdot\boldsymbol{\partial}_{12}+\mathsf{T}'_{\mathbf{q}}(i\epsilon)\cdot(\mathbf{k}-2\mathbf{q})\big]\rho_{\mathbf{k}-\mathbf{q}}(\mathbf{v}_1,t)\rho_{\mathbf{q}}(\mathbf{v}_2,t)\Big\}\end{aligned} \tag{4.82}$$

where $\mathsf{T}_{\mathbf{q}}(z)$ is defined analogously to $\mathsf{T}(z)$ (cf. Eq. 4.6) and where

$$\mathsf{T}'(z)=\frac{\partial\mathsf{T}}{\partial z} \tag{4.83}$$

The $\mathsf{T}'(z)$ contributions come from the commutation of the exponential in (4.81b) through the velocity derivatives of the collision operator (cf. Eqs. 3.35 and 3.41). These commutator terms will be seen to play a determinant role.

Using Eqs. 4.9 through 4.11, A.45, and A.47 to make the quantities T and T' more explicit, and retaining only the dominant terms, we get for (4.82)

$$\begin{aligned}(\partial_t + i\mathbf{k}\cdot\mathbf{v}_1)\rho_{\mathbf{k}}(\mathbf{v}_1,t) &= 2\pi G^2m^2\int d\mathbf{v}_2 \lim_{\epsilon\to 0}\\ &\times\Bigg\{\boldsymbol{\partial}_{12}\cdot\frac{\mathsf{I}v_{12}^2-\mathbf{v}_{12}\mathbf{v}_{12}}{v_{12}^3}\cdot\left[\left(\log\frac{Kv_{12}}{\epsilon}\right)\boldsymbol{\partial}_{12}-\frac{1}{i\epsilon}\mathbf{k}\right]\rho_{\mathbf{k}}(\mathbf{v}_1,t)\varphi(\mathbf{v}_2,t)\\ &+\int d\mathbf{q}\,\boldsymbol{\partial}_{12}\cdot\left[\frac{\mathsf{I}v_{12}^2-\mathbf{v}_{12}\mathbf{v}_{12}}{v_{12}^3}\left(\log\frac{Kv_{12}+i\mathbf{q}\cdot\mathbf{v}_{12}}{\epsilon+i\mathbf{q}\cdot\mathbf{v}_{12}}\right)\cdot\boldsymbol{\partial}_{12}\right.\\ &\left.-\frac{2\mathbf{q}\mathbf{v}_{12}\cdot(\mathbf{k}-2\mathbf{q})}{q^2v_{12}^3}\left(\log\frac{q^2v_{12}^2}{\epsilon|\mathbf{q}\cdot\mathbf{v}_{12}|}\right)\right]\rho_{\mathbf{k}-\mathbf{q}}(\mathbf{v}_1,t)\rho_{\mathbf{q}}(\mathbf{v}_2,t)\Bigg\}\end{aligned} \tag{4.84}$$

It is only through the commutator term that the inhomogeneous collision operator introduces the familiar $\log\epsilon$ divergence. For the homogeneous operator, in addition to the $\log\epsilon$ term, we see that the commutator gives rise to the far stronger ϵ^{-1} divergence. This introduces qualitatively new features into the problem, to which we now turn our attention.

2. *Enhanced Evolution on a Hybrid Time Scale*

We return to the non-Markoffian description of (4.73) and focus attention on the $\psi(z-\mathbf{k}\cdot\mathbf{v}_1)$ contribution that gives the ϵ^{-1} divergence when one attempts to Markoffianize. As long as the corresponding evolution does not drive $\rho_\mathbf{k}$ to a distribution that would make the term small, we may treat it as "superdominant" with respect to the other terms of (4.73) and write

$$(\partial_t + i\mathbf{k}\cdot\mathbf{v}_1)\rho_\mathbf{k}(\mathbf{v}_1,t) \cong \int_0^t d\tau \int \frac{dz}{-2\pi}\exp(-iz\tau)\int d\mathbf{v}_2\psi(z-\mathbf{k}\cdot\mathbf{v}_1) \times \rho_\mathbf{k}(\mathbf{v}_1,t-\tau)\varphi(\mathbf{v}_2,t-\tau) \tag{4.85}$$

The evolution equation is now linear. Moreover, since no ϵ^{-1} terms can appear for $\mathbf{k}=0$ (see Eq. 4.84), the time variation of $\varphi(\mathbf{v}_2)$ may be neglected here. Then, going through a Laplace transformation, we can write the formal solution of (4.85) as

$$\rho_\mathbf{k}(\mathbf{v}_1,t) = \frac{1}{2\pi i}\int dz \frac{\exp(-izt)}{z-\mathbf{k}\cdot\mathbf{v}_1 - i\int d\mathbf{v}_2\psi(z-\mathbf{k}\cdot\mathbf{v}_1)\varphi(\mathbf{v}_2)}\rho_\mathbf{k}(\mathbf{v}_1,0) \tag{4.86}$$

We explicit the collision operator ψ using Eqs. 4.6 through 4.10 [the contribution of (4.11) is negligible here] and introducing the weak-coupling operator ψ_W of (3.51):

$$\psi(z-\mathbf{k}\cdot\mathbf{v}_1) = n_0\psi_W \log\left[\frac{iKv_{12}}{(z-\mathbf{k}\cdot\mathbf{v}_1)}\right] - 2\pi G^2m^2n_0 v_{12}^{-3}\partial_{12}\cdot(\mathsf{I}v_{12}^2 - \mathbf{v}_{12}\mathbf{v}_{12})\cdot\mathbf{k}(z-\mathbf{k}\cdot\mathbf{v}_1)^{-1} \tag{4.87}$$

The second term results from the commutation of the logarithm through the velocity derivatives $\partial_i = \partial/\partial \mathbf{v}_i$ appearing in ψ.

To make an estimate of the time scale on which $\rho_\mathbf{k}(\mathbf{v}_1,t)$ develops, we shall replace ψ by the model

$$\psi(z-\mathbf{k}\cdot\mathbf{v}_1) \to -\nu\log\frac{i\gamma}{z-\mathbf{k}\cdot\mathbf{v}_1} + \frac{\alpha\nu k\bar{v}}{z-\mathbf{k}\cdot\mathbf{v}_1} \tag{4.88}$$

Here ν^{-1} is a positive number of order of the kinetic time scale τ_R' (cf. Eq. 3.7); $\bar{v}$ is a mean speed; $\gamma \cong \langle K v_{12} \rangle \cong \langle K \rangle \bar{v}$ corresponds to a very short time scale since K^{-1} is the short-distance cut-off; finally α is a complex number of order unity. With (4.88), the formal expression (4.86) can be represented by the fully explicit model

$$\rho_{\mathbf{k}}^{(\nu)}(t) = \frac{1}{2\pi i} \int d\zeta \frac{\zeta \exp[-i(\zeta + \mathbf{k}\cdot\mathbf{v}_1)t]}{\zeta^2 + i\zeta\nu \log(i\gamma/\zeta) - i\alpha\nu k\bar{v}} \rho_{\mathbf{k}}^{(\nu)}(0) \tag{4.89}$$

It is not restrictive to choose k such that

$$|\alpha| k\bar{v} \gg \nu \log\left(\frac{\gamma}{\nu}\right) = O(G^2) \tag{4.90}$$

We see that in the denominator of (4.89) the logarithmic term is small; the relevant singularities are poles located at distances from the real axis

$$|\bar{\zeta}_j| = O\left[(\nu k\bar{v})^{1/2}\right] \tag{4.91}$$

and, ignoring the branch cut contributions (cf. Section IV.C),

$$\rho_{\mathbf{k}}^{(\nu)}(t) \cong \sum_j \exp\left[-i\left(\mathbf{k}\cdot\mathbf{v}_1 + \bar{\zeta}_j\right)t\right] \rho_{\mathbf{k}}^{(\nu)}(0) \tag{4.92}$$

The position of the poles (4.91) determines a new hybrid time scale

$$\tau_h = (\nu k\bar{v})^{-1/2} \tag{4.93a}$$

that can be written, in the case where the inhomogeneities $\rho_{\mathbf{k}}$ describe subclusters of Jeans' length L_J so that $k\bar{v}$ is the corresponding crossing time τ_C,

$$\tau_h = (\tau_C \tau_R')^{1/2} \tag{4.93b}$$

Up to a factor of order $\log N_J$, the scale τ_R' is the relaxation time for the whole system. We thus see that τ_h determines an intermediate time scale of evolution. It corresponds, moreover, to a cross-effect between the characteristics of the subclusters and of the system as a whole.

This order of magnitude analysis leaves open such fundamental questions as the sign of the imaginary part of the poles ζ_j. The further elaboration of these very recent results requires an analysis of the properties of the operator (4.87).

However, we can already specify an important point. The appearance of the new time scale is a characteristic property of the infinite model. The collision operator for a finite system of dimensions κ^{-1} would differ from (4.87) only by the substitution of $z + i\kappa v_{12}$ for z (cf. Eq. 4.10). This would change the denominator of (4.89) to

$$\zeta^2 + i\zeta\left(\kappa\bar{v} + \nu\log\frac{i\gamma}{\zeta + i\kappa\bar{v}}\right) - \nu\left(i\alpha k\bar{v} + \kappa\bar{v}\log\frac{i\gamma}{\zeta + i\kappa\bar{v}}\right) \tag{4.94}$$

The dominant term is now $\kappa\bar{v}$; the zeros of (4.94) give rise to the usual relaxation time scale $\nu\log(\gamma/\kappa\bar{v})$, with a strongly damped transient on the scale $(\kappa\bar{v})^{-1}$ in addition.

It is of interest to consider the evolution of the mean energy densities corresponding to the new time scale. For the kinetic energy density

$$\mathcal{E}^K(\mathbf{x}_1, t) = \int d\mathbf{v}_1 \frac{1}{2} m v_1^2 f(\mathbf{x}_1, \mathbf{v}_1, t) \tag{4.95}$$

and with only the superdominant term of (4.87) retained, Eq. 4.85 gives

$$\partial_t \mathcal{E}^K(\mathbf{x}_1, t) + \boldsymbol{\nabla}_1 \cdot \left\langle \frac{1}{2} m v^2 \mathbf{v} \right\rangle = -2\pi G^2 m^3 n_0 \int d\mathbf{v}_1 d\mathbf{v}_2 \varphi(\mathbf{v}_2)$$

$$\times v_{12}^{-3} \mathbf{v}_1 \cdot \left(\mathsf{I} v_{12}^2 - \mathbf{v}_{12}\mathbf{v}_{12}\right) \cdot \boldsymbol{\nabla}_1 \int_0^t d\tau \hat{f}(\mathbf{x}_1, \mathbf{v}_1) \tag{4.96}$$

where

$$\hat{f}(\mathbf{x}_1, \mathbf{v}_1) = \left[\exp - (\tau \mathbf{v}_1 \cdot \boldsymbol{\nabla}_1)\right] f(\mathbf{x}_1, \mathbf{v}_1, t - \tau) \tag{4.97}$$

represents the one-particle distribution at time $t - \tau$ projected to time t according to the collisionless evolution (here represented without the mean field). The $1/\epsilon$ factor in (4.84), resulting from the attempted Markoffianization, shows that the integration with respect to τ is to be replaced by a factor of order t. This gives the compact formula

$$\partial_t \mathcal{E}^K + \boldsymbol{\nabla}_1 \cdot \left\langle \frac{1}{2} m v^2 \mathbf{v} \right\rangle = -\epsilon_h^{-1} \int d\mathbf{v}_1 \mathbf{v}_1 \cdot \mathsf{D}_1 \cdot \boldsymbol{\nabla}_1 \hat{f}\,(\mathbf{x}_1, \mathbf{v}_1) \tag{4.98}$$

where

$$\epsilon_h = \epsilon_h^{-1} O\left(\langle kv \rangle \tau_R'\right), \qquad \epsilon_h = \tau_h^{-1} \tag{4.99}$$

in agreement with (4.93). Here D_1 is the "coefficient of dispersion" of the

Fokker–Planck equation (to within a factor $\log N_J$):

$$(2\pi G^2 m^3 n_0)^{-1}\mathsf{D}_1=\int d\mathbf{v}_2 v_{12}^{-3}(\mathsf{I}v_{12}^2-\mathbf{v}_{12}\mathbf{v}_{12})\varphi(\mathbf{v}_2)=\partial_1\partial_1\int d\mathbf{v}_2 v_{12}\varphi(\mathbf{v}_2) \tag{4.100}$$

One can easily extract corresponding results for the correlational energy density $\mathcal{E}^V(\mathbf{x}_1,t)$. It is noteworthy that the approach is rather different. For the Fourier transform of $\mathcal{E}^V(\mathbf{x}_1,t)$, cf. Eq. 4.69, one has

$$\mathcal{E}_{\mathbf{k}}^V(t)=4\pi^3 m\int d\mathbf{v}_1 d\mathbf{v}_2\int d\mathbf{l}\, V_l \gamma_{\mathbf{l}+\mathbf{k},-\mathbf{l}}(\mathbf{v}_1,\mathbf{v}_2,t) \tag{4.101}$$

From (3.21b) written for γ (but with the mean fields removed), the contribution containing $\varphi(\mathbf{v}_2)$ is readily found to be

$$\left[\mathcal{E}_{\mathbf{k}}^V(t)\right]_0=4\pi^3 m\int d\mathbf{v}_1 d\mathbf{v}_2\int_0^t d\tau\int d\mathbf{l}\, V_l \exp\left[-i(\mathbf{l}\cdot\mathbf{v}_{12}+\mathbf{k}\cdot\mathbf{v}_1)\tau\right]$$
$$\times\, iV_l\mathbf{l}\cdot\partial_{12}\rho_{\mathbf{k}}(\mathbf{v}_1,t-\tau)\varphi(\mathbf{v}_2,t-\tau) \tag{4.102}$$

Carrying out the $\mathbf{l}$ integration (cf. Eq. 4.20) and integrating by parts, one can write the inverse Fourier transform as

$$\left[\mathcal{E}^V(\mathbf{x}_1,t)\right]_0=-4\pi G^2 m^3\int d\mathbf{v}_1 d\mathbf{v}_2 v_{12}^{-1}\varphi(\mathbf{v}_2)\int_0^t d\tau \hat{f}(\mathbf{x}_1,\mathbf{v}_1)$$
$$+\pi G^2 m^3\int d\mathbf{v}_1 d\mathbf{v}_2 v_{12}^{-1}\varphi(\mathbf{v}_2)\mathbf{v}_{12}\cdot\nabla_1\int_0^t d\tau\tau\hat{f}(\mathbf{x}_1,\mathbf{v}_1) \tag{4.103}$$

For the spatial average we recover the homogeneous result (4.50) and (4.53) by differentiating with respect to time. Note that we get an irreversible evolution ($\partial\langle E^V\rangle/\partial t<0$) without taking an asymptotic time limit, and that this obtains rather directly from the BBGKY hierarchy. This comes from the form of V_l^2, which is such that there is no memory kernel $\exp(-\kappa v_{12}\tau)$.

The second term of (4.103) gives an evolution on the hybrid time scale. This is most easily seen by comparing it to the first term. Let t_h be the time at which the second term attains the same order of magnitude as the first at $t=\tau_R'$. Then, since the factor of order $[\langle kv\rangle t_h^2]/\tau_R'$ by which the two terms differ is to be unity, we recover (4.93) with $\tau_h=t_h$. Again a memory kernel $\exp(-\kappa v_{12}\tau)$ would prevent such a result: The factor would simply become $[\langle kv\rangle(\kappa v_{12})^{-2}]/(\kappa v_{12})^{-1}$, of order unity when t is the same in each term.

As mentioned, only a preliminary exploration of the hybrid time scale has been undertaken at the time of writing. It would seem that the superdominant term has little effect upon the energy densities in regions that are neither expanding nor collapsing rapidly. It could, however, destabilize the convective diffusion process for a mean flow having a gradient normal to the streamlines. For expanding or collapsing regions the subclustering process appears respectively to be inhibited or enhanced if one considers the extrapolation to intermediate coupling of the superdominant energy evolution.

3. *Evolution of the Mean Kinetic Energy*

The hybrid time scale does not intervene in the evolution of the velocity distribution (4.74), and we can here simply apply the Markoffianization ansatz as in Section IV.D. We also again consider an initially uncorrelated system, $D=0$. The term containing ψ is the same as in the homogeneous limit, and its contribution $(\partial_t E_K)_0$ to the mean kinetic energy per particle

$$\partial_t E_K = (\partial_t E_K)_0 + (\partial_t E_K)_{\mathbf{q}} \tag{4.104}$$

is given by (4.50). The extra term comes from the $\psi_{\mathbf{q}}$ contribution to (4.74), which after Markoffianization reads:

$$[\partial_t \varphi(\mathbf{v}_1,t)]_{\mathbf{q}} = G^2 m^2 \left(\frac{8\pi^3}{\Omega n_0}\right) \int d\mathbf{v}_2 \oint d\mathbf{q}\, \boldsymbol{\partial}_{12} \cdot \left[\mathsf{T}_{\mathbf{q}}(i\epsilon_0)\cdot \boldsymbol{\partial}_{12} - 2\mathsf{T}'_q(i\epsilon_0)\cdot \mathbf{q}\right] \rho_{-\mathbf{q}}(\mathbf{v}_1,t)\rho_{\mathbf{q}}(\mathbf{v}_2,t) \tag{4.105}$$

(cf. Eqs. 4.82 and 4.83). The memory factor ϵ_0, which will only subsist in a logarithm, is again to be identified with the relaxation time τ_R. Using the definition (4.7) of T, one sees that the extra contribution to the kinetic energy can be written as

$$[\partial_t E_K]_{\mathbf{q}} = G^2 m^3 \left(\frac{8\pi^3}{\Omega n_0}\right) \int d\mathbf{v}_1 d\mathbf{v}_2 \oint d\mathbf{q} (\mathbf{q}\cdot\mathbf{v}_{12} - i\epsilon_0)\,\mathrm{Tr}\,\mathsf{T}'_{\mathbf{q}}(i\epsilon_0) \times \rho_{-\mathbf{q}}(\mathbf{v}_1,t)\rho_{\mathbf{q}}(\mathbf{v}_2,t) \tag{4.106}$$

The quantity $\mathsf{T}'_q(i\epsilon)$ is evaluated in the appendix, the dominant contribution being given by (A.47). Introducing the homogeneous term (4.50), we get for the total kinetic energy per particle

$$\partial_t E_K \cong 4\pi G^2 m^3 n_0 \int d\mathbf{v}_1 d\mathbf{v}_2 v_{12}^{-1} \Bigg\{ \varphi(\mathbf{v}_1,t)\varphi(\mathbf{v}_2,t)$$

$$- \frac{8\pi^3}{\Omega n_0^2} \int d\mathbf{q} \left(\frac{\mathbf{q}\cdot\mathbf{v}_{12}}{q v_{12}} \right)^2 \left(\log \frac{q^2 v_{12}^2}{\epsilon |\mathbf{q}\cdot\mathbf{v}_{12}|} \right) \rho_{-\mathbf{q}}(\mathbf{v}_1,t)\rho_{\mathbf{q}}(\mathbf{v}_2,t) \Bigg\} \qquad (4.107)$$

The inhomogeneous contribution is enhanced with respect to the homogeneous term by a logarithmic factor. Furthermore, the effect would appear to be of opposite sign. In particular, whenever $\rho_{\mathbf{q}}(\mathbf{v}) = n_{\mathbf{q}}\varphi(\mathbf{v})$, the product $\rho_{-\mathbf{q}}(\mathbf{v}_1)\rho_{\mathbf{q}}(\mathbf{v}_2)$ is real and positive; the other part of the integrand is also positive except in a vanishingly small domain.

We suggest that this result is to be interpreted according to the picture of tidal disruption of Spitzer[44]; that is, it should be thought of as a source of relaxation of the amplitude of small scale inhomogeneities that give up their binding energy to modes of larger characteristic lengths.

To be more specific let us consider a situation where a quasistable cluster in a definite location is moving through a field of "clouds" of various masses. Spitzer considers the pair interactions between these external clouds and the stars of the cluster. Such interactions could give a characteristic evolution time for the cluster comparable to τ_R' for star–star encounters, (*a*) if the mass density of such clouds decreases no more slowly with the sampling volume than the inverse of the mean masses of the clouds, and (*b*) if the velocity dispersion of the clouds is comparable to that of the stars, so that the interaction is suitably resonant.

Such conditions can be mutually consistent. For instance, suppose that the number of clouds having the mass of the cluster, and lying within a sphere of radius $L_{J\,(\text{clouds})}$, is the same as the number of stars in the cluster. Then condition (*a*) implies that the mass density is inversely proportional to the 3/2 power[45] of the size:

$$\rho_{(\text{clouds})}/\rho_{(\text{stars})} = \left[L_{J(\text{clouds})}/L_{J(\text{stars})} \right]^{-3/2} \qquad (4.108)$$

so that for successively larger scales

$$\langle v^2 \rangle^{1/2} \propto L_J^{1/4} \qquad (4.109)$$

The slow increase of the velocity dispersion with cloud mass is consistent both with condition (*b*) and with the quasistability of the cluster against fast disruption through coalescence with neighboring clouds.

The "mean" field of the clouds can be taken to be small. It may also be justifiable to ignore the role of the mean field of the cluster in the calculation of tidal energy exchange; for instance, stars could be exchanged rather freely between the cluster and neighboring clouds of comparable mass.

If the source mass m is interpreted as the mass of such nearby small clouds having diameters comparable to the cluster size, then Spitzer's results[44] agree with (4.107) in order of magnitude.

V. HISTORICAL SURVEY

Since the kinetic-theoretical developments of the preceding sections are essentially drawn from the authors' own work, a rapid historical survey of the field seems necessary to establish some degree of perspective.

The subject can be taken as originating in the remarkable monograph by Chandrasekhar[2] (1942), where he developed what with minor modifications (essentially, the displacement of the long-distance cut-off from $n_0^{-1/3}$ to L_J[46]) has become the classical description of stellar encounters. This analysis raised two main problems: The relaxation times estimated for large systems such as our galaxy are too large to account for their apparently relaxed state; and the description of the encounter process involves a sweeping idealization that a priori seems incompatible with the consideration of long-range forces in finite systems.

To reduce the relaxation time to some value lower than the Hubble age of the universe, a wide variety of mechanisms have been suggested. The majority do not involve refinements in the description of the kinetic description itself, but it is appropriate to make a brief review of these first. A reduction of the relaxation time obtains if the mass of the scatterer increases, which suggests that encounters between large subclusters[47] can be of importance. Unlike star clusters, the galaxy is a "dirty" system containing vast quantities of plasma and gas that have relatively short relaxation times.[48] In a rotating system resonances between orbits can also give important effects.[49] The plasma analogy has inspired a number of investigations concerning the possible effect of interactions between stars and density waves in the spirit of quasilinear theory.[50] It was shown by Genkin that in the solution of the Boltzmann equation a cross-effect due to the mean field term can give rise to enhanced diffusion and thus also enhanced relaxation.[51]

The "violent relaxation" mechanism analyzed by Lynden-Bell[4] (1967) and extended by Saslaw[52] (1969, 1970), while being a collisionless process, is nevertheless directly relevant to the observation of relaxed systems. It moreover accounts for the absence of equipartition observed in many systems. The lack of equipartition between stars of different mass species[53] is an important phenomenon that contributes to restrict the applicability of standard equilibrium statistical mechanics.

In Chandrasekhar's work stellar encounters are treated as completed and separate events in a homogeneous and force-free medium, the $1/r$ potential giving then the familiar logarithmic divergence at long distances. Henon[27] (1958) proposed a refinement of this description whereby for distant encounters the asymptotic calculation of the momentum and energy exchange is replaced by a detailed perturbation analysis of the motion along the straight line orbits. This yields transition moments growing as $t \log t$. On physical grounds Henon argues that the coherent interaction between two stars cannot last longer than a crossing time, and thereby recovers the classical results.

Henon's analysis provides in fact a detailed representation of the collision term appearing in the non-Markoffian model of Prigogine and Severne[8] (1966), which, together with connex papers,[37,38] has been discussed here in detail in Section IV. Indeed the characteristic $\log z$ dependence of the collision operator (4.17) can be seen to give a $\log t$ factor upon inverse Laplace transformation. In an analysis conceptually similar to that of Henon's, Ostriker and Davidsen[40] (1968) recover the $\log t$ behavior; on the basis of a very simple model in which encounters are given unit or zero weight according to the value of the impact parameter, these authors propose the self-consistent determination of the relaxation time as given here by (4.44).

In an extension of the master equation formalism, Prigogine and Severne[25] (1968) write a formal kinetic equation for an inhomogeneous weakly-coupled gravitational system in the self-gravitational field. More general but equally formal results are obtained by Gilbert[54] (1968) and Chappell[55] (1968) who retain the collective effects as expressed by the integral of (3.18). These various formal results provide descriptions of the system essentially at the level of (3.25) (where collective interactions are omitted, however).

Chappell's investigation is based upon the Klimontovich formalism,[56] while Gilbert essentially develops a direct perturbation analysis in powers of $1/N$ of the Liouville equation to obtain the approximations (3.11) and (3.18) to the BBGKY hierarchy. The hierarchy is then discussed in terms of the polarization function $f^*(1|2)$ rather than the pair correlation. In a subsequent study of the polarization effect,[23] Gilbert obtains an integral equation for the integrated mass of polarization. For a particular model the polarization is shown to give a mass increase of up to a factor of two and a corresponding reduction of the relaxation time. Polarization effects have also been studied in rotating systems[57] and comparable results have been obtained.

The discussion of encounters in rotating gravitational systems closely parallels that of plasmas in a magnetic field. Wu[58] (1968) discusses the case of a homogeneous system in uniform rotation and using the Klimontovich formalism obtains a collision integral retaining collective effects; the long-distance cut-off necessary for stability ensures the validity of the Markoffian approximation and the obtainment of a "good" collision operator driving the system to Maxwellian equilibrium. Haggerty and Severne[41] (1971) derive the weak-coupling equations for a rotating system that is inhomogeneous along the axis of rotation. Non-Markoffian behavior with systematic increase in kinetic energy is exhibited in the limit of rapid rotation (cf. Eq. 4.51); as for nonrotating systems (cf. Eq. 4.107) the inhomogenities can give rise locally to the opposite effect. This analysis was extended by Haggerty[59] to include transverse inhomogeneities.

A novel approach to the kinetic-theoretical analysis was made by Lerche[60] (1971). He considers the microscopic Poisson equation, in the Klimontovich sense, as the limiting case for infinite propagation velocity of a wave equation. The dynamical problem can then be formulated as that of the interaction between the mass particles and the field oscillators. An expansion in powers of the mass per star (which defines a discreteness parameter) yields a modified BBGKY hierarchy description with coupled equations for the single particle distribution and for the pair correlation.

The force autocorrelation function is a quantity of considerable interest in characterizing the behavior of gravitational systems. For an infinite homogeneous system, Chandrasekhar[61] (1944) showed that this function, which is contained implicitly in the collision terms of kinetic theory, has the slow $1/t$ decay typical of the long-range $1/r$ potential. Cohen and Ahmad[62] (1975) have recently found that this decay steepens considerably for a finite system, and they have obtained a $1/t^5$ law for a homogeneous system with a Maxwellian velocity distribution. Thus, as in the kinetic analysis of Section III, one sees that the anomalies of the gravitational problem tend to disappear for finite systems.

The classical Fokker–Planck equation of Chandrasekhar has found its main field of application in the study of the rate of evaporation of star clusters,[63,30] which in turn is an essential factor in their secular evolution.[31] Larson[5] (1970) developed a global approach to the study of the secular evolution by considering the moment equations of the classical kinetic equation (i.e., Eq. 3.52) truncated only at fifth order. These fluid-dynamical equations are solved numerically, the collision contributions being evaluated for a perturbed Maxwellian distribution. Larson's results, which have been confirmed by calculations using a different truncation procedure,[64] are in quite good agreement with the computer N-body simulations.[29]

The irreversible creation of correlational energy characteristic of the non-Markoffian behavior was shown by Haggerty[65] (1970, 1971) to provide a mechanism for the creation of hierarchical structures in an expanding universe. The role of correlations in gravitational clustering has also been studied by Saslaw[33] (1972) directly from the hierarchy.

The investigations on the nature of quasars have focused interest on the processes occurring in galactic nuclei.[66] In a recent paper[67] Saslaw investigates the collective effects induced by massive objects ejected at large velocity from such a nucleus. He suggests that this external perturbation can provoke subclustering, rather than the tidal disruption as argued in Section IV.E.3. The study of the dense galactic nuclei opens for kinetic theory a new and promising field of investigation.

VI. CONCLUDING REMARKS

There are two main features unique to the gravitational interaction potential that account for nontrivial results even with a weak-coupling approximation. First, the dynamics are complicated by the *long-range, purely attractive potential*. The homogeneous uncorrelated initial state is so unstable that it supplies a stimulus to which even the weakly coupled system responds. It produces correlational energy through a normally closed channel at a rate determined by the pair interactions rather than by the precise details of the initial condition. One can imagine pairs at ever-increasing range beginning to move toward one another.

The second feature unique to the gravitational interactions is the *equality of "charge" and mass* of the particles. This vastly simplifies the treatment of bound subsystems. One can usually treat "tidal" perturbations by distant sources as having small effects on the internal dynamics of a cluster, and similarly the cluster can be approximated for many purposes as reducing to a massive particle at its center of mass. To discuss the creation and lifetime of such clusters, one will think of the system as "infinite;" but on the larger scale for the evolution of the distribution of the centers of mass of the clusters within the containing field, the "finite-system" description assuming quasistability of the units will be more appropriate.

The center-of-mass motion absorbs most of the energy and momentum transferred between clusters, so a *hierarchical structure* can be weakly coupled and quasistable on every scale. Every statistical framework can have a time scale for which the "units" are quasistable while the "system" evolves. The *choice of statistical treatment* of the *same* astronomically observable agglomeration of matter as either an infinite system, a finite system, a manifestation of growing correlations, an unstable composite particle, a quasistable composite particle, a polarization cloud, a coherent inhomogenity, an incoherent density wave, or even an unimportant fluctuation is certainly determined *in part* by such *physical considerations* as the density contrast with its immediate neighborhood. We emphasize rather that an important consideration is *also* the *length and time scales of interest to the observer*. One may want to study the evolution of "forests" rather than "trees." The simplicity of the gravitational hierarchies is that the "forests" interact just like giant "trees," and the same set of statistical descriptions can apply to both.

Such differing descriptions may be complementary but should bear a degree of *mutual consistency*. (For example, the lifetime of trees has a bearing on the evolution of a forest.) That requirement helps to explain the profound similarities we have found between the evolution of gravitational systems under very different physical situations or observational input. For example, the "secular evolution" of a "finite" cluster with a density initially specified as a function of position should bear some resemblance to the average evolution of similar clusters in unspecified positions. Thus the systematic build-up of *correlations* between composite particles may in the intermediate coupling extrapolation be physically identifiable with the growth of *inhomogeneities* due to a Jeans instability. Calculations of the evolution of the correlations may be useful for visualizing either the progression of the Jeans instability beyond the linear growth stage or the secular evolution of finite clusters. The kinetic equations for the pair-correlational growth seem relatively tractable if not of the ideal degree of precision.

Both the theory of the finite system and that of the infinite system are incompletely developed. For finite systems the emphasis of future theoretical work should bear on the *secular phase* involving the change in the density profile after the main evolution of the velocity distributions. For infinite systems the *weak-coupling approximation* imposed a serious constraint on the rigor of extrapolation of results to processes occurring on short time scales. Such an extrapolation to intermediate coupling is rather to be thought of as suggesting *trends* for evolutionary processes. Its main interest lies in the relative simplicity of the results. We have suggested that a *"superdominant"* term occurs; one then has concise evolution equations that are *linear* in the distribution function; approximate solutions involving a new *hybrid time scale* should, therefore, easily be generated. Comparisons with observational data and computer simulation experiments would seem to be a reasonable undertaking.

In all cases the equations do *not* predict an evolution to a well-defined *final state*; rather, the theory indicates precise directions and rates only within *limited length and time scales*. We feel that such complexities and their resolution could be suggestive in a variety of other problems, dealing with complicated reaction mechanisms, in physical chemistry as well as in astrophysics.

APPENDIX

This appendix collects a number of computational results relative to the wavevector and to the inverse Laplace integrations.

Wavevector Integration—The General Case

The integrals to be evaluated are vector and tensor integrals of the form

$$\mathbf{S}^{(n)}(\mathbf{q},z,\kappa,\mathbf{u})=\frac{2}{i\pi}\int d\mathbf{l}\frac{1}{\mathbf{l}\cdot\mathbf{u}-z}\frac{(\mathbf{l})^n}{(l^2+\kappa^2)\left[(\mathbf{l}-\mathbf{q})^2+\kappa^2\right]} \tag{A.1}$$

for $n=1,2$. Decomposing $\mathbf{q}$ into parallel and perpendicular components relative to $\mathbf{u}$,

$$\mathbf{q}=\frac{\mathbf{u}(\mathbf{q}\cdot\mathbf{u})}{u^2}+\mathbf{q}_\perp=q_u\mathbf{1}_u+q_\perp\mathbf{1}_\perp \tag{A.2}$$

we take coordinate axes $OXYZ$ with OZ along $\mathbf{u}$ and OX along $\mathbf{q}_\perp$. One will revert to a general system of reference by writing

$$\mathbf{S}^{(1)}=S_X^{(1)}\mathbf{1}_\perp+S_Z^{(1)}\mathbf{1}_u \tag{A.3}$$

$$\begin{aligned}\mathbf{S}^{(2)}=&\,S_{YY}^{(2)}(\mathsf{I}-\mathbf{1}_u\mathbf{1}_u)+\left(S_{XX}^{(2)}-S_{YY}^{(2)}\right)\mathbf{1}_\perp\mathbf{1}_\perp+S_{ZZ}^{(2)}\mathbf{1}_u\mathbf{1}_u\\&+S_{XZ}\left(\mathbf{1}_\perp\mathbf{1}_u+\mathbf{1}_u\mathbf{1}_\perp\right)\end{aligned} \tag{A.4}$$

In $OXYZ$, we take cylindrical polar coordinates λ,μ,φ, so that

$$\mathbf{S}^{(n)}(\mathbf{q},z,\kappa,\mathbf{u})=-i\int_{-\infty}^{\infty}d\lambda(\lambda u-z)^{-1}\mathsf{U}^{(n)}(\lambda,\mathbf{q},\kappa) \tag{A.5}$$

$$\mathsf{U}^{(n)}=\frac{2}{\pi}\int_0^K d\mu\frac{\mu}{\lambda^2+\mu^2+\kappa^2}\int_0^{2\pi}d\varphi\frac{(\mathbf{1}_X\mu\cos\varphi+\mathbf{1}_Y\mu\sin\varphi+\mathbf{1}_Z\lambda)^n}{(\lambda-q_u)^2+\mu^2+q_\perp^2+\kappa^2-2\mu q_\perp\cos\varphi} \tag{A.6}$$

The short-distance divergence characteristic of the weak-coupling approximation is cut off at $\mu=K$. It is convenient not to introduce the cut-off in the λ integrals, which are well defined. By (3.46) K corresponds to the short-length scale $L_{\min}$, while q and κ (for $\kappa\neq 0$) are scaled by the Jeans' length L_J. To reduce the algebra we shall systematically ignore corrections in

$$\frac{q}{K}=O\left(\frac{L_{\min}}{L_J}\right)=O\left(N_J^{-1}\right), \qquad \frac{\kappa}{K}\lesssim O\left(N_J^{-1}\right) \tag{A.7}$$

noting that these relations are, in fact, valid only on average.

The integrations (A.6) are elementary and yield

$$\mathbf{U}^{(1)}=q_\perp^{-1}\left[L(\lambda)-L(\lambda+q_u)+\left(q^2-2\lambda q_u\right)W\right]\mathbf{1}_X+2\lambda W\mathbf{1}_Z \tag{A.8}$$

$$\mathsf{U}^{(2)}=\left[L(\lambda)-1+\left(\frac{1}{2q_\perp}\right)\left(q^2-2\lambda q_u\right)U_X^{(1)}\right](\mathbf{1}_X\mathbf{1}_X-\mathbf{1}_Y\mathbf{1}_Y)$$

$$+2\left[L(\lambda)-(\lambda^2+\kappa^2)W\right]\mathbf{1}_Y\mathbf{1}_Y+2\lambda^2W\mathbf{1}_Z\mathbf{1}_Z+\lambda U_X^{(1)}(\mathbf{1}_X\mathbf{1}_Z+\mathbf{1}_Z\mathbf{1}_X) \tag{A.9}$$

where

$$L(\lambda)=\log\frac{K^2+(\lambda-q_u)^2}{\kappa^2+(\lambda-q_u)^2} \tag{A.10}$$

$$W(\lambda)=-\frac{1}{\sqrt{\gamma}}\log\frac{(\lambda^2+\kappa^2)\left(\gamma^{1/2}-2\lambda q_u+q^2-2q_\perp^2\right)}{\left(\left[\lambda-q_u\right]^2+q_\perp^2+\kappa^2\right)\left(\gamma^{1/2}-2\lambda q_u+q^2\right)+2q_\perp^2(\lambda^2+\kappa^2)} \tag{A.11}$$

$$\gamma=q^2(2\lambda-q_u)^2+q_\perp^2\left(q^2+4\kappa^2\right) \tag{A.12}$$

The remaining λ integration can only be partially carried through, giving (cf. Ref. 14, p. 137) for $\mathbf{S}^{(1)}(\mathbf{q},z,\kappa,\mathbf{u})$:

$$q_\perp S_X^{(1)}=\Lambda_1+\left(q^2-2q_u\bar{z}\right)I_0-2q_uJ_0 \tag{A.13a}$$

$$S_Y^{(1)}=0 \tag{A.13b}$$

$$S_Z^{(1)}=2(\bar{z}I_0+J_0) \tag{A.13c}$$

and for $\mathbf{S}^{(2)}(\mathbf{q}, z, \kappa, \mathbf{u})$:

$$S_{XX}^{(2)} = -S_{YY}^{(2)} + 2\Lambda_0 - 2(\bar{z}^2 + \kappa^2)I_0 - 2(J_1 + \bar{z}J_0) \tag{A.14a}$$

$$2q_\perp^2 S_{YY}^{(2)} = 2q_\perp^2(\Lambda_0 + \pi u^{-1}) - (q^2 - 2q_u\bar{z})\Lambda_1 + 4q_u q^2 J_0$$
$$- (q^4 - 4q_u q^2\bar{z} + 4q^2\bar{z}^2 + 4\kappa^2 q_\perp^2)I_0 - 4q^2(J_1 + \bar{z}J_0) \tag{A.14b}$$

$$S_{ZZ}^{(2)} = 2\bar{z}^2 I_0 + 2(J_1 + \bar{z}J_0) \tag{A.14c}$$

$$q_\perp S_{XZ}^{(2)} = q_\perp S_{ZX}^{(2)} = \bar{z}\Lambda_1 + \bar{z}(q^2 - 2q_u\bar{z})I_0 + q^2 J_0 - 2q_u(J_1 + \bar{z}J_0) \tag{A.14d}$$

$$S_{YX}^{(2)} = S_{XY}^{(2)} = S_{YZ}^{(2)} = S_{ZY}^{(2)} = 0 \tag{A.14e}$$

with

$$\bar{z} = \frac{z}{u} \tag{A.15}$$

$$\Lambda_0 = \frac{2\pi}{u}\log\frac{\bar{z} - q_u + iK}{\bar{z} - q_u + i\kappa} = \frac{2\pi}{u}\log\frac{z - \mathbf{q}\cdot\mathbf{u} + iKu}{z - \mathbf{q}\cdot\mathbf{u} + i\kappa u} \tag{A.16}$$

$$\Lambda_1 = \Lambda_0 - \frac{2\pi}{u}\log\frac{\bar{z} + iK}{\bar{z} + i\kappa} \cong \frac{2\pi}{u}\log\frac{z + i\kappa u}{z - \mathbf{q}\cdot\mathbf{u} + i\kappa u} \tag{A.17}$$

$$I_0 = (iu)^{-1}\int_{-\infty}^{\infty} d\lambda(\lambda - \bar{z})^{-1}W(\lambda) \tag{A.18}$$

$$J_n = (iu)^{-1}\int_{-\infty}^{\infty} d\lambda\lambda^n W(\lambda) \tag{A.19}$$

Wavevector Integration for $\mathbf{q}_\perp = 0$ and for $\mathbf{q} = 0$

In view of the impossibility of obtaining fully explicit forms for the integrals I_0 and J_n (cf. Eqs. A.18 and A.19), it is of interest to quote the simple results obtaining when $\mathbf{q}_\perp = 0$. One then has

$$I_0 = \frac{\Lambda_1 - \Lambda_2}{q_u(2\bar{z} - q_u)}; \qquad J_n = \frac{\Lambda_2}{2q_u}\left(\frac{q_u}{2}\right)^n, \qquad (n = 0, 1) \tag{A.20}$$

with

$$\Lambda_2 = \frac{2\pi}{u}\log\frac{\mathbf{q}\cdot\mathbf{u} + 2i\kappa u}{-\mathbf{q}\cdot\mathbf{u} + 2i\kappa u} \underset{\kappa\to 0}{\longrightarrow} \frac{2\pi^2}{iu}\operatorname{sgn}(\mathbf{q}\cdot\mathbf{u}) \tag{A.21}$$

This gives the following for $\mathbf{S}^{(1)}(q_u \mathbf{1}_u, z, \kappa, \mathbf{u})$:

$$\mathbf{S}^{(1)} = S_Z^{(1)} \mathbf{1}_u = (2\bar{z}q_u - q_u^2)^{-1}(2\bar{z}\Lambda_1 - q_u\Lambda_2)\mathbf{1}_u \tag{A.22}$$

and for $\mathsf{S}^{(2)}(q_u \mathbf{1}_u, z, \kappa, \mathbf{u})$:

$$\mathsf{S}^{(2)} = S_{YY}^{(2)}(\mathsf{I} - \mathbf{1}_u\mathbf{1}_u) + S_{ZZ}^{(2)}\mathbf{1}_u\mathbf{1}_u \tag{A.23}$$

$$S_{YY}^{(2)} = \Lambda_0 + (q_u^2 - 2\bar{z}q_u)^{-1}\left[(\bar{z}^2 + \kappa^2)\Lambda_1 - \left(\frac{1}{4}q_u^2 + \kappa^2\right)\Lambda_2\right] \tag{A.24a}$$

$$S_{ZZ}^{(2)} = (2\bar{z}q_u - q_u^2)^{-1}\left(2\bar{z}^2\Lambda_1 - \frac{1}{2}q_u^2\Lambda_2\right), \tag{A.24b}$$

where $\bar{z}$, Λ_0, Λ_1, are defined by (A.15) through (A.17).

Let us here also quote the expressions for the homogeneous case $\mathbf{q}=0$:

$$\mathbf{S}^{(1)} = 2\pi\left[(z + i\kappa u)^{-1} - (z + iKu)^{-1}\right]\mathbf{1}_u \tag{A.25}$$

$$\mathsf{S}^{(2)} = \frac{2\pi}{u}\left(\log\frac{z + iKu}{z + i\kappa u} - \frac{1}{2}\frac{iKu}{z + iKu}\right)(\mathsf{I} - \mathbf{1}_u\mathbf{1}_u)$$

$$+ 2\pi i\left[\frac{K}{(z + iKu)} - \frac{\kappa}{(z + i\kappa u)}\right]\mathbf{1}_u\mathbf{1}_u \tag{A.26}$$

Collision Operator for a Finite System

From the definitions (3.38), (3.41), (3.42), and (A.1), one has, with $\mathsf{S}^{(n)} = \mathsf{S}^{(n)}(\mathbf{q}, z, \kappa, \mathbf{v}_{12})$

$$\tilde{\psi}_{\mathbf{q}}(z) = G^2m^2\boldsymbol{\partial}_{12}\cdot\left\{\left[\mathsf{S}^{(2)} - \mathbf{q}\mathbf{S}^{(1)}\right]\cdot\boldsymbol{\partial}_{12} + \left[\frac{\partial(\mathsf{S}^{(2)} - \mathbf{q}\mathbf{S}^{(1)})}{\partial z}\right]\cdot\boldsymbol{\nabla}_{12}\right\} \tag{A.27}$$

The contribution that will be dominant asymptotically is that containing $\log K$, that is, Λ_0 (cf. Eq. A.16); we shall single it out to write

$$\tilde{\psi}_{\mathbf{q}}(z) = 2\pi G^2m^2\boldsymbol{\partial}_{12}\cdot\left(\log\frac{z - \mathbf{q}\cdot\mathbf{v}_{12} + iKv_{12}}{z - \mathbf{q}\cdot\mathbf{v}_{12} + i\kappa v_{12}}\right)\frac{\mathsf{I}v_{12}^2 - \mathbf{v}_{12}\mathbf{v}_{12}}{v_{12}^3}\cdot\boldsymbol{\partial}_{12} + \hat{\psi}_{\mathbf{q}}(z) \tag{A.28}$$

where we have used Eqs. A.14b and A.4 and where $\hat{\psi}_{\mathbf{q}}(z)$ is defined by comparison with (A.27).

The time dependence of the inverse Laplace transform $\tilde{\Psi}_{\mathbf{q}}(\tau)$ can in part be explicited, using the result

$$\int \frac{dz}{-2\pi} \exp(-iz\tau) \log \frac{z-a+ib}{z-a+ib'} = \exp(-ia\tau) \frac{\exp(-b'\tau) - \exp(-b\tau)}{\tau} \tag{A.29}$$

which obtains by an elementary contour integration. Thus the first term in (A.28) decays at least as fast as $\tau^{-1}\exp(-\kappa v_{12}\tau)$. One can verify that the contributions from $\hat{\psi}_{\mathbf{q}}(z)$ also decay on the crossing time scale $\tau_C = (\kappa\langle v_{12}\rangle)^{-1}$. This is immediate for the contributions containing Λ_1 (cf. Eq. A.17). As for $I_0 = I_0(z)$, from (A.11) and (A.18) one verifies that $I_0(z)$ has no singularities on the real axis [which is no longer true if $\mathbf{q}_\perp = 0$ (see Eq. A.20) or if $\kappa = 0$ (see Eq. A.37)]. Then, since the only times that can appear in I_0 are $(\kappa v_{12})^{-1}$ and $(qv_{12})^{-1} \lesssim \tau_C$, one sees that $\tilde{\Psi}_{\mathbf{q}}(\tau)$ necessarily decays on the scale

$$\tilde{\Psi}_{\mathbf{q}}(\tau) \to 0, \quad \tau \gg \tau_C \qquad (\kappa \neq 0) \tag{A.30}$$

This property continues to hold if one considers a homogeneous model situation as in Section III.F, for which $\mathbf{q} = 0$ and $\boldsymbol{\nabla}_{12} = 0$ so that $\tilde{\psi}_{\mathbf{q}}$ reduces to (cf. Eq. A.27)

$$\psi_0(z) = G^2 m^2 \boldsymbol{\partial}_{12} \cdot \mathsf{T}(z,\kappa) \cdot \boldsymbol{\partial}_{12} \tag{A.31}$$

with

$$\mathsf{T}(z,\kappa) = \mathsf{S}^{(2)}(\mathbf{q} = 0, z, \kappa, \mathbf{v}_{12}) \tag{A.32}$$

given by (A.26). Let us also note for reference the explicit expression obtaining for $z = 0$:

$$\psi_0(z=0) = 2\pi G^2 m^2 \left[-\frac{1}{2} + \log\left(\frac{K}{\kappa}\right) \right] v_{12}^{-3} \boldsymbol{\partial}_{12} \cdot (\mathsf{I} v_{12}^2 - \mathbf{v}_{12}\mathbf{v}_{12}) \cdot \boldsymbol{\partial}_{12} \tag{A.33}$$

Destruction Term for a Finite System

For initial correlations having the same functional form as in equilibrium (cf. Eq. 4.18), the defining relation (3.23) gives

$$D(\mathbf{x}_1, \mathbf{v}_1, t) = -\sigma^{-2} \int d\mathbf{v}_2 d\mathbf{x}_2 \Theta_{12} \exp[-(\mathcal{L}_1 + \mathcal{L}_2)t] V^{12} f(\mathbf{x}_1, \mathbf{v}_1, 0) f(\mathbf{x}_2, \mathbf{v}_2, 0) \tag{A.34}$$

In the p.e. approximation (compare the passage between Eqs. 3.25 and 3.34), and with the interaction potential given by (3.38), one gets:

$$D(\mathbf{x}_1,\mathbf{v}_1,t) = -\sigma^{-2}\int d\mathbf{v}_2 d\mathbf{x}_2 \int d\mathbf{r}\,\delta(\mathbf{r}-\mathbf{x}_1+\mathbf{x}_2)\Theta_{12}\exp\left(-t\mathbf{v}_{12}\cdot\frac{\partial}{\partial\mathbf{r}}\right)V^{12}(r)$$

$$\times\exp[-(\mathfrak{L}_1+\mathfrak{L}_2)t]f(\mathbf{x}_1,\mathbf{v}_1,0)f(\mathbf{x}_2,\mathbf{v}_2,0)$$

$$= G^2m^2(16\pi^4 i\sigma^2)^{-1}\int d\mathbf{v}_2 d\mathbf{x}_2 \int dz\exp(-izt)\int d\mathbf{q}\exp[i\mathbf{q}\cdot(\mathbf{x}_1-\mathbf{x}_2)]\boldsymbol{\partial}_{12}$$

$$\cdot\mathbf{S}^{(1)}(-\mathbf{q},z-\mathbf{q}\cdot\mathbf{v}_{12},\kappa,\mathbf{v}_{12})\exp[-(\mathfrak{L}_1+\mathfrak{L}_2)t]f(\mathbf{x}_1,\mathbf{v}_1,0)f(\mathbf{x}_2,\mathbf{v}_2,0) \quad (A.35)$$

The decay of the destruction term is determined by the inverse Laplace transform of $\mathbf{S}^{(1)}(z-\mathbf{q}\cdot\mathbf{v}_{12})$, that is, of $\Lambda_1(z-\mathbf{q}\cdot\mathbf{v}_{12})$ and of $I_0(z-\mathbf{q}\cdot\mathbf{v}_{12})$ (cf. Eq. A.13). The arguments developed previously for the collision term apply here also, and as in (A.30) one has

$$D(\mathbf{x}_1,\mathbf{v}_1,t)\to 0, \quad t\gg\tau_C \qquad (\kappa\neq 0) \qquad (A.36)$$

Characteristic Integral for the Infinite Inhomogeneous Model

From the foregoing results for a finite system, one passes to the case of an infinite system by setting $\kappa=0$. The characteristic integrals $\mathbf{S}^{(n)}(z,\kappa=0)$ now have logarithmic singularities on the real axis in $z=0$ and $z=\mathbf{q}\cdot\mathbf{u}$. These are introduced not only by the quantities Λ_0 and Λ_1 of (A.16) and (A.17) but also by (A.18). More specifically, it can be verified that $W(\lambda, \kappa=0)$ (cf. Eq. A.11) has singularities only in the points $\lambda=0$ (for the numerator) and $\lambda=q_u$ (for the denominator). In $I_0(z,\ \kappa=0)$, (A.18), these give rise to logarithmic singularities in $\bar{z}=0$ and $\bar{z}=q_u$, as can be exhibited by considering the special case $\mathbf{q}_\perp=0$ for which

$$I_0(\mathbf{q}_\perp=0,\kappa=0)=\frac{2\pi u}{(2z-\mathbf{q}\cdot\mathbf{u})\mathbf{q}\cdot\mathbf{u}}\left[\log\frac{z}{z-\mathbf{q}\cdot\mathbf{u}}+\pi i\,\mathrm{sgn}(\mathbf{q}\cdot\mathbf{u})\right] \quad (A.37)$$

(cf. Eqs. A.17, A.20, and A.21).

The destruction term (4.79) introduces the quantity $\mathbf{S}_\mathbf{q}(z-\mathbf{k}\cdot\mathbf{v}_1)$ that (cf. Eq. A.35) coincides with

$$\mathbf{S}_\mathbf{q}(z-\mathbf{k}\cdot\mathbf{v}_1)\equiv\mathbf{S}^{(1)}(\mathbf{q},\ z-\mathbf{k}\cdot\mathbf{v}_1,\ \kappa=0,\ \mathbf{v}_{12}) \qquad (A.38)$$

and is thus given by (A.3) and (A.13). The singularities of $\mathbf{S}_\mathbf{q}(z')$ are logarithmic, as in (A.37), and thus by (A.29) decrease with time as $1/t$.

The tensor integral $\mathsf{T}_{\mathbf{q}}(z)$ appearing in the collision term of (4.82) is by definition

$$\mathsf{T}_{\mathbf{q}}(z) = \mathsf{S}^{(2)}(\mathbf{q}, z, \kappa = 0, \mathbf{v}_{12}) - \mathbf{q}\mathbf{S}^{(1)}(\mathbf{q}, z, \kappa = 0, \mathbf{v}_{12}) \tag{A.39}$$

In the asymptotic limit $z = i\epsilon \to +i0$, the dominant contributions (i.e., containing $\log\epsilon$ or $\log K$) to the components of $\mathsf{T}_{\mathbf{q}}(z)$ are, from (A.13) through (A.19), seen to be

$$(\mathsf{T}_{\mathbf{q}})_{XX} = \tilde{\Lambda}_0 - \left(1 - \frac{q^2}{2q_\perp^2}\right)(\tilde{\Lambda}_1 + q^2\tilde{I}_0) + O(1) \tag{A.40a}$$

$$(\mathsf{T}_{\mathbf{q}})_{YY} = \tilde{\Lambda}_0 - \left(\frac{q^2}{2q_\perp^2}\right)(\tilde{\Lambda}_1 + q^2\tilde{I}_0) + O(1) \tag{A.40b}$$

with

$$\tilde{\Lambda}_0 = \frac{2\pi}{v_{12}}\log\frac{\mathbf{q}\cdot\mathbf{v}_{12} - iKv_{12}}{\mathbf{q}\cdot\mathbf{v}_{12} - i\epsilon}, \qquad \tilde{\Lambda}_1 = \frac{2\pi}{v_{12}}\log\frac{-i\epsilon}{\mathbf{q}\cdot\mathbf{v}_{12} - i\epsilon} \tag{A.41}$$

$$\tilde{I}_0 = -i\int_{-\infty}^{\infty} d\lambda(\lambda v_{12} - i\epsilon)^{-1} W(\lambda, \kappa = 0) \tag{A.42}$$

The dominant contribution to $\tilde{I}_0$ comes from the neighborhood of $\lambda = 0$; using Eqs. A.11 and A.12 one gets

$$\tilde{I}_0 = \tilde{I}_0^{(0)} + O(1) \tag{A.43}$$

$$\tilde{I}_0^{(0)} = \frac{i}{q^2}\int_{-\infty}^{\infty} d\lambda \frac{1}{\lambda v_{12} - i\epsilon}\log\frac{\lambda^2(\mathbf{q}\cdot\mathbf{v}_{12})^2}{q^4 v_{12}^2} = \frac{2\pi}{q^2 v_{12}}\log\frac{q^2 v_{12}^2}{\epsilon|\mathbf{q}\cdot\mathbf{v}_{12}|} \tag{A.44}$$

The compensations between $\tilde{\Lambda}_1$ and $q^2\tilde{I}_0$ finally leaves us with

$$\mathsf{T}_{\mathbf{q}}(i\epsilon) \cong 2\pi v_{12}^{-3}(\mathsf{I}v_{12}^2 - \mathbf{v}_{12}\mathbf{v}_{12})\log\frac{\mathbf{q}\cdot\mathbf{v}_{12} - iKv_{12}}{\mathbf{q}\cdot\mathbf{v}_{12} - i\epsilon} \tag{A.45}$$

The derivative $\mathsf{T}'_{\mathbf{q}} = \partial\mathsf{T}_q/\partial i\epsilon$ introduces the quantity $\partial\tilde{I}_0/\partial i\epsilon$ that, proceeding as for (A.44), can be written as

$$\frac{\partial\tilde{I}_0}{\partial i\epsilon} \cong \frac{\partial\tilde{I}_0^{(0)}}{\partial i\epsilon} + \frac{2\mathbf{q}\cdot\mathbf{v}_{12}}{q^2 v_{12}^2}\tilde{I}_0^{(0)} \tag{A.46}$$

At this dominant order there are numerous compensations in $\mathsf{T}'_q(i\epsilon)$; one verifies easily that $\mathsf{T}'_{\mathbf{q}}(i\epsilon)$ reduces to

$$\mathsf{T}'_{\mathbf{q}}(i\epsilon) \cong -\mathbf{q}\mathbf{v}_{12}\frac{1}{v_{12}}\left(\frac{\partial S_Z^{(1)}(\kappa=0)}{\partial z}\right)_{z=i\epsilon} = -\mathbf{q}\mathbf{v}_{12}\frac{4\pi}{q^2v_{12}^3}\log\frac{q^2v_{12}^2}{\epsilon|\mathbf{q}\cdot\mathbf{v}_{12}|} \quad \text{(A.47)}$$

It may be noted that such asymptotic results can be obtained directly from the defining relations without going through all the complication of the wavevector integration (Eqs. A.5 through A.19). Thus, proceeding from (A.1), and expanding the integrand for small $\mathbf{l}$ one gets, in agreement with (A.47),

$$\left(\frac{\partial \mathbf{S}^{(1)}(\kappa=0)}{\partial z}\right)_{z=i\epsilon} = \frac{2}{\pi i q^2 v_{12}^2}\int d\lambda \int d\mathbf{l}_\perp \frac{\mathbf{l}(1+2q^{-2}\mathbf{q}\cdot\mathbf{l}+\cdots)}{(l_\perp^2+\lambda^2)(\lambda-i\epsilon/v_{12})^2}$$

$$= \frac{4\mathbf{v}_{12}}{q^2v_{12}^3}\int_{-\infty}^{\infty} d\lambda \frac{(\lambda^2\epsilon/v_{12})\log\left[O(q^2)/\lambda^2\right]}{\left[\lambda^2+(\epsilon/v_{12})^2\right]^2}$$

$$= \frac{2\pi\mathbf{v}_{12}}{q^2v_{12}^3}\log\frac{O(q^2v_{12}^2)}{\epsilon^2} + O\left(\left[\frac{\epsilon}{qv_{12}}\right]\right) + O\left(\frac{\epsilon}{qv_{12}}\log\frac{qv_{12}}{\epsilon}\right) \quad \text{(A.48)}$$

Finally, let us note the complete asymptotic forms for the case $\mathbf{q}_\perp = 0$, as derived from (A.39) and (A.22) through (A.24):

$$\mathsf{T}'_{\mathbf{q}_\perp=0}(i\epsilon) = -\mathbf{q}\mathbf{v}_{12}\left(\frac{4\pi}{q^2v_{12}^3}\right)\left\{\log\left[\frac{(i\epsilon-\mathbf{q}\cdot\mathbf{v}_{12})}{i\epsilon}\right] - 1 - \pi i\,\text{sgn}(\mathbf{q}\cdot\mathbf{v}_{12})\right\}$$

$$+\left(\frac{2\pi}{v_{12}^3}\right)\left\{\left[\mathsf{I}v_{12}^2 - \mathbf{v}_{12}\mathbf{v}_{12}\right]\left[\mathbf{q}\cdot\mathbf{v}_{12}\right]^{-1}\left[1+\tfrac{1}{2}\pi i\,\text{sgn}(\mathbf{q}\cdot\mathbf{v}_{12})\right] - \mathbf{v}_{12}\mathbf{v}_{12}\left(\frac{\pi i}{|\mathbf{q}\cdot\mathbf{v}_{12}|}\right)\right\}$$

(A.49)

References

Principal abbreviations used: *Ap. J* = *Astrophysical Journal*; *Astr. J* = *Astronomical Journal*; *Astr.* and *Ap.* = *Astronomy and Astrophysics*; *Ap.* and *Sp. Sc.* = *Astrophysics and Space Science*; *M. N.* = *Monthly Notices of the Royal Astronomical Sosiety* (G. B.).

1. J. Jeans, *Astronomy and Cosmogony*, Cambridge Univ. Press, London, 1929.
2. S. Chandrasekhar, *Principles of Stellar Dynamics*, Univ. of Chicago Press, Chicago, 1942; Dover, New York, 1960.
3. I. King, *Astr. J.*, **67**, 471 (1962).

4. D. Lynden-Bell, *M.N.*, **136**, 101 (1967).
5. R. B. Larson, *M.N.*, **147**, 323; **150**, 93 (1970).
6. M. Henon, *Astr. and Ap.*, **2**, 151 (1969).
7. D. Lynden-Bell and R. Wood, *M.N.*, **138**, 495 (1968).
8. I. Prigogine and G. Severne, *Physica*, **32**, 1376 (1966).
9. R. H. Miller, *Advances in Chemical Physics*, Vol. 26, edited by I. Prigogine and S. Rice, Wiley, New York, 1974.
10. M. Henon, in *Dynamical Structure and Evolution of Stellar Systems*, lectures by M. Contopoulos, M. Henon and D. Lynden-Bell, edited by L. Martinet and L. Mayor, Geneva Observatory, 1973.
11. For a good discussion see, for example, K. Ogorodnikov, *Dynamics of Stellar Systems*, Pergamon, New York, 1965.
12. See, for example, Ref. 2, Section V.5.8.
13. See, for example, M. Lecar and L. Cohen, *Ap. and Sp. Sc.*, **13**, 397 (1971).
14. R. Balescu, *Statistical Mechanics of Charged Particles*, Wiley-Interscience, New York, 1963.
15. G. Severne and A. Kuszell, *Ap. and Sp. Sc.*, **32**, 447 (1975).
16. D. Lynden-Bell and N. Sanitt, *M.N.*, **143**, 167 (1969). R. Kulsrud and J. Mark, *Ap. J.*, **160**, 471 (1970).
17. D. Lynden-Bell, *M.N.*, **124**, 279 (1962).
18. E. P. Lee, *Ap. J.*, **148**, 185 (1967). [See also C. S. Wu, *Phys. Fluids*, **11**, 545 (1969).]
19. A. Toomre, *Ap. J.*, **139**, 1217 (1964).
20. C. C. Lin and F. H. Shu, *Ap. J.*, **140**, 646 (1964). A. Kalnajs, *Ap. J.*, **166**, 275 (1971).
21. S. Chandrasekhar, *Hydrodynamic and Hydromagnetic Stability*, Oxford Univ. Press, Oxford, 1961, Chap. XIII.
22. For a discussion of the thermodynamical limit for gravitational systems, see P. Hertel and W. Thirring, *Comm. Math. Phys.*, **24**, 22 (1971).
23. I. Gilbert, *Ap. J.*, **159**, 239 (1970).
24. I. Prigogine, *Non-equilibrium Statistical Mechanics*, Wiley-Interscience, New York, 1962.
25. I. Prigogine and G. Severne, *Bull. Astron.*, **3**, 273 (1968).
26. G. Severne, *Physica*, **31**, 877 (1965). H. T. Davis, *Advances in Chemical Physics*, Vol. 24, edited by I. Prigogine and S. Rice, Wiley, New York, 1973.
27. M. Henon, *Annales d'Astrophys.*, **21**, 186 (1958).
28. M. Henon, in *Dynamics of Stellar Systems* (I.A.U. Symposium N° 69 Besancon 1974), edited by A. Hayli, Reidel, Holland, 1975.
29. S. Aarseth, M. Henon, and R. Wielen, *Astr. and Ap.*, **37**, 183 (1974).
30. I. King, *Astr. J.*, **65**, 122 (1960) and papers referred to therein.
31. R. Michie, *Ap. J.*, **133**, 781 (1961); I. King, *Astr. J.*, **71**, 64 (1966).
32. D. Montgomery and C. Nielsen, *Phys. Fluids*, **13**, 1405 (1970); J.-Y. Hsu, G. Joyce, and D. Montgomery, *J. Plasma Phys.*, **12**, 27 (1974); M. Yoo and B. Abraham-Shrauner, *Physica*, **68**, 133 (1973).
33. W. Saslaw, *Ap. J.*, **177**, 17 (1972).
34. P. J. E. Peebles, *Comm. Ap. and Sp. Phys.*, **4**, 53 (1972); W. Press and P. Schechter, *Ap. J.*, **187**, 425 (1974).
35. M. Haggerty and G. Janin, *Astr. and Ap.*, **36**, 415 (1974).
36. P. Hodge and R. Michie, *Astr. J.*, **74**, 587 (1969).
37. M. Haggerty and G. Severne, *Bull. Acad. Roy. Belgique Cl. Sc.*, **60**, 226 (1974); M. Haggerty and G. Severne, *Nature*, **249**, 537 (1974).
38. G. Severne, *Physica*, **61**, 307 (1972).
39. For an elementary example, see Ref. 8, Section 3.

40. J. Ostriker and A. Davidsen, *Ap. J.*, **151**, 679 (1968).
41. M. Haggerty and G. Severne, *Physica*, **51**, 461 (1971).
42. T. Guo and W. Guo, *Bull. Acad. Roy. Belgique Cl. Sc.*, **60**, 1490 (1975); *Physica*, **79A**, 120 (1975).
43. R. H. Miller, *Ap. J.*, **180**, 759 (1973).
44. L. Spitzer, *Ap. J.*, **127**, 17 (1958).
45. The observational exponent 1.7 has been suggested by G. de Vancouleurs, *Science*, **167**, 1203 (1970); for theoretical discussions see Ref. 35 and papers quoted therein.
46. R. Cohen, L. Spitzer, and P. Routly, *Phys. Rev.*, **80**, 230 (1950).
47. L. Spitzer and M. Schwarzschild, *Ap. J.*, **114**, 385 (1951).
48. L. Marochnik, *Sov. Astr.*, **12**, 371 (1968); **12**, 1000 (1969).
49. I. Genkin, *Sov. Astr.*, **12**, 858 (1969).
50. L. Marochnik, *Sov. Astr.*, **14**, 36 (1970); S. Pomagaev, *Sov. Astr.*, **13**, 635 (1970) and papers quoted therein.
51. I. Genkin, *Sov. Phys. Dokl.*, **16**, 261 (1971).
52. W. Saslaw, *M.N.*, **143**, 437 (1969); **150**, 299 (1970).
53. L. Spitzer, *Ap. J. Lett.*, **158**, L139 (1969); W. Saslaw and D. De Young, *Ap. J.*, **170**, 423 (1971).
54. I. Gilbert, *Ap. J.*, **152**, 1043 (1968).
55. W. Chappell, *J. Math. Phys.*, **9**, 1921 (1968).
56. Y. Klimontovich, *The Statistical Theory of Non-equilibrium Processes in a Plasma*, Pergamon, New York, 1967.
57. R. Thorne, *Ap. J.*, **151**, 671 (1968).
58. C. S. Wu, *Phys. Fluids*, **11**, 316 (1968).
59. M. Haggerty, *Physica*, **51**, 477 (1971).
60. I. Lerche, *Ap. J.*, **166**, 207 (1971).
61. S. Chandrasekhar, *Ap. J.*, **99**, 47 (1944).
62. L. Cohen and A. Ahmad, *Ap. J.*, **197**, 667 (1975).
63. S. Chandrasekhar, *Ap. J.*, **98**, 54 (1943).
64. Y. Talpaert and F. Lefèvre, *Bull. Acad. Roy. Belg. Cl. Sc.*, **58**, 759 (1972); R. Coutrez, *Comm. Obs. Roy. Belg.*, N^0 15 (1950).
65. M. Haggerty, *Physica*, **50**, 391 (1970); *Ap. J.*, **166**, 257 (1971).
66. W. Saslaw, *Pub. Astr. Soc. Pacific*, **85**, 5 (1973).
67. W. Saslaw, *Ap. J.*, **195**, 773 (1975).

MAGNETIC CIRCULAR DICHROISM

P. J. STEPHENS

Department of Chemistry,
University of Southern California,
Los Angeles, California 90007

CONTENTS

I. INTRODUCTION

Circular dichroism—differential absorption of left and right circularly polarized light—is induced in all matter by a uniform longitudinal magnetic field. This is the phenomenon of magnetic circular dichroism (MCD). Magnetic circular dichroism is the absorptive counterpart of optical rotation induced by a magnetic field—magnetic optical rotation (MOR). The latter, more commonly known as the Faraday effect, has been studied for many years.[1] On the other hand, MCD has a more limited history and has been pursued intensively only recently,[1,2] However, the utility of MCD is already clear, and applications in chemistry, physics, and biophysics are currently increasing rapidly.[2]

Our objective is to present an introduction to the theory of MCD, demonstrating the techniques used in its analysis and the kinds of information made available thereby. Our approach is illustrative. Examples are chosen for their simplicity and specific experimental systems are not treated. In particular, we limit our discussion to electronic transitions of impurities in crystalline solids, with the further assumptions that the impurities are rigid and dilute and their absorption is localized. Such systems have been a principal focus of recent MCD studies.

In Section II basic formulas for the absorption and differential absorption of circularly polarized light are derived. In Section III we calculate the MCD of various simple model systems, in which it is possible to obtain analytic forms for the MCD dispersion. In Section IV we discuss the application to MCD of the method of moments. Throughout Sections III and IV only allowed electronic transitions are considered. In Section V the discussion is extended to forbidden, vibration-induced transitions.

II. BASIC THEORY

In an MCD experiment the sample is placed in a uniform magnetic field $\boldsymbol{H}$. Its MCD is the difference in absorption of left and right circularly polarized (CP) light, propagating parallel to $\boldsymbol{H}$, that is induced by the magnetic field.

To relate the macroscopic MCD measurement to microscopic properties of the absorbing impurity centers in the crystalline sample, we employ standard semiclassical radiation absorption theory, treating the radiation classically and the absorbing matter quantum-mechanically. First, we solve Maxwell's equations to obtain the electromagnetic fields of the light, propagating through the sample. These are to be CP—otherwise CD has no meaning—and such waves are only solutions of Maxwell's equations when, for propagation vector $\mathbf{k}$, the medium is isotropic in the xy-plane. $\mathbf{E}$, $\mathbf{H}$, $\mathbf{D}$, and $\mathbf{B}$ all lie in the xy-plane and $\mathbf{D}=\hat{\epsilon}\mathbf{E}$, $\mathbf{B}=\hat{\mu}\mathbf{H}$, therefore, where $\hat{\epsilon}$ and $\hat{\mu}$ are the complex dielectric constant and magnetic permeability in the xy-plane, respectively. Then follows

$$\begin{aligned}\mathbf{E}_{\pm} &= E_0^{\circ} \exp\left[\frac{i\mathcal{E}(t-\hat{n}_{\pm}z/\mathbf{c})}{\hbar}\right](\mathbf{i}\pm i\mathbf{j}) \\ \mathbf{H}_{\pm} &= \hat{n}_{\pm} E_0^{\circ} \exp\left[\frac{i\mathcal{E}(t-\hat{n}_{\pm}z/\mathbf{c})}{\hbar}\right](\mp i\mathbf{i}+\mathbf{j})\end{aligned} \tag{1}$$

where + and − denote right and left CP light respectively, $\hat{n}=n-ik$ is the complex refractive index, $\mathcal{E}=h\nu$ is the photon energy, and we have made the approximation $\hat{\mu}=1$. The intensity $I(z)$ of the light at z is the energy passing through unit area in unit time and is given by the Poynting vector

$(\mathbf{c}/4\pi)(\mathbf{E}\times\mathbf{H})$. From (1)

$$
\begin{aligned}
I_{\pm}(z) &= \left(\frac{\mathbf{c}}{4\pi}\right) n_{\pm} E_0^{\circ 2} \exp\left[-\frac{2\mathcal{E} k_{\pm} z}{\hbar \mathbf{c}}\right] \\
&= I_{\pm}(0) \exp\left[-\frac{2\mathcal{E} k_{\pm} z}{\hbar \mathbf{c}}\right]
\end{aligned} \tag{2}
$$

whence

$$
k_{\pm} = \frac{h}{n_{\pm} \mathcal{E} |\mathbf{E}_{\pm}(z)|^2} \left[-\frac{\partial I_{\pm}}{\partial z}(z)\right] \tag{3}
$$

Now $-\partial I/\partial z$ is the energy absorbed per unit time per unit volume at z and can be related to the transition probabilities of the absorbing centers there. Let us consider the eigenstates responsible for absorption of a center to be as shown in Fig. 1. If the probability of absorption per unit time at energy $\mathcal{E}$ due to the transition $a \to j$ is $P_{a\to j}$ and N_a is the number of centers per unit volume in state a

$$
-\frac{\partial I}{\partial z}(z) = \sum_{a,j} N_a P_{a\to j}(z) \mathcal{E} \tag{4}
$$

whence

$$
k_{\pm} = \frac{h}{n_{\pm} |\mathbf{E}_{\pm}(z)|^2} \sum_{a,j} N_a P_{a\to j}^{\pm}(z) \tag{5}
$$

The calculation of $P_{a\to j}$ is standard: The Hamiltonian of the center plus radiation is

$$
\mathcal{H} = \mathcal{H}_0 + \mathcal{H}_1 \tag{6}
$$

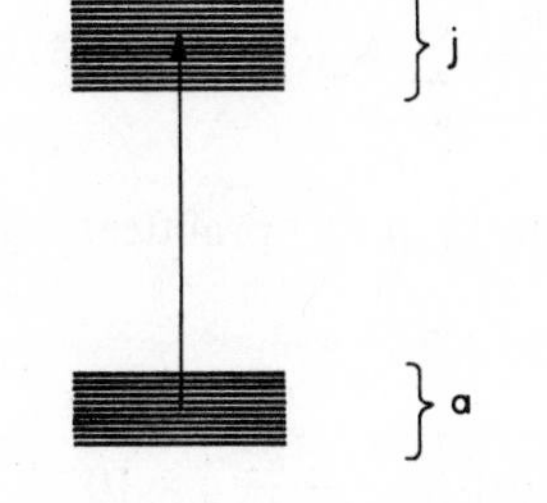

Fig. 1. Ground and excited eigenstates, a and j, of the absorbing center. It is assumed that only the ground states a are populated.

where $\mathcal{H}_0$ and $\mathcal{H}_1$ are the unperturbed Hamiltonian and the interaction with the radiation, respectively. We here limit our treatment to electric dipole absorption and approximate $\mathcal{H}_1$ immediately by

$$\mathcal{H}_1 = -\mathbf{m} \cdot \mathbf{e} \tag{7}$$

where $\mathbf{m} = \sum_i e_i \mathbf{r}_i$ is the electric dipole operator and $\mathbf{e}$ is the electric field due to the light wave at the center, assumed uniform. $\mathbf{e}$ is related to the macroscopic $\mathbf{E}$ field by a dielectric factor (e.g., the Lorentz effective field correction); for our purposes we make only the assumption that $\mathbf{e}$ is proportional to $\mathbf{E}$:

$$\mathbf{e} = \alpha \mathbf{E} \tag{8}$$

Then

$$\mathcal{H}_1^{\mp}(z) = -\sqrt{2}\,\alpha |\mathbf{E}_{\pm}(z)| \operatorname{Re}\left\{ m_{\pm} \exp\left[\frac{i\mathcal{E}t}{\hbar} \right] \right\} \tag{9}$$

where $m_{\pm} = (1/\sqrt{2}\,)(m_x \pm i m_y)$. Time-dependent perturbation theory gives

$$P_{a\to j} = \frac{1}{t} \left| \frac{1}{\hbar} \int_0^t \exp\left(i\mathcal{E}_{ja}t'/\hbar\right) \langle j | \mathcal{H}_1 | a \rangle dt \right|^2 \tag{10}$$

where $|a\rangle$ and $|j\rangle$ are eigenstates of $\mathcal{H}_0$ of energies $\mathcal{E}_a$ and $\mathcal{E}_j$, respectively, and $\mathcal{E}_{ja} = \mathcal{E}_j - \mathcal{E}_a$. With (9), Eq. 10 leads to

$$P_{a\to j}^{\pm}(z) = \frac{\alpha^2 \pi}{\hbar} |E_{\pm}(z)|^2 |\langle a | m_{\pm} | j \rangle|^2 \delta(\mathcal{E}_{ja} - \mathcal{E}) \tag{11}$$

and, hence,

$$k_{\pm} = \frac{2\pi^2\alpha^2}{n_{\pm}} \sum_{a,j} N_a |\langle a | m_{\pm} | j \rangle|^2 \delta(\mathcal{E}_{ja} - \mathcal{E}) \tag{12}$$

We now transform Eq. 12 into "practical" units. Equation 2 can be alternatively written

$$\frac{I(z)}{I(0)} = \exp\left\{ -\frac{2\mathcal{E}kz}{\hbar c} \right\} = 10^{-A} = 10^{-\epsilon c z} \tag{13}$$

where A is absorbance (= optical density), ϵ is molar extinction coefficient, and c is concentration in moles per liter of absorbing centers, whence

$$\frac{A_{\pm}}{\mathcal{E}} = \left(\frac{\epsilon_{\pm}}{\mathcal{E}}\right)cz$$
$$= \gamma_{\pm}\left\{\sum_{a,j}\left(\frac{N_a}{N}\right)|\langle a|m_{\pm}|j\rangle|^2\delta(\mathcal{E}_{ja}-\mathcal{E})\right\}cz \tag{14}$$
$$\gamma_{\pm} = \frac{\mathfrak{N}\pi^2\alpha^2\log_{10}e}{250\hbar c n_{\pm}}$$

where $\mathfrak{N}$ is Avogadro's number and (N_a/N) is the fractional population of state a.

The absorbance due to the transition $a \to j$ is thus proportional to the concentration of a (N_a), the appropriate CP transition moment squared ($|\langle a|m_{\pm}|j\rangle|^2$) and pathlength ($z$). It has a delta function dispersion ($\delta(\mathcal{E}_{ja}-\mathcal{E})$) and is independent of light intensity. Several assumptions have led to these results. The linear optics approximations, embodied in (1) and (7), lead to intensity-independent A; Eq. 14 does not allow for nonlinear phenomena occurring at high-light intensities. The assumption of negligible interaction between absorbing centers leads to Beer's law ($A \propto N_a$); Eq. 14 does not apply to cooperatively absorbing systems. The electric dipole approximation leads to the simple form of the transition moment; Eq. 14 does not allow for magnetic dipole or electric quadrupole transition mechanisms. The neglect of radiative lifetimes produces infinitely sharp transitions.

Within the domain of its validity, Eq. 14 leads further to the corresponding basic formula for CD. We define

$$\Delta A = A_- - A_+$$
$$\Delta\epsilon = \epsilon_- - \epsilon_+ \tag{15}$$

and thus, further approximating $n_+ = n_- = n, \gamma_+ = \gamma_- = \gamma$,

$$\frac{\Delta A}{\mathcal{E}} = \left(\frac{\Delta\epsilon}{\mathcal{E}}\right)cz$$
$$= \gamma\sum_{a,j}\left(\frac{N_a}{N}\right)\left[|\langle a|m_-|j\rangle|^2 - |\langle a|m_+|j\rangle|^2\right]\delta\{(\mathcal{E}_{ja}-\mathcal{E})\}cz \tag{16}$$

Equations 14 and 16 relate the macroscopic CP absorption and CD observables to the eigenstates of $\mathcal{H}_0$—the absorbing center. More specifically, writing

$$\mathcal{H}_0 = \mathcal{H}_0^\circ + \mathcal{H}_0'(H) \tag{17}$$

where $\mathcal{H}_0^\circ$ is the zero-field Hamiltonian and $\mathcal{H}_0'$ is the magnetic field perturbation, the zero-field absorption, $A_\pm^\circ$, and CD, ΔA°, are obtained using eigenstates of $\mathcal{H}_0^\circ$ and the absorption, $A_\pm$, and CD, ΔA, in the presence of the field require eigenstates of $\mathcal{H}_0^\circ + \mathcal{H}_0'$. Now, $\mathcal{H}_0^\circ$ is real, and its eigenfunctions can be chosen to be real, whence

$$|\langle a|m_+|j\rangle|^2 = |\langle a|m_-|j\rangle|^2 \tag{18}$$

and

$$A_+^\circ = A_-^\circ; \; \Delta A^\circ = 0 \tag{19}$$

In the electric dipole approximation, therefore, natural CD does not exist, all CD is MCD, and we can define the zero-field absorption (ZFA) interchangeably by

$$A^\circ = A_+^\circ = A_-^\circ = \frac{1}{2}(A_+^\circ + A_-^\circ). \tag{20}$$

III. DISPERSION CALCULATIONS

In this section we present calculations of MCD in which the entire dispersion—the $\mathcal{E}$-dependence—is obtained. As indicated earlier, we restrict discussion to electronic transitions of dilute crystalline solutions of rigid localized absorbing centers. In discussing the vibrational-electronic (vibronic) states of such systems in this and following sections, a number of concepts and equations will be drawn upon repeatedly. For convenience, these are collected in appendix I.

A. A Rigid Shift Model

We first discuss the ZFA and MCD of absorbing centers with the following additional properties:

(1) the Born–Oppenheimer (BO) approximation is obeyed in ground and excited states, and

(2) the Franck–Condon (FC) approximation is adequate for matrix elements of electronic operators.

We consider the electronic transition $A \rightarrow J$, where A and J may be degenerate. The zero-field eigenstates within the BO approximation can be written

$$
\begin{aligned}
|A_\alpha a\rangle &= \psi_{A_\alpha}(\mathbf{r},\mathbf{R})\chi_a(\mathbf{R}) \quad (\alpha = 1 \text{ to } d_A) \\
|J_\lambda j\rangle &= \psi_{J_\lambda}(\mathbf{r},\mathbf{R})\chi_j(\mathbf{R}) \quad (\lambda = 1 \text{ to } d_J) \\
\mathcal{H}_0^\circ|A_\alpha a\rangle &= \mathcal{E}_a|A_\alpha a\rangle, \qquad \mathcal{H}_0^\circ|J_\lambda j\rangle = \mathcal{E}_j|J_\lambda j\rangle
\end{aligned}
\tag{21}
$$

where $\mathbf{r}$ and $\mathbf{R}$ denote electronic and nuclear coordinates, respectively, and d_A and d_J are the degeneracies of A and J. The ground state vibrational functions χ_a have significant amplitude only for $\mathbf{R}$ in the region of the equilibrium configuration $\mathbf{R}_0$ if nuclear motion is small. The transition matrix elements can then be simplified by the FC approximation:

$$
\begin{aligned}
\langle A_\alpha a|m_\pm|J_\lambda j\rangle &= \langle A_\alpha a|m_\pm^e + m_\pm^n|J_\lambda j\rangle \\
&= \langle a|\langle A_\alpha|m_\pm^e|J_\lambda\rangle + m_\pm^n\langle A_\alpha|J_\lambda\rangle|j\rangle \\
&= \langle a|j\rangle\langle A_\alpha|m_\pm^e|J_\lambda\rangle^\circ
\end{aligned}
\tag{22}
$$

where e and n denote electronic and nuclear terms, respectively, and the zero superscript indicates that the electronic matrix element is evaluated at $\mathbf{R}=\mathbf{R}_0$. Substituting in (14), we obtain

$$
\frac{A_\pm^\circ}{\mathcal{E}} = \gamma\left\{\frac{1}{d_A}\sum_{\alpha,\lambda}|\langle A_\alpha|m_\pm^e|J_\lambda\rangle^0|^2\right\}\left\{\sum_{a,j}\frac{N_a}{N}|\langle a|j\rangle|^2\delta(\mathcal{E}_{ja}-\mathcal{E})\right\}cz \tag{23}
$$

where N_a is the total occupancy of vibrational level $a(N_a = d_A N_{A_\alpha a})$. Integrating the absorption over the whole band of vibronic transitions, making the assumption that γ is constant over this interval, gives

$$
\int\frac{A_\pm^\circ}{\mathcal{E}}\,d\mathcal{E} = \gamma\sum_{\alpha,\lambda}\frac{1}{d_A}|\langle A_\alpha|m_\pm^e|J_\lambda\rangle^0|^2 cz \tag{24}
$$

since

$$
\sum_j|\langle a|j\rangle|^2 = 1; \qquad \sum_a\left(\frac{N_a}{N}\right) = 1 \tag{25}
$$

Equation 23 can thus be written

$$
\frac{A_\pm^\circ}{\mathcal{E}} = \gamma D_0 f(\mathcal{E})cz \tag{26}
$$

where

$$D_0 = \frac{1}{d_A} \sum_{\alpha,\lambda} |\langle A_\alpha | m^e_\pm | J_\lambda \rangle^\circ|^2$$

$$= \frac{1}{2d_A} \sum_{\alpha,\lambda} \left[|\langle A_\alpha | m^e_+ | J_\lambda \rangle^\circ|^2 + |\langle A_\alpha | m^e_- | J_\lambda \rangle^\circ|^2 \right] \quad (27)$$

and

$$f(\mathcal{E}) = \sum_{a,j} \left(\frac{N_a}{N} \right) |\langle a | j \rangle|^2 \delta(\mathcal{E}_{ja} - \mathcal{E})$$

$$\int_0^\infty f(\mathcal{E})\, d\mathcal{E} = 1 \quad (28)$$

The ZFA given by (26) has a shape originating entirely in the ground and excited *vibrational* functions and an integrated intensity dependent only on the equilibrium ($\mathbf{R}_0$) *electronic* functions. Notice also that the shape is temperature(T)-dependent through the populations (N_a/N), but the integrated intensity is T-independent.

We turn now to the CP absorption and MCD in the presence of an applied field, $\boldsymbol{H} = H\mathbf{k}$. The magnetic field perturbation $\mathcal{H}_0'$ is given to first order in H by

$$\mathcal{H}_0' = -\sum_i \frac{e}{2mc}(l_{i_z} + 2s_{i_z})H$$

$$\equiv -\mu_z H \equiv \beta(L_z + 2S_z)H \quad (29)$$

where i sums over all electrons (but not nuclei), e and m are the electron charge and mass, $\boldsymbol{\mu}$ is the electronic magnetic moment operator, $\hbar\mathbf{L}$ and $\hbar\mathbf{S}$ are total electronic orbital and spin angular momentum operators and $\beta = |e|\hbar/2mc$ is the electronic Bohr magneton. Interaction with the nuclei is ignored since nuclear magnetons (gyromagnetic factors) are at least three orders of magnitude smaller than the electronic magneton. Within the ground and excited state manifolds $\mathcal{H}_0'$ is diagonal in the FC approximation:

$$\begin{aligned} \langle A_\alpha a | \mathcal{H}_0' | A_{\alpha'} a' \rangle &= -\langle A_\alpha | \mu_z | A_\alpha \rangle^\circ H \delta_{\alpha\alpha'} \delta_{aa'} \\ \langle J_\lambda j | \mathcal{H}_0' | J_{\lambda'} j' \rangle &= -\langle J_\lambda | \mu_z | J_\lambda \rangle^\circ H \delta_{\lambda\lambda'} \delta_{jj'} \end{aligned} \quad (30)$$

as long as the $\mathbf{R}_0$ electronic wavefunctions are chosen to diagonalize μ_z. Hence, neglecting mixing of different electronic states by $\mathcal{H}_0'$, the wavefunctions in the presence of $\mathcal{H}_0'$ are just $|A_\alpha a\rangle$ and $|J_\lambda j\rangle$ with energies

$$\begin{aligned} \mathcal{E}'_{A_\alpha a} &= \mathcal{E}_a - \langle A_\alpha | \mu_z | A_\alpha \rangle^\circ H \\ \mathcal{E}'_{J_\lambda j} &= \mathcal{E}_j - \langle J_\lambda | \mu_z | J_\lambda \rangle^\circ H \end{aligned} \tag{31}$$

To this approximation the Zeeman splitting of each vibronic state is independent of vibrational level, and is identical to the pure electronic splitting at $\mathbf{R}_0$. To a better approximation we must include intermixing of different electronic states. Perturbation theory gives

$$\begin{aligned} |J_\lambda j\rangle' &= |J_\lambda j\rangle - \sum_{\substack{K_\kappa k \\ (K \neq J)}} |K_\kappa k\rangle \frac{\langle K_\kappa | \mu_z | J_\lambda \rangle^\circ H \langle k | j \rangle}{\mathcal{E}_{jk}} \\ |A_\alpha a\rangle' &= |A_\alpha a\rangle - \sum_{\substack{K_\kappa k \\ (K \neq A)}} |K_\kappa k\rangle \frac{\langle K_\kappa | \mu_z | A_\alpha \rangle^\circ H \langle k | a \rangle}{\mathcal{E}_{ak}} \end{aligned} \tag{32}$$

where primes denote wavefunctions in the presence of $\mathcal{H}_0'$, we assume that the intervals $\mathcal{E}_{jk}$ and $\mathcal{E}_{ak}$ are large compared to Zeeman energies and again use the FC approximation. If we further assume that vibrational levels k contributing to the sums in (32) are such that we can approximate

$$\begin{aligned} \mathcal{E}_{jk} &= W_J^\circ - W_K^\circ \\ \mathcal{E}_{ak} &= W_A^\circ - W_K^\circ \end{aligned} \tag{33}$$

where W_K° is the $\mathbf{R}_0$ energy of state K, Eq. 32 reduces to

$$\begin{aligned} |A_\alpha a\rangle' &= |A_\alpha a\rangle - \sum_{K_\kappa \neq A} |K_\kappa a\rangle \frac{\langle K_\kappa | \mu_z | A_\alpha \rangle^\circ H}{W_A^\circ - W_K^\circ} \\ |J_\lambda j\rangle' &= |J_\lambda j\rangle - \sum_{K_\kappa \neq J} |K_\kappa j\rangle \frac{\langle K_\kappa | \mu_z | J_\lambda \rangle^\circ H}{W_J^\circ - W_K^\circ} \end{aligned} \tag{34}$$

To first order in H these wavefunction perturbations do not contribute to the energies, which remain as in (31). However, the transition moments are perturbed to

$$\langle A_\alpha a|m_\pm|J_\lambda j\rangle' = \Bigg[\langle A_\alpha|m_\pm^e|J_\lambda\rangle^\circ + \Bigg\{\sum_{K_\kappa\neq J}\frac{\langle A_\alpha|m_\pm^e|K_\kappa\rangle^\circ\langle K_\kappa|\mu_z|J_\lambda\rangle^\circ}{W_K^\circ - W_J^\circ} + \sum_{K_\kappa\neq A}\frac{\langle A_\alpha|\mu_z|K_\kappa\rangle^\circ\langle K_\kappa|m_\pm^e|J_\lambda\rangle^\circ}{W_K^\circ - W_A^\circ}\Bigg\}H\Bigg]\langle a|j\rangle$$

$$\equiv[\langle A_\alpha|m_\pm^e|J_\lambda\rangle^\circ + \langle A_\alpha|m_\pm^e|J_\lambda\rangle' H]\langle a|j\rangle \tag{35}$$

The Zeeman splittings of the ground states also lead to population changes:

$$\frac{N'_{A_\alpha a}}{N} = \frac{\exp\{-\mathcal{E}'_{A_\alpha a}/kT\}}{\sum_{\alpha,a}\exp\{-\mathcal{E}'_{A_\alpha a}/kT\}}$$

$$= \frac{\exp\{-\mathcal{E}_a/kT\}\exp\{\langle A_\alpha|\mu_z|A_\alpha\rangle^\circ H/kT\}}{\sum_{\alpha,a}\exp\{-\mathcal{E}_a/kT\}\exp\{\langle A_\alpha|\mu_z|A_\alpha\rangle^\circ H/kT\}} \tag{36}$$

At "high" temperatures where Zeeman energies are small compared to kT

$$\exp\left\{\langle A_\alpha|\mu_z|A_\alpha\rangle^\circ \frac{H}{kT}\right\} \approx 1 + \frac{\langle A_\alpha|\mu_z|A_\alpha\rangle^\circ H}{kT} \tag{37}$$

and

$$\frac{N'_{A_\alpha a}}{N} = \frac{\exp\{-\mathcal{E}_a/kT\}}{\sum_a \exp\{-\mathcal{E}_a/kT\}}\frac{1}{d_A}\left[1 + \frac{\langle A_\alpha|\mu_z|A_\alpha\rangle^\circ H}{kT}\right]$$

$$= \left(\frac{N'_{A_\alpha a}}{N}\right)\left[1 + \frac{\langle A_\alpha|\mu_z|A_\alpha\rangle^\circ H}{kT}\right] \tag{38}$$

using the property of Zeeman splittings that the center of gravity is retained

$$\sum_\alpha \langle A_\alpha|\mu_z|A_\alpha\rangle^\circ H = 0 \tag{39}$$

Thus, the fractional population change of $|A_\alpha a\rangle$ is independent of vibrational level and the same as would obtain in a purely electronic system clamped to $\mathbf{R}_0$.

Having evaluated the effects of the magnetic field on transition energies and moments and on ground state populations, we can now derive the CP absorption and MCD. From (14)

$$\frac{A_\pm}{\mathcal{E}} = \gamma\left\{\sum_{\substack{\alpha,a\\ \lambda,j}}\left(\frac{N'_{A_\alpha a}}{N}\right)|\langle A_\alpha a|m_\pm|J_\lambda j\rangle'|^2\delta\left(\mathcal{E}'_{J_\lambda j;\,A_\alpha a}-\mathcal{E}\right)\right\}cz$$

$$=\gamma\left\{\sum_{\alpha,\lambda}\frac{1}{d_A}\left[1+\frac{\langle A_\alpha|\mu_z|A_\alpha\rangle^\circ H}{kT}\right]\right.$$

$$\left.\times|[\langle A_\alpha|m^e_\pm|J_\lambda\rangle^\circ+\langle A_\alpha|m^e_\pm|J_\lambda\rangle' H]|^2 f'_{\alpha\lambda}(\mathcal{E})\right\}cz \tag{40}$$

Here,

$$f'_{\alpha\lambda}(\mathcal{E})=\sum_{a,j}\left(\frac{N_a}{N}\right)|\langle a|j\rangle|^2\delta\left(\mathcal{E}_{ja}-[\langle J_\lambda|\mu_z|J_\lambda\rangle^\circ-\langle A_\alpha|\mu_z|A_\alpha\rangle^0]H-\mathcal{E}\right) \tag{41}$$

is a function identical in shape to $f(\mathcal{E})$ but shifted rigidly along the $\mathcal{E}$-axis by the $A_\alpha\rightarrow J_\lambda$ Zeeman shift $-[\langle J_\lambda|\mu_z|J_\lambda\rangle^\circ-\langle A_\alpha|\mu_z|A_\alpha\rangle^\circ]H$. For a broad band the Zeeman shift is a very small perturbation and the shifted function can be expressed in terms of the unshifted function by a Taylor expansion:

$$f'_{\alpha\lambda}(\mathcal{E})=f(\mathcal{E})+[\langle J_\lambda|\mu_z|J_\lambda\rangle^\circ-\langle A_\alpha|\mu_z|A_\alpha\rangle^\circ]H\left(\frac{\partial f}{\partial\mathcal{E}}\right) \tag{42}$$

Collecting terms of zeroth and first order in H in (40) then gives

$$\frac{A_\pm}{\mathcal{E}}=\frac{A^\circ_\pm}{\mathcal{E}}+\gamma\left\{\sum_{\alpha,\lambda}\frac{1}{d_A}|\langle A_\alpha|m^e_\pm|J_\lambda\rangle^\circ|^2[\langle J_\lambda|\mu_z|J_\lambda\rangle^\circ-\langle A_\alpha|\mu_z|A_\alpha\rangle^\circ]\left(\frac{\partial f}{\partial\mathcal{E}}\right)\right.$$

$$+\sum_{\alpha,\lambda}\frac{2}{d_A}\operatorname{Re}[\langle A_\alpha|m^e_\pm|J_\lambda\rangle^\circ\langle A_\alpha|m^e_\pm|J_\lambda\rangle'^*]f$$

$$\left.+\sum_{\alpha,\lambda}\frac{1}{d_A}\frac{\langle A_\alpha|\mu_z|A_\alpha\rangle^\circ|\langle A_\alpha|m^e_\pm|J_\lambda\rangle^\circ|^2 f}{kT}\right\}Hcz \tag{43}$$

whence

$$\frac{\Delta A}{\mathcal{E}} = +\gamma\left\{\mathcal{A}_1\left(-\frac{\partial f}{\partial \mathcal{E}}\right)+\left(\mathcal{B}_0+\frac{\mathcal{C}_0}{kT}\right)f\right\}(\beta H)cz \tag{44}$$

where

$$\mathcal{A}_1 = +\frac{1}{d_A}\sum_{\alpha,\lambda}\left[|\langle A_\alpha|m_-^e|J_\lambda\rangle^\circ|^2-|\langle A_\alpha|m_+^e|J_\lambda\rangle^\circ|^2\right]$$

$$\times\left[\langle J_\lambda|L_z+2S_z|J_\lambda\rangle^\circ-\langle A_\alpha|L_z+2S_z|A_\alpha\rangle^\circ\right]$$

$$\mathcal{B}_0 = -\frac{2}{d_A}\sum_{\alpha,\lambda}\mathrm{Re}\left\{\sum_{K_\kappa\neq J}\left[\langle A_\alpha|m_-^e|J_\lambda\rangle^\circ\langle K_\kappa|m_+^e|A_\alpha\rangle^\circ\right.\right.$$

$$\left.-\langle A_\alpha|m_+^e|J_\lambda\rangle^\circ\langle K_\kappa|m_-^e|A_\alpha\rangle^\circ\right]$$

$$\times\frac{\langle J_\lambda|L_z+2S_z|K_\kappa\rangle^\circ}{W_K^\circ-W_J^\circ} \tag{45}$$

$$+\sum_{K_\kappa\neq A}\left[\langle A_\alpha|m_-^e|J_\lambda\rangle^\circ\langle J_\lambda|m_+^e|K_\kappa\rangle^\circ-\langle A_\alpha|m_+^e|J_\lambda\rangle^\circ\langle J_\lambda|m_-^e|K_\kappa\rangle^\circ\right]$$

$$\left.\times\frac{\langle K_\kappa|L_z+2S_z|A_\alpha\rangle^\circ}{W_K^\circ-W_A^\circ}\right\}$$

$$\mathcal{C}_0 = -\frac{1}{d_A}\sum_{\alpha,\lambda}\left[|\langle A_\alpha|m_-^e|J_\lambda\rangle^\circ|^2-|\langle A_\alpha|m_+^e|J_\lambda\rangle^\circ|^2\right]$$

$$\times\langle A_\alpha|L_z+2S_z|A_\alpha\rangle^\circ$$

Equation 40 shows the manner in which the CP absorption is modified by the applied magnetic field. Each component transition $A_\alpha \rightarrow J_\lambda$ contributes absorption that is identical in shape to that at zero field. The energy of the band is modified by the Zeeman shift, and the intensity is changed by the ground state Zeeman effect and by intermixing of the zero field electronic wavefunctions. Each of these effects is identical to that obtained by clamping the nuclei to the configuration $\mathbf{R}_0$. The shift of the absorption in the applied field without change in shape leads to the naming of this situation as "rigid shift".

The total CP absorption and MCD is the sum of the contributions of all Zeeman components $A_\alpha \rightarrow J_\lambda$. Equations 43 and 45 show that under the conditions of "high" temperature and large band width the changes in intensity and frequency of the Zeeman components of the transition due to the magnetic field contribute additively and linearly in H to the change in CP absorption and to the MCD. Further, the contributions of the Zeeman splitting, of the intensity change due to intermixing of electronic states and of the intensity change due to ground state population redistribution are separable physically through their dependence on either $\mathcal{E}$ or T. The three terms in the MCD are referred to as $\mathcal{A}$, $\mathcal{B}$, and $\mathcal{C}$ terms; their relative magnitudes determine the overall form of the MCD.

The existence of $\mathcal{A}$, $\mathcal{B}$, and $\mathcal{C}$ terms depends on the ground and excited electronic states. $\mathcal{A}$ terms require either A or J to be degenerate. $\mathcal{C}$ terms require A to be degenerate. $\mathcal{B}$ terms exist under any conditions of ground and excited state degeneracy. When all terms exist, in order of magnitude we can put

$$\mathcal{A}_1 : \mathcal{B}_0 : \mathcal{C}_0 \approx 1 : \frac{1}{\Delta W} : 1 \tag{46}$$

where ΔW is the order of magnitude of an electronic energy gap. Since

$$\left(\frac{\partial f}{\partial \mathcal{E}} \right)_{\text{max}} \approx \frac{(f)_{\text{max}}}{\Gamma} \tag{47}$$

where Γ is the width of f, the maximum contributions of $\mathcal{A}$, $\mathcal{B}$, and $\mathcal{C}$ terms to ΔA are

$$\mathcal{A} : \mathcal{B} : \mathcal{C} \approx \frac{1}{\Gamma} : \frac{1}{\Delta W} : \frac{1}{kT} \tag{48}$$

Narrow bands, closely spaced electronic states, and low temperatures thus favor $\mathcal{A}$, $\mathcal{B}$, and $\mathcal{C}$ terms, respectively. For a typical band at room temperature we might put $\Gamma = 10^3$ cm^{-1}, $\Delta W = 10^4$ cm^{-1} and $kT = 200$ cm^{-1} when the MCD terms are in the ratio

$$\mathcal{A} : \mathcal{B} : \mathcal{C} \approx 10 : 1 : 50 \tag{49}$$

Figure 2 illustrates the form of the MCD obtained from (44) in a variety of circumstances.

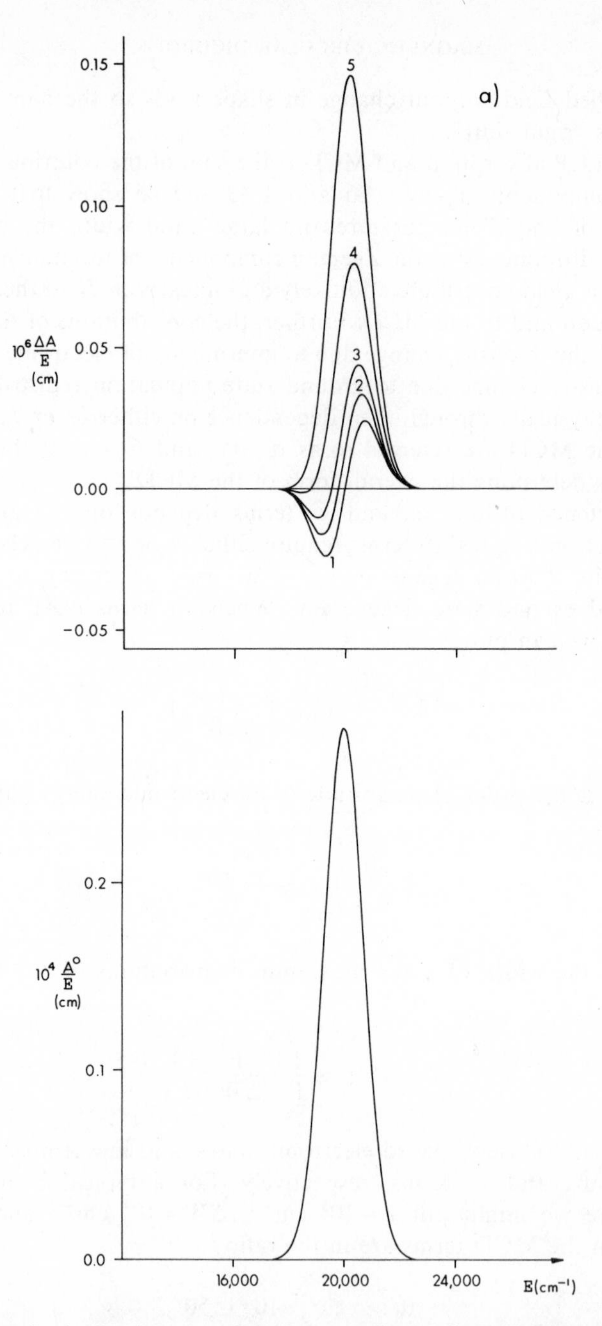

Fig. 2. Magnetic circular dichroism dispersion according to the rigid shift model. In *a*) the effect of varying the relative magnitudes of the $\mathcal{A}$ and $(\mathcal{B} + \mathcal{C})$ terms is illustrated. The ZFA is obtained from (26) with $D_0 = 1$, $\gamma cz = 5 \times 10^{-2}$, and $f = \Delta^{-1}\pi^{-1/2}\exp[-(\mathcal{E} - \mathcal{E}^\circ)^2/\Delta^2]$, $\mathcal{E}^\circ = 20{,}000$ cm^{-1}, $\Delta = 1000$ cm^{-1}. The MCD is calculated from (44) with $\mathcal{A}_1 = 1$, $\beta H = 1$ cm^{-1}, and $(\mathcal{B}_0 + \mathcal{C}_0/kT)$ equal to 1) 0, 2) 5×10^{-4}, 3) 10^{-3}, 4) 2.5×10^{-3}, and 5) 5×10^{-3}.

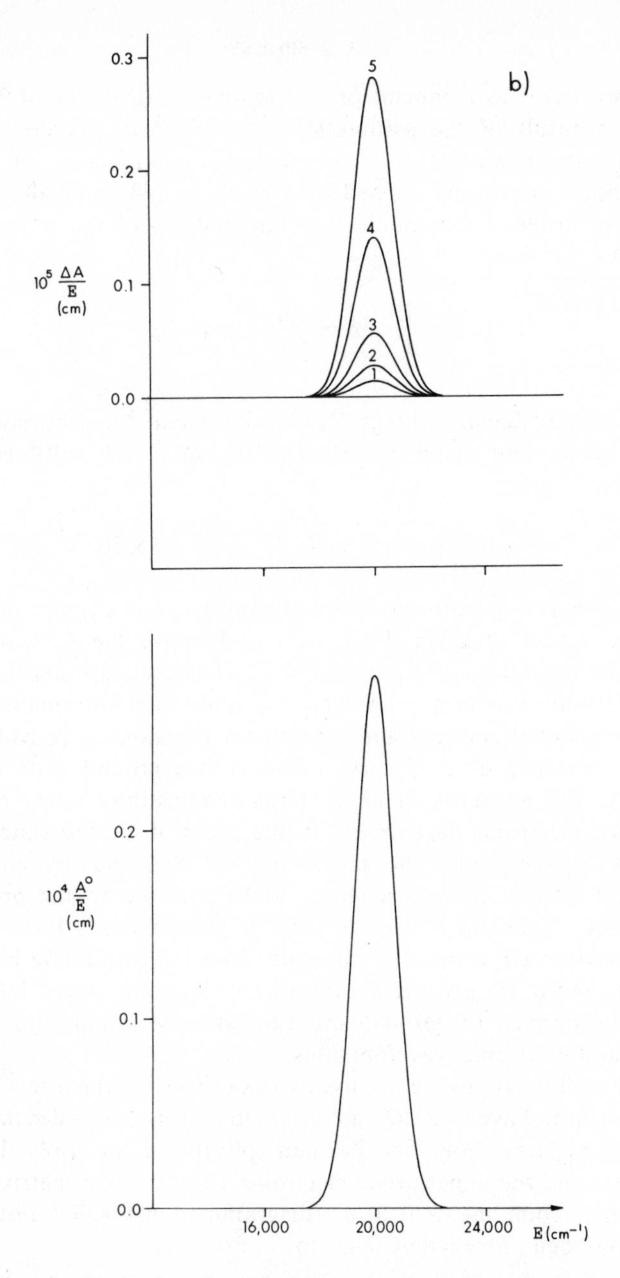

Fig. 2 (Continued). In *b*) the effect of varying the temperature T on the $\mathcal{C}$ terms is illustrated. The ZFA is identical to Fig. 2*a*. The MCD is calculated with $\mathcal{A}_1 = \mathcal{B}_0 = 0$, $\mathcal{C}_0 = 1$ and kT equal to 1) 200 cm^{-1}, 2) 100 cm^{-1}, 3) 50 cm^{-1}, 4) 20 cm^{-1}, and 5) 10 cm^{-1}.

It is also useful to comment on the relative magnitudes of MCD and ZFA. As a result of the assumptions that Zeeman energies are small compared to band width Γ, kT, and electronic energy gaps, the change in CP absorption due to the applied field given by (43) is small, and hence $\Delta A \ll A^\circ$. In order of magnitude the contributions of the $\mathcal{A}$, $\mathcal{B}$, and $\mathcal{C}$ terms to $\Delta A / A^\circ$ are

$$\mathcal{A} \approx \frac{Z}{\Gamma};\ \mathcal{B} \approx \frac{Z}{\Delta W};\ \mathcal{C} \approx \frac{Z}{kT} \tag{50}$$

where Z denotes Zeeman energy. Taking $Z = 1\ \mathrm{cm}^{-1}$ appropriate to fields of $\sim 10^4$ gauss, and again putting $\Gamma = 10^3\ \mathrm{cm}^{-1}$, $\Delta W = 10^4\ \mathrm{cm}^{-1}$, and $kT = 200\ \mathrm{cm}^{-1}$ gives

$$\mathcal{A} \approx 10^{-3}; \qquad \mathcal{B} \approx 10^{-4}; \qquad \mathcal{C} \approx 5 \times 10^{-3} \tag{51}$$

We are now in a position to assess the information content of MCD in this simple model situation. First, over and above the ZFA, the MCD provides just the values of $\mathcal{A}_1$, $\mathcal{B}_0$, and $\mathcal{C}_0$. These in turn depend entirely on $\mathbf{R}_0$ electronic function properties. No additional information on the electronic potential surfaces and vibrational functions is provided therefore. The existence of a $\mathcal{C}$ term demonstrates ground state electronic degeneracy. The existence of an $\mathcal{A}$ term demonstrates either ground or excited state electronic degeneracy. In the event of ground state degeneracy alone, $\mathcal{A}_1 = \mathcal{C}_0$, and the satisfaction of this equality enables the existence of excited state degeneracy to be assessed in the presence of ground state degeneracy. The $\mathcal{A}$ and $\mathcal{C}$ terms depend on magnetic moments and on CP transition moments. These in turn relate both to the symmetries and to the specific natures of the electronic states. Information can thus be derived on ground and excited state symmetries, magnetic moments and electronic wavefunctions.

As further illustrations we discuss two examples in which the absorbing center is taken to have exact O_h site symmetry.[3] First, consider the $\mathcal{A}$ term of an $A_{1g} \rightarrow T_{1u}$ transition. The Zeeman splitting of the triply degenerate excited state and the unperturbed electronic CP transition matrix elements to the Zeeman components at $\mathbf{R}_0$ are diagrammed in Fig. 3. Substituting in (27) and (45) then immediately leads to

$$\begin{aligned} \mathcal{A}_1 &= 2gm^2 \qquad \mathcal{D}_0 = m^2 \\ \frac{\mathcal{A}_1}{\mathcal{D}_0} &= 2g \end{aligned} \tag{52}$$

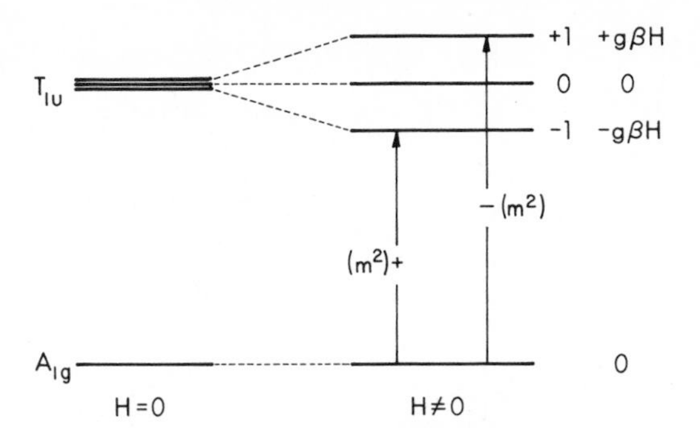

Fig. 3. The Zeeman effect in a $A_{1g} \rightarrow T_{1u}$ transition. The $T_{1u}+1$, 0, and -1 functions diagonalize $(L_z+2S_z)\,\beta H$ with expectation values indicated at the right. Excitations for which $|\langle A_{1g}|m_{\pm}^{e}|T_{1u}\lambda\rangle^{\circ}|^2$ is nonzero are indicated; the sign denotes the $m_{\pm}^{e}$ component involved; the magnitude in parenthesis is the value of the squared transition moment.

Determination of the $\mathcal{A}_1$ value from the MCD, together with the $\mathcal{D}_0$ value from the ZFA, then gives the excited state g value (magnitude and sign). Second, consider the $\mathcal{C}$ terms in the transitions $T_{1u} \rightarrow A_{1g}$ and $T_{1u} \rightarrow E_g$. The $\mathbf{R}_0$ Zeeman splitting and CP transition matrix elements are given in Fig. 4. Thence,

$$
\begin{array}{ll}
T_{1u} \rightarrow A_{1g}: & T_{1u} \rightarrow E_g: \\
\mathcal{C}_0 = \dfrac{2}{3}\, g m_1^2 & \mathcal{C}_0 = -\dfrac{4}{3}\, g m_2^2 \\
\mathcal{D}_0 = \dfrac{1}{3} m_1^2 & \mathcal{D}_0 = \dfrac{4}{3} m_2^2 \\
\dfrac{\mathcal{C}_0}{\mathcal{D}_0} = 2g & \dfrac{\mathcal{C}_0}{\mathcal{D}_0} = -g
\end{array}
\tag{53}
$$

The signs of the $\mathcal{C}$ terms depend on the excited state symmetries, as well as on the sign of the ground state g value. Knowing the ground state g value, the MCD allows excited state symmetry assignment; conversely, if the excited state symmetry is known, the MCD provides the ground state g value (magnitude and sign).

B. Overlapping Bands

We now consider the more complex situation in which the ZFA and MCD comprise overlapping bands due to two transitions. This can arise when there exist either two near-degenerate excited states or an excited

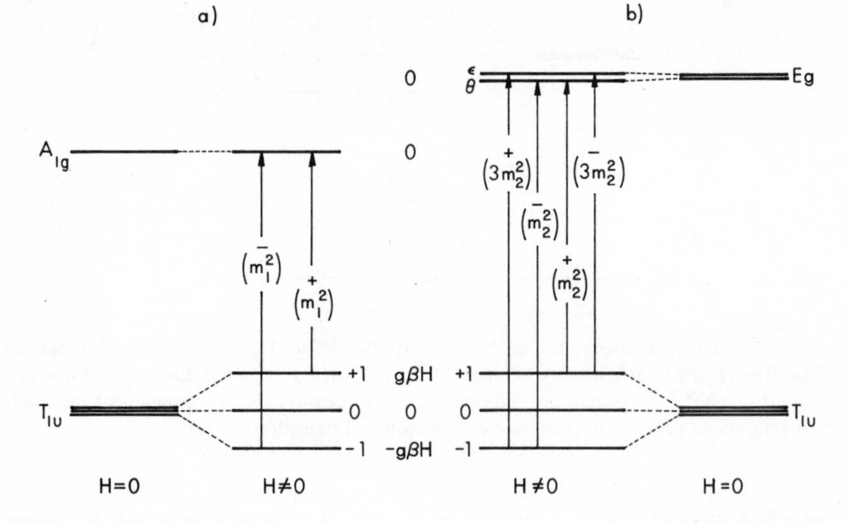

Fig. 4. The Zeeman effect in *a*) $T_{1u} \rightarrow A_{1g}$ and *b*) $T_{1u} \rightarrow E_g$ transitions. The nomenclature follows that in Fig. 3. Note that the two E_g components are separated slightly only to allow their individual transition moments to be shown and are *not* split by the magnetic field.

state close to the ground state, and when the separation of the near-degenerate states is comparable to or much less than the absorption band width. In the second case the low-lying excited state must also be thermally occupied, and hence within kT of the ground state. In either situation we make the highly simplifying assumption that near-degenerate states have parallel potential surfaces.

Consider first the excited state near-degeneracy and suppose there to be two excited states J^1 and J^2 in place of the one J state of Section III.A. We adopt the same assumptions as before, and add the requirement that J^1 and J^2 have parallel potential surfaces:

$$W_{J^2}(\mathbf{R}) - W_{J^1}(\mathbf{R}) = \Delta W_{J^2J^1} \tag{54}$$

where $\Delta W_{J^2J^1}$ is independent of $\mathbf{R}$. Then the zero-field eigenfunctions are

$$\begin{aligned} |A_\alpha a\rangle &= \psi_{A_\alpha}\chi_a \qquad (\alpha = 1 \text{ to } d_A) \\ |J^1_{\lambda_1} j_1\rangle &= \psi_{J^1_{\lambda_1}}\chi_{j_1} \qquad (\lambda_1 = 1 \text{ to } d_{J^1}) \\ |J^2_{\lambda_2} j_2\rangle &= \psi_{J^2_{\lambda_2}}\chi_{j_2} \qquad (\lambda_2 = 1 \text{ to } d_{J^2}) \end{aligned} \tag{55}$$

in which the sets of vibrational functions χ_{j_1} and χ_{j_2} are identical due to the identicality of the associated potential surfaces. The ZFA is easily shown to be a simple extension of (26);

$$\frac{A_{\pm}^{\circ}}{\mathcal{E}} = \gamma\{\mathcal{D}_0^1 f_1 + \mathcal{D}_0^2 f_2\}cz \tag{56}$$

where $\mathcal{D}_0^i$ and f_i $(i=1,2)$, referring to transitions $A \rightarrow J^i$, are defined as in (27) and (28) and f_2 is identical in shape to f_1 but shifted $\Delta W_{J^2J^1}$ higher in energy:

$$f_2(\mathcal{E}) = f_1(\mathcal{E} - \Delta W_{J^2J^1}) \tag{57}$$

The matrix of $\mathcal{H}_0'$ within the excited state manifold is now

$$\begin{aligned} \langle J_{\lambda_i}^i j_i | \mathcal{H}_0' | J_{\lambda_i'}^i j_i' \rangle &= -\langle J_{\lambda_i}^i | \mu_z | J_{\lambda_i'}^i \rangle^{\circ} H \delta_{\lambda_i \lambda_i'} \delta_{j_i j_i'} \\ \langle J_{\lambda_i}^i j j_i | \mathcal{H}_0' | J_{\lambda_{i'}}^{i'} j_{i'} \rangle &= -\langle J_{\lambda_i}^i | \mu_z | J_{\lambda_{i'}}^{i'} \rangle^{\circ} H \delta_{j_i j_{i'}} \qquad (i' \neq i) \end{aligned} \tag{58}$$

using the FC approximation and the identicality of the j_1 and j_2 vibrational functions and diagonalizing $\langle J_{\lambda_i}^i | \mu_z | J_{\lambda_i'}^i \rangle^{\circ}$. Vibrational states of J^1 and J^2 are thus only coupled by $\mathcal{H}_0'$ when having the same vibrational function. Within our model such states are always separated by $\Delta W_{J^2J^1}$ and as long as this is much greater than Zeeman energies the mixing of J^1 and J^2 can be treated by perturbation theory. Then, including as before the interaction with non-near-degenerate electronic states we obtain, in the presence of the field,

$$\begin{aligned} |A_\alpha a\rangle' &= |A_\alpha a\rangle - \sum_{K_\kappa \neq A} |K_\kappa a\rangle \frac{\langle K_\kappa | \mu_z | A_\alpha \rangle^{\circ} H}{W_A^{\circ} - W_K^{\circ}} \\ |J_{\lambda_i}^i j_i\rangle' &= |J_{\lambda_i}^i j_i\rangle - \sum_{\substack{\lambda_{i'} \\ (i' \neq i)}} |J_{\lambda_{i'}}^{i'} j_i\rangle \frac{\langle J_{\lambda_{i'}}^{i'} | \mu_z | J_{\lambda_i}^i \rangle^{\circ} H}{\Delta W_{J^i J^{i'}}} \\ &\quad - \sum_{K_\kappa \neq J^i} |K_\kappa j_i\rangle \frac{\langle K_\kappa | \mu_z | J_{\lambda_i}^i \rangle^{\circ} H}{W_{J^i}^{\circ} - W_K^{\circ}} \\ \mathcal{E}_{A_\alpha a}' &= \mathcal{E}_a - \langle A_\alpha | \mu_z | A_\alpha \rangle^{\circ} H \\ \mathcal{E}_{J_{\lambda_i}^i j_i}' &= \mathcal{E}_{ij_i} - \langle J_{\lambda_i}^i | \mu_z | J_{\lambda_i}^i \rangle^{\circ} H \end{aligned} \tag{59}$$

The influence of J^2 on J^1 is formally the same as if J^2 were distant from

J^1—and vice versa—and the MCD calculation therefore follows the same path as in Section III.A, with the result

$$\frac{\Delta A}{\mathcal{E}} = \gamma\left\{ \mathcal{A}_1^1\left(-\frac{\partial f_1}{\partial \mathcal{E}}\right) + \mathcal{A}_1^2\left(-\frac{\partial f_2}{\partial \mathcal{E}}\right) + \left[\mathcal{B}_0^1 + \frac{\mathcal{C}_0^1}{kT}\right] f_1 + \left[\mathcal{B}_0^2 + \frac{\mathcal{C}_0^2}{kT}\right] f_2 \right\}(\beta H)cz \tag{60}$$

$\mathcal{A}_1^i$, $\mathcal{B}_0^i$, and $\mathcal{C}_0^i$ for $A \to J^i$ being parameters as defined in (45). As the states J^1 and J^2 approach in energy and their associated absorption bands overlap the MCD thus remains a simple additive function of the individual contributions.

In general, the analysis of the ZFA and MCD of the two overlapping $A \to J^1$ and $A \to J^2$ bands leads to the $\mathcal{A}_1$, $\mathcal{B}_0$, $\mathcal{C}_0$, and $\mathcal{D}_0$ parameters for each and the energy gap $\Delta W_{J^2J^1}$. The information content of the $\mathcal{A}_1$, $\mathcal{B}_0$, and $\mathcal{C}_0$ parameters is as discussed in Section III.A. In some cases $\Delta W_{J^2J^1}$ will be obtainable from the ZFA alone. However, in other cases, particularly when the bands are completely unresolved ($\Delta W_{J^2J^1} \ll \Gamma$, the band width), this will not be possible and the extraction of $\Delta W_{J^2J^1}$ depends on the sensitivity of the MCD to its value. This in turn varies according to the relative signs and magnitudes of the component terms. Consider, for example, the $\mathcal{C}$ terms, which give

$$\frac{\Delta A}{\mathcal{E}} = \gamma\left[\mathcal{C}_0^1 f_1 + \mathcal{C}_0^2 f_2\right]\left(\frac{\beta H}{kT}\right)cz \tag{61}$$

and the situation when $\Delta W_{J^2J^1} \ll \Gamma$. Then to a first approximation

$$f_2 = f_1 - \Delta W_{J^2J^1}\left(\frac{\partial f_1}{\partial \mathcal{E}}\right) \tag{62}$$

and

$$\frac{\Delta A}{\mathcal{E}} = \gamma\left\{\left[\mathcal{C}_0^1 + \mathcal{C}_0^2\right] f_1 + \mathcal{C}_0^2 \Delta W_{J^2J^1}\left(-\frac{\partial f_1}{\partial \mathcal{E}}\right)\right\}\left(\frac{\beta H}{kT}\right)cz \tag{63}$$

If $\mathcal{C}_0^1$ and $\mathcal{C}_0^2$ are comparable in magnitude and of the same sign $[\mathcal{C}_0^1 + \mathcal{C}_0^2] f_1 \gg \mathcal{C}_0^2 \Delta W_{J^2J^1}(\partial f_1/\partial \mathcal{E})$ and the MCD is independent of $\Delta W_{J^2J^1}$ to the same extent as the ZFA:

$$\frac{A^\circ}{\mathcal{E}} = \gamma[\mathcal{D}_0^1 f_1 + \mathcal{D}_0^2 f_2]cz$$
$$\approx \gamma[\mathcal{D}_0^1 + \mathcal{D}_0^2] f_1 cz \tag{64}$$

On the other hand, if $\mathcal{C}_0^1$ and $\mathcal{C}_0^2$ are opposite in sign, $\mathcal{C}_0^2 \Delta W_{J^2J^1}(\partial f_1/\partial \mathcal{E})$ can be comparable to or larger than $(\mathcal{C}_0^1 + \mathcal{C}_0^2) f_1$, and the MCD is then very sensitive to $\Delta W_{J^2J^1}$. In the most extreme case, when $\mathcal{C}_0^1 = -\mathcal{C}_0^2$,

$$\frac{\Delta A}{\mathcal{E}} = \gamma \mathcal{C}_0^2 \Delta W_{J^2J^1}\left(-\frac{\partial f_1}{\partial \mathcal{E}}\right)\left(\frac{\beta H}{kT}\right)cz \tag{65}$$

and the MCD is a "pseudo-$\mathcal{A}$" term with the dispersion of the derivative of the ZFA and proportional in magnitude to $\Delta W_{J^2J^1}$.

It is also interesting to examine the behavior of the $\mathcal{B}$ terms arising from the coupling of J^1 and J^2 since these are inversely proportional to $\Delta W_{J^2J^1}$. For simplicity let us further suppose that A, J^1, and J^2 are all nondegenerate, when this contribution to the MCD is

$$\frac{\Delta A}{\mathcal{E}} = \gamma\{\mathcal{B}_0^1(J^2) f_1 + \mathcal{B}_0^2(J^1) f_2\}(\beta H)cz \tag{66}$$

where

$$\mathcal{B}_0^1(J^2) = -2\,\mathrm{Re}\left\{[\langle A|m_-^e|J^1\rangle^\circ\langle J^2|m_+^e|A\rangle^\circ - \langle A|m_+^e|J^1\rangle^\circ\langle J^2|m_-^e|A\rangle^\circ]\frac{\langle J^1|L_z+2S_z|J^2\rangle^\circ}{\Delta W_{J^2J^1}}\right\}$$
$$= -\mathcal{B}_0^2(J^1) \tag{67}$$

Then

$$\frac{\Delta A}{\mathcal{E}} = \gamma\{\mathcal{B}_0^1(J^2)[f_1 - f_2]\}(\beta H)cz \tag{68}$$

The contributions of $\mathcal{B}_0^1(J^2)$ and $\mathcal{B}_0^2(J^1)$ are of opposite sign and result in a net MCD changing in sign. When $\Delta W_{J^2J^1}$ is much smaller than the band width, Eq. 68 becomes

$$\frac{\Delta A}{\mathcal{E}} = \gamma \mathcal{B}_0^1(J^2)\Delta W_{J^2J^1}\left(\frac{\partial f_1}{\partial \mathcal{E}}\right)(\beta H)cz$$

$$= \gamma \mathcal{Q}_1'\left(-\frac{\partial f_1}{\partial \mathcal{E}}\right)(\beta H)cz \tag{69}$$

where

$$\mathcal{Q}_1' = 2\,\mathrm{Re}\Big\{\big[\langle A|m_-^e|J^1\rangle^\circ\langle J^2|m_+^e|A\rangle^\circ - \langle A|m_+^e|J^1\rangle^\circ\langle J^2|m_-^e|A\rangle^\circ\big] \times \langle J^1|L_z+2S_z|J^2\rangle^\circ\Big\} \tag{70}$$

In this limit the ZFA is given by (64). As $\Delta W_{J^2J^1}$ decreases, therefore, the oppositely-signed $\mathcal{B}$ terms produce a "pseudo-$\mathcal{Q}$" term, with the dispersion of the derivative of the ZFA and independent of $\Delta W_{J^2J^1}$. This implies that the MCD is then identical to that when $\Delta W_{J^2J^1}=0$ and J^1 and J^2 are exactly degenerate, and this result is easily obtained. Writing

$$\mathcal{Q}_1' = \mathrm{Re}\Big\{\sum_{i,i'}\big[\langle A|m_-^e|J^i\rangle^\circ\langle J^{i'}|m_+^e|A\rangle^\circ - \langle A|m_+^e|J^i\rangle^\circ\langle J^{i'}|m_-^e|A\rangle^\circ\big] \times \langle J^i|L_z+2S_z|J^{i'}\rangle^\circ\Big\} \qquad (i,i'=1,2) \tag{71}$$

and diagonalizing $\langle J^i|L_z+2S_z|J^{i'}\rangle^\circ$ through the unitary transformation

$$\psi^\circ{}_{J^i} = \sum_\lambda \psi^\circ{}_{J_\lambda} U_{\lambda i} \qquad (\lambda=1,2) \tag{72}$$

gives

$$\mathcal{Q}_1' = \sum_\lambda \big[|\langle A|m_-^e|J_\lambda\rangle^\circ|^2 - |\langle A|m_+^e|J_\lambda\rangle^\circ|^2\big] \times \langle J_\lambda|L_z+2S_z|J_\lambda\rangle^\circ = \mathcal{Q}_1(A\to J) \tag{73}$$

Likewise

$$\mathcal{D}_0^1 + \mathcal{D}_0^2 = \mathcal{D}_0(A\to J) \tag{74}$$

Thus, in this case when $\Delta W_{J^2J^1} \ll \Gamma$ the MCD can be evaluated either via Eqs. 64, 69, and 70 or from (26), (27), (44), and (45) treating J^1 and J^2 as components of the degenerate state J.

Illustrations of (56) and (60) for various exemplary situations are given in Fig. 5 to further clarify the foregoing discussion.

We now turn to ground state near-degeneracy and suppose there to be two thermally populated ground states A^1 and A^2 in place of the one earlier assumed, returning to the situation of one excited state J. We again require A^1 and A^2 to have parallel potential surfaces so that

$$W_{A^2}(\mathbf{R}) - W_{A^1}(\mathbf{R}) = \Delta W_{A^2A^1} \tag{75}$$

where $\Delta W_{A^2A^1}$ is independent of $\mathbf{R}$ and positive. The zero-field eigenfunctions are then

$$\begin{aligned} |A^1_{\alpha_1}a_1\rangle &= \psi_{A^1_{\alpha 1}}\chi_{a_1} \qquad (\alpha_1 = 1 \text{ to } d_{A^1}) \\ |A^2_{\alpha_2}a_2\rangle &= \psi^2_{A_{\alpha 2}}\chi_{a_2} \qquad (\alpha_2 = 1 \text{ to } d_{A^2}) \\ |J_\lambda j\rangle &= \psi_{J_\lambda}\chi_j \qquad (\lambda = 1 \text{ to } d_J) \end{aligned} \tag{76}$$

with χ_{a_1} and χ_{a_2} spanning identical sets of vibrational functions. The zero-field absorption spectrum is obtained as before and is found to be

$$\frac{A^\circ_\pm}{\mathcal{E}} = \gamma\{\delta_1 \mathfrak{D}^1_0 f_1 + \delta_2 \mathfrak{D}^2_0 f_2\}cz \tag{77}$$

where δ_1 and δ_2 are the fractional populations of A^1 and A^2:

$$\begin{aligned} \delta_1 &= \frac{d_{A^1}}{d_{A^1} + d_{A^2}\exp\{-\Delta W_{A^2A^1}/kT\}} \\ \delta_2 &= \frac{d_{A^2}\exp\{-\Delta W_{A^2A^1}/kT\}}{d_{A^1} + d_{A^2}\exp\{-\Delta W_{A^2A^1}/kT\}} \\ &\delta_1 + \delta_2 = 1 \end{aligned} \tag{78}$$

$\mathfrak{D}^i_0$ and $f_i (i = 1,2)$ refer to the transitions $A^i \to J$ and

$$f_2(\mathcal{E}) = f_1(\mathcal{E} + \Delta W_{A^2A^1}) \tag{79}$$

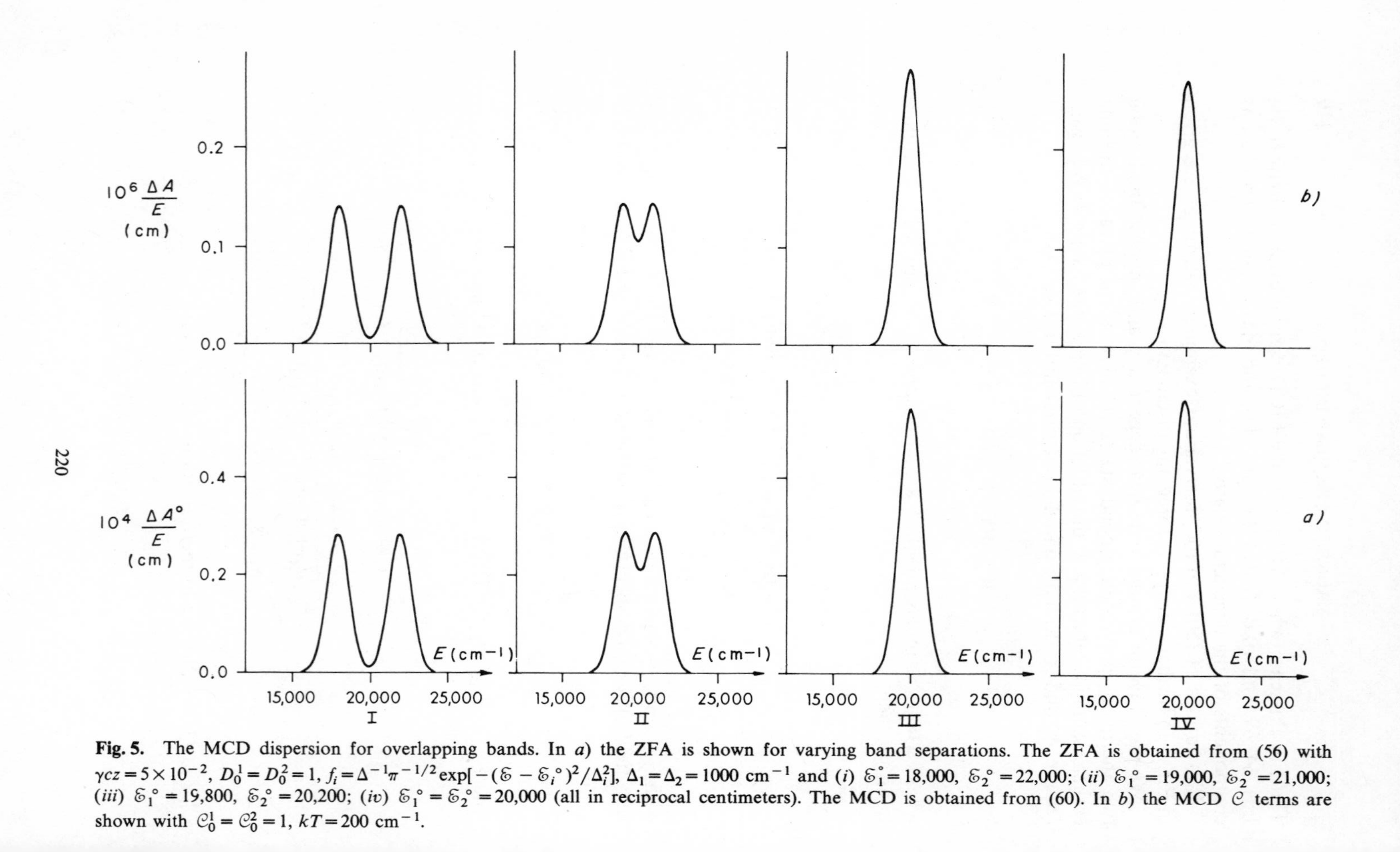

Fig. 5. The MCD dispersion for overlapping bands. In *a*) the ZFA is shown for varying band separations. The ZFA is obtained from (56) with $\gamma cz = 5\times10^{-2}$, $D_0^1 = D_0^2 = 1$, $f_i = \Delta^{-1}\pi^{-1/2}\exp[-(\mathcal{E} - \mathcal{E}_i^\circ)^2/\Delta_i^2]$, $\Delta_1 = \Delta_2 = 1000\ \text{cm}^{-1}$ and (*i*) $\mathcal{E}_1^\circ = 18{,}000$, $\mathcal{E}_2^\circ = 22{,}000$; (*ii*) $\mathcal{E}_1^\circ = 19{,}000$, $\mathcal{E}_2^\circ = 21{,}000$; (*iii*) $\mathcal{E}_1^\circ = 19{,}800$, $\mathcal{E}_2^\circ = 20{,}200$; (*iv*) $\mathcal{E}_1^\circ = \mathcal{E}_2^\circ = 20{,}000$ (all in reciprocal centimeters). The MCD is obtained from (60). In *b*) the MCD $\mathcal{C}$ terms are shown with $\mathcal{C}_0^1 = \mathcal{C}_0^2 = 1$, $kT = 200\ \text{cm}^{-1}$.

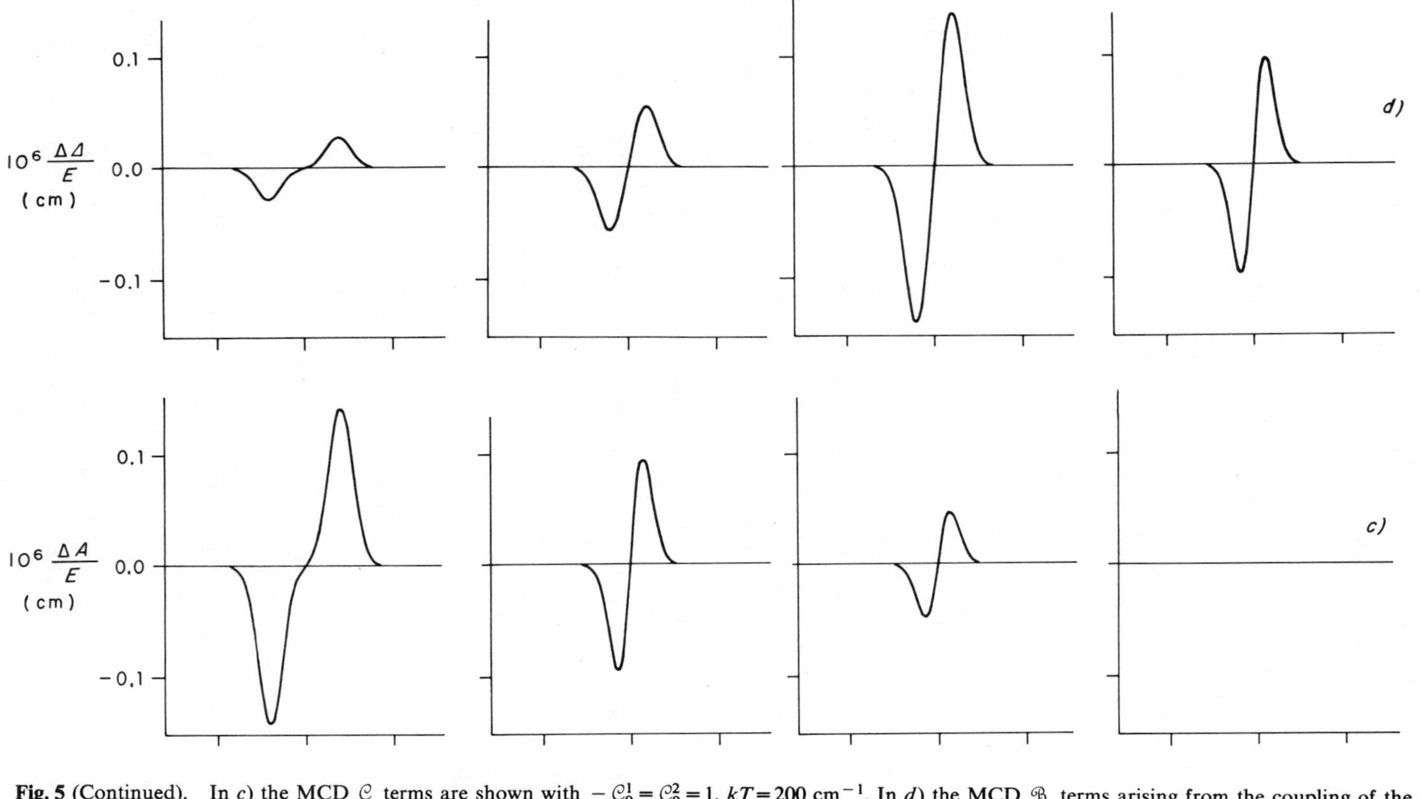

Fig. 5 (Continued). In *c*) the MCD $\mathcal{C}$ terms are shown with $-\mathcal{C}_0^1 = \mathcal{C}_0^2 = 1$, $kT = 200\ \text{cm}^{-1}$. In *d*) the MCD $\mathcal{B}$ terms arising from the coupling of the two excited states are shown. Here, $-\mathcal{B}_0^1 = \mathcal{B}_0^2 = 0.4/(\mathcal{E}_2^\circ - \mathcal{E}_1^\circ)$.

Likewise, the MCD becomes

$$\frac{\Delta A}{\mathcal{E}} = \gamma \left\{ \delta_1 \left[\mathcal{A}_1^1 \left(-\frac{\partial f_1}{\partial \mathcal{E}} \right) + \left(\mathcal{B}_0^1 + \frac{\mathcal{C}_0^1}{kT} \right) f_1 \right] + \delta_2 \left[\mathcal{A}_1^2 \left(\frac{-\partial f_2}{\partial \mathcal{E}} \right) + \left(\mathcal{B}_0^2 + \frac{\mathcal{C}_0^2}{kT} \right) f_2 \right] \right\} (\beta H) cz \quad (80)$$

and is a simple additive function of the MCD of the individual transitions $A^1 \to J$ and $A^2 \to J$.

As with excited state near-degeneracy the information content of the ZFA and MCD consists in the $\mathcal{A}_1$, $\mathcal{B}_0$, $\mathcal{C}_0$, and $\mathcal{D}_0$ parameters of the two transitions and the energy gap between the near-degenerate states. In the case of ground state near-degeneracy, the dependence on $\Delta W_{A^2A^1}$ occurs not only via the dispersion shift between f_1 and f_2 and the $\mathcal{B}$ terms due to coupling of A^1 and A^2, but also through the population factors δ_1 and δ_2. The sensitivity of ZFA and MCD to $\Delta W_{A^2A^1}$ can vary markedly. If $\mathcal{D}_0^1 = \mathcal{D}_0^2$ and $\Delta W_{A^2A^1} \ll \Gamma$, the ZFA is insensitive to $\Delta W_{A^2A^1}$. At the other extreme, if $\mathcal{D}_0^1 = 0$, the ZFA is proportional to α_2 and is very sensitive to $\Delta W_{A^2A^1}$ when $\Delta W_{A^2A^1}$ is comparable to or less than kT. Similar features are exhibited by the MCD, although in a way not necessarily correlated with the ZFA. For example, if $\mathcal{C}_0^1 = 0$ when $\mathcal{D}_0^1 = \mathcal{D}_0^2$ and $\Delta W_{A^2A^1} \ll \Gamma$, the MCD $\mathcal{C}$ terms are proportional to α_2 while the ZFA is insensitive to $\Delta W_{A^2A^1}$. Thus, the MCD may permit the determination of $\Delta W_{A^2A^1}$ when the ZFA does not.

The behavior of the $\mathcal{B}$ terms due to interaction of A^1 and A^2 as $\Delta W_{A^2A^1}$ decreases is worth further examination. For simplicity let us again assume A^1, A^2, and J to be nondegenerate. This contribution to the MCD is then

$$\frac{\Delta A}{\mathcal{E}} = \gamma \mathcal{B}_0^1(A^2) [\delta_1 f_1 - \delta_2 f_2] (\beta H) cz \quad (81)$$

since

$$\begin{aligned} \mathcal{B}_0^1(A^2) &= -\mathcal{B}_0^2(A^1) \\ &= -2\,\mathrm{Re}\{ \langle A^1 | L_z + 2S_z | A^2 \rangle^\circ [\langle A^2 | m_-^e | J \rangle^\circ \langle J | m_+^e | A^1 \rangle^\circ \\ &\quad - \langle A^2 | m_+^e | J \rangle^\circ \langle J | m_-^e | A^1 \rangle^\circ] / \Delta W_{A^2A^1} \} \end{aligned} \quad (82)$$

As $\Delta W_{A^2A^1}$ diminishes, the $\mathcal{B}$ terms increase and δ_1 and δ_2 become equal, as also do f_1 and f_2. When $\Delta W_{A^2A^1}$ becomes simultaneously much smaller than the band width and kT

$$\delta_1 = \left[2 - \frac{\Delta W_{A^2A^1}}{kT}\right]^{-1} \approx \frac{1}{2} + \frac{\Delta W_{A^2A^1}}{4kT}$$

$$\delta_2 \approx \frac{1}{2} - \frac{\Delta W_{A^2A^1}}{4kT} \tag{83}$$

$$f_2 = f_1 + \Delta W_{A^2A^1} \frac{\partial f_1}{\partial \mathcal{E}}$$

and

$$\frac{\Delta A}{\mathcal{E}} = \gamma \left\{ \mathcal{A}_1' \left(-\frac{\partial f_1}{\partial \mathcal{E}} \right) + \frac{\mathcal{C}_0'}{kT} f_1 \right\} (\beta H) cz \tag{84}$$

where

$$\mathcal{A}_1' = \mathcal{C}_0' = -\mathrm{Re}\{\langle A^1|L_z + 2S_z|A^2\rangle^\circ \times [\langle A^2|m_-^e|J\rangle^\circ \langle J|m_+^e|A^1\rangle^\circ - \langle A^2|m_+^e|J\rangle^\circ \langle J|m_-^e|A^1\rangle^\circ]\} \tag{85}$$

At the same time Eq. 77 reduces to

$$\frac{A^\circ}{\mathcal{E}} = \frac{\gamma}{2}\{\mathcal{D}_0^1 + \mathcal{D}_0^2\} f_1 cz \equiv \gamma \mathcal{D}_0' f_1 cz \tag{86}$$

At "high" temperatures ($\Delta W_{A^2A^1} \ll kT$) the MCD due to these $\mathcal{B}$ terms is thus the sum of "pseudo-$\mathcal{A}$" and "pseudo-$\mathcal{C}$" terms, both independent of $\Delta W_{A^2A^1}$. By diagonalizing $\langle A^1|L_z + 2S_z|A^2\rangle^\circ$ it is easy to show that $\mathcal{A}_1'$ and $\mathcal{C}_0'$ given by (85) are identical to the $\mathcal{A}$ and $\mathcal{C}$ parameters obtained assuming A^1 and A^2 to be exactly degenerate ($\Delta W_{A^2A^1} = 0$). At small values of $\Delta W_{A^2A^1}$ and "high" temperatures, therefore, the MCD behaves as if $\Delta W_{A^2A^1} = 0$.

C. Saturation

The discussion so far has assumed "high" temperatures where kT is much greater than Zeeman energies. We now return to the simple model of Section III.A and examine the consequences of entering into the "low" temperature region where this condition no longer obtains.

According to Eq. 36 the redistribution of population among the ground state Zeeman levels increases as H increases and T decreases. At "high" temperatures the population change is linear in H/T (Eq. 38). As the Zeeman energy becomes comparable to and greater than kT the population change varies increasingly slowly with H/T. Eventually, as the Zeeman energy becomes very much greater than kT only the lowest level is populated, and further increase in H/T causes no further population change. At this point the system is saturated.

Quantitatively, at "low" temperatures Eq. 38 is replaced by

$$\frac{N'_{A_\alpha a}}{N} = \frac{\exp\{\langle A_\alpha|\mu_z|A_\alpha\rangle^\circ H/kT\}}{\sum_\alpha \exp\{\langle A_\alpha|\mu_z|A_\alpha\rangle^\circ H/kT\}}\left(\frac{N_a}{N}\right) \tag{87}$$

At saturation if $\alpha=\alpha_1$ is the component of lowest energy ($-\langle A_{\alpha_1}|\mu_z|A_{\alpha_1}\rangle^\circ H$ is most negative),

$$\frac{N'_{A_\alpha a}}{N} = \frac{N_a}{N}\delta_{\alpha\alpha_1} \tag{88}$$

Use of (87) in place of (38) leads to

$$\frac{A_\pm}{\mathcal{E}} = \gamma\left\{\sum_{\alpha,\lambda}\frac{\exp\{\langle A_\alpha|\mu_z|A_\alpha\rangle^\circ H/kT\}}{\sum_\alpha \exp\{\langle A_\alpha|\mu_z|A_\alpha\rangle^\circ H/kT\}}\right.$$

$$\left.\times|\langle A_\alpha|m^e_\pm|J_\lambda\rangle^\circ + \langle A_\alpha|m^e_\pm|J_\lambda\rangle' H|^2 f'_{\alpha\lambda}(\mathcal{E})\right\}cz \tag{89}$$

The change in transition probabilities and band shape due to the field remain fractionally small, while the changes in population of the ground state Zeeman levels is large. To a good approximation therefore Eq. 89 becomes

$$\frac{A_\pm}{\mathcal{E}} = \gamma\left\{\sum_{\alpha,\lambda}\frac{\exp\{\langle A_\alpha|\mu_z|A_\alpha\rangle^\circ H/kT\}}{\sum_\alpha \exp\{\langle A_\alpha|\mu_z|A_\alpha\rangle^\circ H/kT\}}\right.$$

$$\left.\times|\langle A_\alpha|m^e_\pm|J_\lambda\rangle^\circ|^2 f\right\}cz \tag{90}$$

whence

$$\frac{\Delta A}{\mathcal{E}} = \gamma\left\{\sum_{\alpha,\lambda} \frac{\exp\{\langle A_\alpha|\mu_z|A_\alpha\rangle^\circ H/kT\}}{\sum_\alpha \exp\{\langle A_\alpha|\mu_z|A_\alpha\rangle^\circ H/kT\}} \times\left[|\langle A_\alpha|m_-^e|J_\lambda\rangle^\circ|^2 - |\langle A_\alpha|m_+^e|J_\lambda\rangle^\circ|^2\right] f\right\} cz \tag{91}$$

In the saturation limit Eqs. 90 and 91 reduce to

$$\frac{A_\pm}{\mathcal{E}} = \gamma\left\{\sum_\lambda |\langle A_{\alpha_1}|m_\pm^e|J_\lambda\rangle^\circ|^2\right\} fcz$$

$$\frac{\Delta A}{\mathcal{E}} = \gamma\left\{\sum_\lambda \left[|\langle A_{\alpha_1}|m_-^e|J_\lambda\rangle^\circ|^2 - |\langle A_{\alpha_1}|m_+^e|J_\lambda\rangle^\circ|^2\right]\right\} fcz \tag{92}$$

which are independent of H and T.

The simplest example of saturation behavior is provided by a ground state Kramers doublet. Writing

$$-\langle A_\alpha|\mu_z|A_\alpha\rangle^\circ H = g\beta H M_\alpha \begin{pmatrix} M_1 = -\frac{1}{2} \\ M_2 = +\frac{1}{2} \end{pmatrix} \tag{93}$$

leads to

$$\frac{\exp\{\langle A_\alpha|\mu_z|A_\alpha\rangle^\circ H/kT\}}{\sum_\alpha \exp\{\langle A_\alpha|\mu_z|A_\alpha\rangle^\circ H/kT\}} = \frac{\exp(g\beta H/2kT)\delta_{\alpha 1} + \exp(-g\beta H/2kT)\delta_{\alpha 2}}{\exp(g\beta H/2kT) + \exp(-g\beta H/2kT)} \tag{94}$$

Also, we must have

$$\sum_\lambda \left[|\langle A_{\alpha_1}|m_-^e|J_\lambda\rangle^\circ|^2 - |\langle A_{\alpha_1}|m_+^e|J_\lambda\rangle^\circ|^2\right] = -\sum_\lambda \left[|\langle A_{\alpha_2}|m_-^e|J_\lambda\rangle^\circ|^2 - |\langle A_{\alpha_2}|m_+^e|J_\lambda\rangle^\circ|^2\right] \tag{95}$$

in order that ΔA vanish at $H=0$. The MCD is then given by

$$\frac{\Delta A}{\mathcal{E}} = \gamma \tanh\left(\frac{g\beta H}{2kT}\right)\left[\sum_{\lambda} |\langle A_{\alpha_1}|m_-^e|J_\lambda\rangle^\circ|^2 - |\langle A_{\alpha_1}|m_+^e|J_\lambda\rangle^\circ|^2\right] f \quad (96)$$

and depends on H and T as tanh $(g\beta H/2kT)$, as illustrated in Fig. 6. In this specific case the saturation behavior (i.e., the H/T dependence) of the MCD is independent of the nature of the electronic transition and depends only on the ground state g-value. In general, for ground state degeneracies greater than two, this will not be the case and the specific selection rules will modulate the form of saturation.

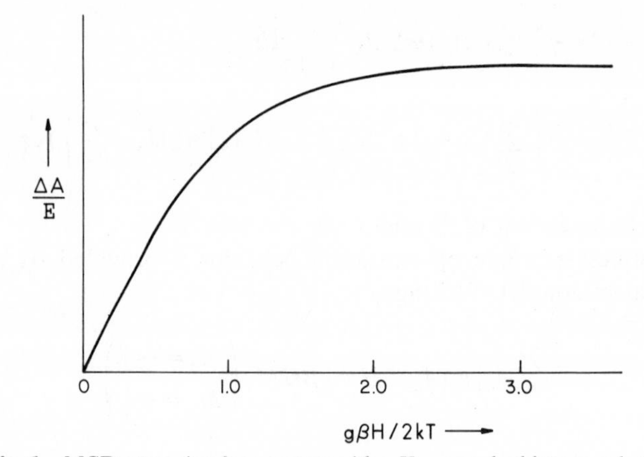

Fig. 6. MCD saturation for a system with a Kramers doublet ground state.

D. Conclusions

The calculations presented previously involve many simplifying approximations. Some of these can be removed without great difficulty. However, the removal of others causes major problems, which we now discuss.

Returning to Section III.A, the most drastic assumption made therein is that the BO approximation holds for ground and excited electronic states. The Jahn–Teller theorem states that all spatial-symmetry-derived electronic degeneracy is broken by at least some nuclear displacements and hence that the potential surfaces emanating from a degenerate state do not remain exactly degenerate. The BO approximation then breaks down: It is no longer possible to associate a vibronic state with only one electronic state. To an equivalent level of approximation—the generalized BO approximation—the vibronic states associated with the electronic states ψ_{K_κ}, exactly degenerate at $\mathbf{R}_0$, must be written

$$\Psi_{Kk} = \sum_{\kappa=1}^{d_K} \psi_{K_\kappa} \chi_{K_\kappa k} \tag{97}$$

where the $\chi_{K_\kappa k}$ vibrational functions are determined by d_K coupled equations generalizing equation 21. If the ground and excited vibronic states are assumed to have the form of (97), the calculation of both ZFA and MCD become very much more difficult. In particular, the diagonality of $\mathcal{H}_0'$ in (30) is not maintained and the magnetic field perturbation scrambles different vibronic levels of the same electronic state. As a result the calculation of the Zeeman effect within an electronic state is dependent on the details of the Jahn–Teller phenomenon. Furthermore, in general it cannot even be handled by perturbation theory, since there is no reason why matrix elements of $\mathcal{H}_0'$ should exist only between states separated by energies much greater than Zeeman energies. Diagonalization of $\mathcal{H}_0'$ is thus an infinite continuum problem.

The following alternatives then exist:

(1) There is a good reason for the Jahn–Teller effect to be zero, when the BO approximation can be retained. The JT effect is exactly zero for either nondegenerate states or Kramers doublets, and BO breakdown is not a problem therefore for singlet–singlet or Kramers doublet–Kramers doublet transitions of even or odd electron systems respectively. In the case of any other degeneracy, however, BO breakdown can only be legitimately ignored if a special reason exists for the JT effect being accidentally zero.

(2) The JT effect is included explicitly in the vibronic wavefunctions of a degenerate electronic state and in the ZFA and MCD calculations.

These problems are further enhanced in the case of near-degeneracy discussed in Section III.B. Even without breakdown of the BO approximation the diagonalization of $\mathcal{H}_0'$ within the near-degenerate vibronic manifolds becomes complex when the assumption of parallel potential surfaces is relaxed. Thus, for example, the off-diagonal terms of (58) become

$$\langle J^1_{\lambda_1} j_1 | \mathcal{H}_0' | J^2_{\lambda_2} j_2 \rangle = -\langle J^1_{\lambda_1} | \mu_z | J^2_{\lambda_2} \rangle^\circ H \langle j_1 | j_2 \rangle \tag{98}$$

and any one vibronic level of J^1 interacts with all levels of J^2. Again perturbation theory breaks down and the diagonalization of $\mathcal{H}_0'$ is an infinite continuum problem, depending on the specific potential surfaces of the near-degenerate states. Breakdown of the BO approximation further compounds the problem, and in general will now also couple the near-degenerate electronic states. Thus, for example, in place of (55) the excited levels become

$$|Jj\rangle = \sum_{i,\lambda_i} \psi_{J^i_{\lambda_i}} \chi_{i\lambda_i j} \tag{99}$$

It is clear that as the model adopted for a system becomes increasingly sophisticated the calculation of its absorption and MCD becomes rapidly more difficult, and eventually prohibitive. Furthermore, the difficulties associated with JT effects and BO breakdown are prevalent in precisely those situations of ground and excited state degeneracy leading to the most interesting and potentially informative MCD spectra. They are thus of central, rather than peripheral, importance.

In this situation it is of interest to look for alternative methods of analysis that enable more complex models to be treated without the difficulty of the dispersion calculations. The method of moments is such an approach and will now be discussed.

IV. MOMENTS

The moments of a spectrum are averaged properties. Specifically, the nth moments of ZFA and MCD bands are defined by

$$\begin{aligned} \langle A^\circ_\pm \rangle_n^{\mathcal{E}^\circ} &= \int \frac{A^\circ_\pm}{\mathcal{E}} (\mathcal{E} - \mathcal{E}^\circ)^n d\mathcal{E} \\ \langle \Delta A \rangle_n^{\mathcal{E}^\circ} &= \int \frac{\Delta A}{\mathcal{E}} (\mathcal{E} - \mathcal{E}^\circ)^n d\mathcal{E} \end{aligned} \tag{100}$$

where $\mathcal{E}^\circ$ is some chosen energy about which the moments are taken. We note immediately that moments can only be evaluated for finite bands, whose absorption is zero on either side of the band. The zeroth moments of $A^\circ_\pm$ and ΔA are simply the integrated areas of $A^\circ_\pm/\mathcal{E}$ and $\Delta A/\mathcal{E}$. Throughout our discussion we choose $\mathcal{E}^\circ$ to be the average energy of the ZFA, $\bar{\mathcal{E}}$, defined by

$$\begin{aligned} \langle A^\circ_\pm \rangle_1^{\bar{\mathcal{E}}} &= 0 \\ \bar{\mathcal{E}} &= \frac{\int A^\circ_\pm d\mathcal{E}}{\int \frac{A^\circ_\pm d\mathcal{E}}{\mathcal{E}}} \end{aligned} \tag{101}$$

Higher moments emphasize different aspects of the band shapes. For example, $\langle A^\circ_\pm \rangle_2^{\bar{\mathcal{E}}} / \langle A^\circ_\pm \rangle_0$ is a measure of the ZFA width; $\langle A^\circ_\pm \rangle_3^{\bar{\mathcal{E}}}$ is sensitive to its skewness. As n increases the outer wings of the band contribute increasingly heavily. It is important to note that an individual moment of $A_{\pm}^{\circ}$ or ΔA is a single number describing an averaged property of the full dispersion curve. In fact, knowledge of all moments (up to $n=\infty$) is required to reconstruct the actual spectrum. However, as we shall later show, the reduced information content of a moment at the same time reduces the complexity of its calculation enabling individual moments to be calculated when the calculation of the entire spectrum is excessively difficult.

Substitution of (14) and (16) into (100) results in the explicit theoretical expressions for the moments:

$$\langle A_{\pm}\rangle_n^{\bar{\mathcal{E}}} = \gamma\left\{\sum_{a,j}\left(\frac{N_a}{N}\right)|\langle a|m_{\pm}|j\rangle|^2\left(\mathcal{E}_{ja}-\bar{\mathcal{E}}\right)^n\right\}cz$$
$$\langle \Delta A\rangle_n^{\bar{\mathcal{E}}} = \gamma\left\{\sum_{a,j}\left(\frac{N_a}{N}\right)\left[|\langle a|m_{-}|j\rangle|^2-|\langle a|m_{+}|j\rangle|^2\right]\left(\mathcal{E}_{ja}-\bar{\mathcal{E}}\right)^n\right\}cz \tag{102}$$

where a and j run over all states contributing to the band. These equations enable us to demonstrate immediately the crucial property of a moment that simplifies its calculation. Consider, for example, the zeroth and first moments of $A_{\pm}$ and ΔA. Since a and j are eigenstates of $\mathcal{H}_0$,

$$\begin{aligned}\mathcal{E}_{ja} &= \langle j|\mathcal{H}_0|j\rangle - \langle a|\mathcal{H}_0|a\rangle \\ &= \sum_{j'}\left[\langle j|\mathcal{H}_0|j'\rangle - \langle a|\mathcal{H}_0|a\rangle\delta_{jj'}\right]\end{aligned} \tag{103}$$

and Eq. 102 can be written

$$\langle A_{\pm}\rangle_0 = \gamma\left\{\sum_{a,j}\frac{N_a}{N}\langle a|m_{\pm}|j\rangle\langle j|m_{\mp}|a\rangle\right\}cz$$
$$\langle A_{\pm}\rangle_1^{\bar{\mathcal{E}}} = \gamma\left\{\sum_{a,j,j'}\frac{N_a}{N}\langle a|m_{\pm}|j\rangle\left[\langle j|\mathcal{H}_0|j'\rangle - \langle a|\mathcal{H}_0|a\rangle\delta_{jj'} - \bar{\mathcal{E}}\delta_{jj'}\right]\langle j'|m_{\mp}|a\rangle\right\}cz$$
$$\langle \Delta A\rangle_0 = \gamma\left\{\sum_{a,j}\frac{N_a}{N}\left[\langle a|m_{-}|j\rangle\langle j|m_{+}|a\rangle - \langle a|m_{+}|j\rangle\langle j|m_{-}|a\rangle\right]\right\}cz$$
$$\langle \Delta A\rangle_1^{\bar{\mathcal{E}}} = \gamma\left\{\sum_{a,j,j'}\frac{N_a}{N}\left[\langle a|m_{-}|j\rangle\langle j'|m_{+}|a\rangle - \langle a|m_{+}|j\rangle\langle j'|m_{-}|a\rangle\right] \times\left[\langle j|\mathcal{H}_0|j'\rangle - \langle a|\mathcal{H}_0|a\rangle\delta_{jj'} - \bar{\mathcal{E}}\delta_{jj'}\right]\right\}cz \tag{104}$$

These expressions have the fundamental property of being invariant to a unitary transformation on the excited manifold j. That is, on substituting

$$|j\rangle = \sum_{j^\circ} |j^\circ\rangle U_{j^\circ j} \tag{105}$$

where $U_{j^\circ j}$ is any unitary transformation, Eq. 104 retains an identical form, with j and j' everywhere replaced by j° and $j^{\circ\prime}$. Suppose, then, that one wishes to approximate the eigenfunctions j of the system by an expansion in some zeroth-order set j° and the relation between j and j° functions is of the form of (105) where $U_{j^\circ j}$ is obtained by diagonalization of the matrix $\langle j^\circ|\mathcal{H}_0|j^{\circ\prime}\rangle$. In the calculation of the dispersion of $A_\pm$ or ΔA, one requires the j functions and hence the explicit diagonalization of $\langle j^\circ|\mathcal{H}_0|j^{\circ\prime}\rangle$. For the moments of (104) only the j° functions and selected $\langle j^\circ|\mathcal{H}_0|j^{\circ\prime}\rangle$ matrix elements are needed, and the complete diagonalization is unnecessary. The saving of labor can make the difference in the tractability of the calculation, and therein lies the importance of the moments. All moments can be written in a form invariant to a unitary transforamtion on the excited state manifold. The form becomes increasingly cumbersome as n increases and the lower the moment the easier its calculation.

It is important to realize that the invariance property of moments exists only with respect to unitary tranformations on the excited state manifold, and cannot be extended to the ground state manifold because of the presence of the population weighting factors (N_a/N).

A. The Rigid Shift Model

To introduce the application of the method of moments we will reconsider the system discussed in Section III.A. In contrast to the more complex systems to be discussed later, this system allows the dispersion and moment approaches to be directly compared.

The earlier calculation of the dispersion of $A^\circ_\pm$ and ΔA resulted in (26) through (28), (44) and (45). We note again that the parameters $\mathcal{A}_1$, $\mathcal{B}_0$, $\mathcal{C}_0$, and $\mathcal{D}_0$ involve only $\mathbf{R}_0$ electronic wavefunctions, whereas the band shape function $f(\mathcal{E})$ relates to the ground and excited vibrational states. The calculation of f thus requires both the explicit forms of the ground and excited state potential surfaces, and the solution of the nuclear motion problems thereon. From the dispersion equations we obtain the zeroth and first moments

$$\langle A^\circ_\pm\rangle_0=\gamma\mathcal{D}_0 cz$$

$$\langle A_\pm\rangle_1^{\bar{\mathcal{E}}}=0;\qquad \bar{\mathcal{E}}=\int f\mathcal{E}\,d\mathcal{E}$$

$$\langle \Delta A\rangle_0=\gamma\left\{\mathcal{B}_0+\frac{\mathcal{C}_0}{kT}\right\}(\beta H)cz \tag{106}$$

$$\langle \Delta A\rangle_1^{\bar{\mathcal{E}}}=\gamma\mathcal{A}_1(\mathcal{B}H)cz$$

where $\bar{\mathcal{E}}$ is the average energy of $A^\circ_\pm/\mathcal{E}$, about which $\langle A^\circ_\pm\rangle_1^{\bar{\mathcal{E}}}=0$. $\langle A^\circ_\pm\rangle_0$, $\langle\Delta A\rangle_0$, and $\langle\Delta A\rangle_1^{\bar{\mathcal{E}}}$ are independent of the form of f and depend only on the $\mathcal{A}_1$, $\mathcal{B}_0$, $\mathcal{C}_0$, and $\mathcal{D}_0$ parameters. The value of $\bar{\mathcal{E}}$ depends on f, however; its explicit derivation requires first the calculation of f.

We will now show that Eq. 106 can be obtained more simply, and an explicit expression for $\bar{\mathcal{E}}$ derived by direct moment analysis without the calculation of f and particularly without the solution of the excited state nuclear motion problem. Consider first the ZFA whose zeroth and first moments are given by (104), a and j being zero-field eigenstates of $\mathcal{H}_0^\circ$. The excited states of (21) were used in the dispersion calculation. Let us expand χ_j in terms of the complete set of ground state vibrational functions χ_a:

$$\chi_j=\sum_a\chi_a\langle a|j\rangle \tag{107}$$

The $|J_\lambda j\rangle$ set then relates to the set of functions $|J_\lambda a\rangle\equiv\psi_{J_\lambda}(\mathbf{r},\mathbf{R})\chi_a(\mathbf{R})$ by a unitary transformation:

$$\begin{aligned}|J_\lambda j\rangle&=\sum_{\lambda',a}|J_{\lambda'}a\rangle U^{(1)}_{\lambda'a;\lambda j}\\ U^{(1)}_{\lambda'a;\lambda j}&=\delta_{\lambda'\lambda}\langle a|j\rangle\end{aligned} \tag{108}$$

and the $|J_\lambda a\rangle$ basis can be used in the calculation of any quantity invariant to a unitary transformation on the eigenfunctions $|J_\lambda j\rangle$. The $|J_\lambda a\rangle$ functions do not diagonalize $\mathcal{H}_0^\circ$, of course; instead

$$\begin{aligned}\langle J_\lambda a|\mathcal{H}_0^\circ|J_{\lambda'}a'\rangle&=\langle a|\langle\psi_{J_\lambda}|\mathcal{H}_{el}|\psi_{J_{\lambda'}}\rangle+T_n\delta_{\lambda\lambda'}|a'\rangle\\ &=\langle a|W_J+T_n|a'\rangle\delta_{\lambda\lambda'}\end{aligned} \tag{109}$$

within the BO approximation. This contrasts with the $|A_\alpha a\rangle$ set, within which $\mathcal{H}_0^\circ$ is diagonal:

$$\langle A_\alpha a|\mathcal{H}_0^\circ|A_{\alpha'}a'\rangle = \langle a|W_A + T_n|a\rangle\delta_{\alpha\alpha'}\delta_{aa'} \tag{110}$$

The transition moments between states of the $|A_\alpha a\rangle$ and $|J_\lambda a\rangle$ sets simplify in the FC approximation to

$$\langle A_\alpha a|m_\pm|J_\lambda a'\rangle = \langle A_\alpha|m_\pm^e|J_\lambda\rangle^\circ\delta_{aa'} \tag{111}$$

Then, using the $|J_\lambda a\rangle$ set in (104) and invoking Eqs. 109 through 111

$$\langle A_\pm^\circ\rangle_0 = \gamma\mathfrak{D}_0 cz$$

$$\langle A_\pm^\circ\rangle_1^{\bar{\mathcal{E}}} = \gamma\left\{\mathfrak{D}_0\left[\sum_a \frac{N_a}{N}\langle a|W_{JA}|a\rangle - \bar{\mathcal{E}}\right]\right\}cz$$

$$\equiv \gamma\mathfrak{D}_0\left[\overline{W_{JA}} - \bar{\mathcal{E}}\right]cz \tag{112}$$

where $\overline{W_{JA}}$ is the thermal average of W_{JA}

$$\overline{W_{JA}} = \sum_a \frac{N_a}{N}\langle a|W_{JA}|a\rangle \tag{113}$$

Choosing $\bar{\mathcal{E}}$ to make $\langle A_\pm^\circ\rangle_1^{\bar{\mathcal{E}}} = 0$ then gives

$$\bar{\mathcal{E}} = \overline{W_{JA}} \tag{114}$$

Equation 112 simply duplicates the result for $\langle A_\pm^\circ\rangle_0$ in (106). Equation 114, however, provides a more explicit expression for $\bar{\mathcal{E}}$ involving the difference in potential surfaces of A and J and the ground state vibrational functions.

Turning now to the MCD, the excited states used in the dispersion calculation are given in (34). These can again be written in terms of a simpler set, $|J_\lambda a\rangle'$, using Eq. 107:

$$|J_\lambda j\rangle' = \sum_{\lambda',a}|J_{\lambda'}a\rangle' U^{(1)}_{\lambda'a;\lambda j} \tag{115}$$

where $U^{(1)}$ is given by (108) and

$$|J_\lambda a\rangle' = |J_\lambda a\rangle - \sum_{K_\kappa\neq J}|K_\kappa a\rangle\frac{\langle K_\kappa|\mu_z|J_\lambda\rangle^\circ H}{W_J^\circ - W_K^\circ} \tag{116}$$

and the $|J_\lambda a\rangle'$ set can be used for moment calculations. The matrices of $\mathcal{H}_0 = \mathcal{H}_0^\circ + \mathcal{H}_0'$ within the $|J_\lambda a\rangle'$ and $|A_\alpha a\rangle'$ manifolds, to first order in H and within the BO and FC approximations, are

$$\begin{aligned}\langle J_\lambda a|\mathcal{H}_0|J_{\lambda'}a'\rangle' &= \langle J_\lambda a|\mathcal{H}_0^\circ|J_{\lambda'}a'\rangle \\ &\quad + \langle J_\lambda a|\mathcal{H}_0'|J_{\lambda'}a'\rangle \\ &= \langle a|W_J + T_n|a'\rangle\delta_{\lambda\lambda'} - \langle J_\lambda|\mu_z|J_\lambda\rangle^0 H\delta_{\lambda\lambda'}\delta_{aa'} \\ \langle A_\alpha a|\mathcal{H}_0|A_{\alpha'}a'\rangle' &= \langle a|W_A + T_n|a\rangle\delta_{\alpha\alpha'}\delta_{aa'} \\ &\quad - \langle A_\alpha|\mu_z|A_\alpha\rangle^0 H\delta_{\alpha\alpha'}\delta_{aa'}\end{aligned} \tag{117}$$

since $\langle J_\lambda|\mu_z|J_{\lambda'}\rangle^\circ$ and $\langle A_\alpha|\mu_z|A_{\alpha'}\rangle^\circ$ are assumed to be diagonalized. The transition moments from $|A_\alpha a\rangle'$ to $|J_\lambda a\rangle'$ are

$$\langle A_\alpha a|m_\pm|J_\lambda a'\rangle' = [\langle A_\alpha|m_\pm^e|J_\lambda\rangle^\circ + \langle A_\alpha|m_\pm^e|J_\lambda\rangle' H]\delta_{aa'} \tag{118}$$

Substituting in (104) and proceeding further as in Section III.A then gives

$$\begin{aligned}\langle\Delta A\rangle_0 &= \gamma\left\{\mathcal{B}_0 + \frac{\mathcal{C}_0}{kT}\right\}(\beta H)cz \\ \langle\Delta A\rangle_1 &= \gamma\mathcal{A}_1(\beta H)cz\end{aligned} \tag{119}$$

–the results already given in (106).

In the derivation of (106), the excited state eigenfunctions $|J_\lambda j\rangle$ and $|J_\lambda j\rangle'$ are used. In calculating the moments of (112), (114), and (119), on the other hand, the alternative basis sets $|J_\lambda a\rangle$ and $|J_\lambda a\rangle'$ are employed, and the unitary transformation $U^{(1)}$ connecting them to the eigenfunctions is not required. The determination of $U^{(1)}$ is tantamount to the solution of the nuclear motion problem in the excited state; this must be carried out in dispersion calculations but is circumvented in moment analysis.

Having shown that moment analysis is easier than the dispersion calculation, let us now examine the relative information content of the moment and dispersion equations. Equation 106 shows that $\mathcal{A}_1$, $\mathcal{B}_0$, $\mathcal{C}_0$, and $\mathcal{D}_0$ can be obtained from $\langle A_\pm^\circ\rangle_0$, $\langle\Delta A\rangle_0$ and $\langle\Delta A\rangle_1^{\mathcal{E}}$, since the $\mathcal{B}$ and $\mathcal{C}$ terms in $\langle\Delta A\rangle_0$ are separable via their different T-dependence. The only difference in information content thus lies with the band shape function $f(\mathcal{E})$. Moment analysis up to moments with $n=1$ yields only $\overline{W_{JA}}$, an averaged property of the ground and excited state potential surfaces. The dispersion of $f(\mathcal{E})$, on the other hand, depends on the explicit form of these surfaces. The information content of the moments is thus less than that of the full dispersion form. Increasing information is yielded by increasingly higher moments and the difference in information content diminishes as the number of moments calculated increases.

The situation we have found here holds generally. Moment analysis is easier than dispersion calculation since it does not require explicit excited state eigenfunctions, and difficult aspects of their calculation can be avoided. On the other hand, the information content of a limited number of moments is less than that of the entire dispersion. Moment analysis thus offers advantage whenever the information of interest is contained in the moments calculated, the advantage increasing the lower the moments needed.

B. Overlapping Bands

In Section III.B we discussed the case of overlapping bands arising from two near-degenerate excited states J^1 and J^2. The potential surfaces of J^1 and J^2 were there made to be parallel in order for the dispersion calculation to be feasible. We now reexamine such systems using the method of moments, showing in particular that nonparallel potential surfaces can be handled without extra difficulty.

It is initially useful to obtain the moments predicted by the earlier dispersion calculation. From (56) and (60) we get:

$$\langle A_{\pm}^{0}\rangle_0 = \gamma\{\mathfrak{D}_0^1 + \mathfrak{D}_0^2\}cz$$

$$\langle A_{\pm}^{0}\rangle_1^{\bar{\mathcal{E}}} = \gamma\{\mathfrak{D}_0^1(\bar{\mathcal{E}}_1 - \bar{\mathcal{E}}) + \mathfrak{D}_0^2(\bar{\mathcal{E}}_2 - \bar{\mathcal{E}})\}cz \tag{120}$$

$$\langle \Delta A\rangle_0 = \gamma\left\{\left(\mathfrak{B}_0^1 + \frac{\mathcal{C}_0^1}{kT}\right) + \left(\mathfrak{B}_0^2 + \frac{\mathcal{C}_0^2}{kT}\right)\right\}(\beta H)cz$$

$$\langle \Delta A\rangle_1^{\bar{\mathcal{E}}} = \gamma\left\{\mathcal{A}_1^1 + \mathcal{A}_1^2 + \left(\mathfrak{B}_0^1 + \frac{\mathcal{C}_0^1}{kT}\right)(\bar{\mathcal{E}}_1 - \bar{\mathcal{E}}) + \left(\mathfrak{B}_0^2 + \frac{\mathcal{C}_0^2}{kT}\right)(\bar{\mathcal{E}}_2 - \bar{\mathcal{E}})\right\}(\beta H)cz$$

where

$$\int f_i(\mathcal{E} - \bar{\mathcal{E}}_i)\,d\mathcal{E} = 0 \tag{121}$$

and from (57)

$$\bar{\mathcal{E}}_2 = \bar{\mathcal{E}}_1 + \Delta W_{J^2 J^1} \tag{122}$$

For $\langle A^\circ_\pm \rangle_1^{\bar{\mathcal{E}}} = 0$

$$\bar{\mathcal{E}} = \frac{\mathfrak{D}_0^1 \bar{\mathcal{E}}_1 + \mathfrak{D}_0^2 \bar{\mathcal{E}}_2}{\mathfrak{D}_0^1 + \mathfrak{D}_0^2} \tag{123}$$

$\langle A^\circ_\pm \rangle_0$, $\langle \Delta A \rangle_0$, and the $\mathcal{A}$ terms of $\langle \Delta A \rangle_1^{\bar{\mathcal{E}}}$ are independent of the band shape function and the excited state splitting $\Delta W_{J^2 J^1}$. $\bar{\mathcal{E}}$ and the $\mathfrak{B}$ and $\mathcal{C}$ terms of $\langle \Delta A \rangle_1^{\bar{\mathcal{E}}}$ depend on $f_1(\mathcal{E})$ and $\Delta W_{J^2 J^1}$.

It is convenient to modify eq. 120 slightly, as follows:

$$\mathfrak{B}_0^1 = \mathfrak{B}_0^1(J^2) + \mathfrak{B}_0^1(K)$$

$$\mathfrak{B}_0^2 = \mathfrak{B}_0^2(J^1) + \mathfrak{B}_0^2(K) \tag{124}$$

where $\mathfrak{B}_0^1(J^2)$ and $\mathfrak{B}_0^2(J^1)$ are the $\mathfrak{B}$ terms originating in the interaction of J^1 and J^2. Then

$$\mathfrak{B}_0^1(J^2) + \mathfrak{B}_0^2(J^1) = 0 \tag{125}$$

and

$$\mathfrak{B}_0^1(J^2)(\bar{\mathcal{E}}_1 - \bar{\mathcal{E}}) + \mathfrak{B}_0^2(J^1)(\bar{\mathcal{E}}_2 - \bar{\mathcal{E}})$$

$$= \mathfrak{B}_0^1(J^2)(\bar{\mathcal{E}}_1 - \bar{\mathcal{E}}_2) = \frac{1}{2}\left[\mathfrak{B}_0^2(J^1) - \mathfrak{B}_0^1(J^2)\right]\Delta W_{J^2 J^1}$$

$$= \frac{1}{d_A} \sum_{\substack{\alpha, i, \lambda_i \\ i', \lambda_{i'} \\ (i' \neq i)}} \left\{\left[\langle A_\alpha | m^e_- | J^i_{\lambda_i} \rangle^\circ \langle J^{i'}_{\lambda_{i'}} | m^e_+ | A_\alpha \rangle^\circ - \langle A_\alpha | m^e_+ | J^i_{\lambda_i} \rangle^\circ \langle J^{i'}_{\lambda_{i'}} | m^e_- | A_\alpha \rangle^\circ\right]\right.$$

$$\left. \times \langle J^i_{\lambda_i} | L_z + 2S_z | J^{i'}_{\lambda_{i'}} \rangle^\circ \right\}$$

$$\equiv \mathcal{A}_1' \tag{126}$$

The contribution of $\mathfrak{B}_0^1(J^2)$ and $\mathfrak{B}_0^2(J^1)$ to $\langle \Delta A \rangle_1^{\bar{\mathcal{E}}}$ is thus independent of $f_1(\mathcal{E})$ and $\Delta W_{J^2 J^1}$, like the $\mathcal{A}$ terms, and $\langle \Delta A \rangle_0$ and $\langle \Delta A \rangle_1^{\bar{\mathcal{E}}}$ can be rewritten

$$\langle \Delta A \rangle_0 = \gamma \left\{ \left(\mathcal{B}_0^1(K) + \frac{\mathcal{C}_0^1}{kT} \right) + \left(\mathcal{B}_0^2(K) + \frac{\mathcal{C}_0^2}{kT} \right) \right\} (\beta H) cz$$

$$\langle \Delta A \rangle_1^{\bar{\mathcal{E}}} = \gamma \left\{ \mathcal{A}_1^1 + \mathcal{A}_1^2 + \mathcal{A}_1' + \left(\mathcal{B}_0^1(K) + \frac{\mathcal{C}_0^1}{kT} \right) (\bar{\mathcal{E}}_1 - \bar{\mathcal{E}}) \right.$$

$$\left. + \left(\mathcal{B}_0^2(K) + \frac{\mathcal{C}_0^2}{kT} \right) (\bar{\mathcal{E}}_2 - \bar{\mathcal{E}}) \right\} (\beta H) cz \tag{127}$$

We return now to the direct calculation of moments. For the ZFA we start from the functions of (55) but abolish Eq. 54 and therefore no longer make the assumption that the χ_{j_1} and χ_{j_2} sets of vibrational functions are the same. Following Eq. 107 we can expand the $|J^i_{\lambda_i} j_i\rangle$ set ($i = 1,2$) of functions in terms of the $|J^i_{\lambda_i} a\rangle$ set:

$$|J^i_{\lambda_i} j_i\rangle = \sum_{i', \lambda_{i'}, a} |J^{i'}_{\lambda_{i'}} a\rangle U^{(2)}_{i'\lambda_{i'} a; i\lambda_i j_i} \qquad (i, i' = 1,2)$$

$$U^{(2)}_{i'\lambda_{i'} a; i\lambda_i j_i} = \delta_{ii'} \delta_{\lambda_i \lambda_{i'}} \langle a | j_i \rangle \tag{128}$$

and therefore use the $|J^i_{\lambda_i} a\rangle$ basis in moment calculations. The necessary matrix elements of $\mathcal{H}^0_0$ and transition moments are

$$\left\langle J^i_{\lambda_i} a \right| \mathcal{H}^0_0 \left| J^{i'}_{\lambda_{i'}} a' \right\rangle = \langle a | W_{J^i} + T_n | a' \rangle \delta_{ii'} \delta_{\lambda_i \lambda_{i'}}$$

$$\left\langle A_\alpha a \right| \mathcal{H}^0_0 \left| A_\alpha a \right\rangle = \langle a | W_A + T_n | a \rangle \tag{129}$$

$$\left\langle A_\alpha a \right| m_\pm \left| J^i_{\lambda_i} a' \right\rangle = \left\langle A_\alpha \right| m^e_\pm \left| J^i_{\lambda_i} \right\rangle^\circ \delta_{aa'}$$

Substituting in (104) then gives

$$\langle A^\circ_\pm \rangle_0 = \gamma \{ \mathcal{D}^1_0 + \mathcal{D}^2_0 \} cz$$

$$\bar{\mathcal{E}} = \frac{\mathcal{D}^1_0 \overline{W_{J^1 A}} + \mathcal{D}^2_0 \overline{W_{J^2 A}}}{\mathcal{D}^1_0 + \mathcal{D}^2_0} \tag{130}$$

where $\overline{W_{J^i A}}$ is defined as in (113). Equation 130 formally replicates Eqs. 120 and 123, with $\overline{W_{J^i A}}$ replacing $\bar{\mathcal{E}}_i$. It is, however, more general since there is now no restriction on the potential surfaces W_{J^1} and W_{J^2} and hence on the band shapes of the two transitions. In particular Eq. 122 no longer holds; instead

$$\overline{W_{J^2A}} - \overline{W_{J^1A}} = \overline{W_{J^2J^1}} \tag{131}$$

where $W_{J^2J^1} = W_{J^2} - W_{J^1}$ and is now a function of **R**.

As we pointed out earlier, when the potential surfaces of J^1 and J^2 are not parallel, the matrix of $\mathcal{H}_0'$ within the $|J^i_{\lambda_i} j_i\rangle$ manifold is not diagonal, and its evaluation requires the χ_{j_i} functions. However, formally we can write the transformation to a set of functions $|j°\rangle$ that diagonalize $\mathcal{H}_0'$:

$$|j^0\rangle = \sum_{i,\lambda_i,j_i} |J^i_{\lambda_i} j_i\rangle U^{(3)}_{i\lambda_i j_i; j^0} \tag{132}$$

where $U^{(3)}$ is H-dependent. Allowing for the further mixing with other electronic states by perturbation theory then gives, for the wavefunctions $|j\rangle$ in the presence of the field,

$$|j\rangle = |j^0\rangle - \sum_{K_\kappa k} |K_\kappa k\rangle \frac{\langle K_\kappa k | \mu_z | j^0\rangle H}{\mathcal{E}_{j^0} - \mathcal{E}_k}$$

$$\approx \sum_{i,\lambda_i,j_i} \left\{ |J^i_{\lambda_i} j_i\rangle - \sum_{K_\kappa} |K_\kappa j_i\rangle \frac{\langle K_\kappa | \mu_z | J^i_{\lambda_i}\rangle^\circ H}{W^0_{J^i} - W^0_K} \right\} U^{(3)}_{i\lambda_i j_i, j^0}$$

$$\equiv \sum_{i,\lambda_i,j_i} |J^i_{\lambda_i} j_i\rangle' U^{(3)}_{i\lambda_i j_i; j^0} \tag{133}$$

that is, the eigenfunctions of $\mathcal{H}_0$ are related by a unitary transformation to the simpler functions $|J^i_{\lambda_i} j_i\rangle'$. Furthermore, again expanding the χ_{j_i} functions in terms of the χ_a set, as in (128), gives

$$|J^i_{\lambda_i} j_i\rangle' = \sum_{i',\lambda_{i'} a} |J^{i'}_{\lambda_{i'}} a\rangle' U^{(2)}_{i'\lambda_{i'} a; i\lambda_i j_i}$$

$$|j\rangle = \sum_{i',\lambda_{i'},a} |J^{i'}_{\lambda_{i'}} a\rangle' U^{(4)}_{i'\lambda_{i'} a; j} \tag{134}$$

$$U^{(4)} = U^{(2)} U^{(3)}$$

The $|J^i_{\lambda_i} a\rangle'$ basis can thus be used in the MCD moment calculation. The requisite matrix elements of $\mathcal{H}_0$ and transition moments are

$$\langle J^i_{\lambda_i} a|\mathcal{H}_0|J^{i'}_{\lambda_{i'}} a'\rangle' = \langle a|W_{J^i} + T_n|a'\rangle \delta_{ii'}\delta_{\lambda_i\lambda_{i'}} - \langle J^i_{\lambda_i}|\mu_z|J^{i'}_{\lambda_{i'}}\rangle^\circ H\delta_{aa'}$$

$$\langle A_\alpha a|\mathcal{H}_0|A_\alpha a\rangle' = \langle a|W_A + T_n|a\rangle - \langle A_\alpha|\mu_z|A_\alpha\rangle^\circ H \tag{135}$$

$$\langle A_\alpha a|m^e_\pm|J^i_\lambda a'\rangle' = \left[\langle A_\alpha|m^e_\pm|J^i_{\lambda_i}\rangle^\circ + \langle A_\alpha|m^e_\pm|J^i_{\lambda_i}\rangle' H\right]\delta_{aa'}$$

where we further assume $\langle J^i_{\lambda_i}|\mu_z|J^i_{\lambda_{i'}}\rangle^\circ$ to be diagonal and the only off-diagonal elements to involve $i \neq i'$. Substituting in (104) then leads to

$$\langle \Delta A\rangle_0 = \gamma\left\{\left(\mathcal{B}^1_0(K) + \frac{\mathcal{C}^1_0}{kT}\right) + \left(\mathcal{B}^2_0(K) + \frac{\mathcal{C}^2_0}{kT}\right)\right\}(\beta H)cz$$

$$\langle \Delta A\rangle^{\overline{\mathcal{E}}}_1 = \gamma\left\{\mathcal{A}^1_1 + \mathcal{A}^2_1 + \mathcal{A}'_1 + \left(\mathcal{B}^1_0(K) + \frac{\mathcal{C}^1_0}{kT}\right)\left(\overline{W_{J^1A}} - \overline{\mathcal{E}}\right) + \left(\mathcal{B}^2_0(K) + \frac{\mathcal{C}^2_0}{kT}\right)\left(\overline{W_{J^2A}} - \overline{\mathcal{E}}\right)\right\}(\beta H)cz \tag{136}$$

Equation 136 is formally identical with (127), with $\overline{\mathcal{E}}_i$ replaced by $\overline{W_{J^iA}}$ and $\overline{\mathcal{E}}$ defined by (130). Again, however, we emphasize that it is obtained with no restriction on the relative potential surfaces of J^1 and J^2.

In this example, moment analysis has avoided not only the calculation of excited state vibrational functions, but also the diagonalization of the Zeeman perturbation within the excited state vibronic manifold, both of which are required in a dispersion calculation. Because of this it is possible to obtain results for any J^1 and J^2 potential surfaces with no increase in difficulty over the special case of parallel potential surfaces.

Let us now examine the information content of (130) and (136). $\langle A^\circ_\pm\rangle_0$ provides $\mathcal{D}^1_0 + \mathcal{D}^2_0$ and $\langle \Delta A\rangle_0$ as a function of T yields $\mathcal{B}^1_0(K) + \mathcal{B}^2_0(K)$ and $\mathcal{C}^1_0 + \mathcal{C}^2_0$, independently of the excited state potential surfaces. $\langle \Delta A\rangle^{\overline{\mathcal{E}}}_1$ provides the two quantities

$$\{\mathcal{A}^1_1 + \mathcal{A}^2_1 + \mathcal{A}'_1\} + \left\{\mathcal{B}^1_0(K)\left(\overline{W_{J^1A}} - \overline{\mathcal{E}}\right) + \mathcal{B}^2_0(K)\left(\overline{W_{J^2A}} - \overline{\mathcal{E}}\right)\right\} \tag{137}$$

and

$$\mathcal{C}^1_0\left(\overline{W_{J^1A}} - \overline{\mathcal{E}}\right) + \mathcal{C}^2_0\left(\overline{W_{J^2A}} - \overline{\mathcal{E}}\right) \tag{138}$$

which both depend on $\overline{W_{J^1A}}$ and $\overline{W_{J^2A}}$, and hence on the J^1 and J^2 potential surfaces. If the $\mathcal{B}$ terms in (137) can be neglected, $\mathcal{A}_1^1 + \mathcal{A}_1^2 + \mathcal{A}_1'$ is obtained and does not depend on $\overline{W_{J^iA}}$ and $\overline{\mathcal{E}}$. The moments do not immediately provide $\mathcal{A}_1$, $\mathcal{B}_0$, $\mathcal{C}_0$, and $\mathcal{D}_0$ parameters for the individual $A \to J^1$ and $A \to J^2$ transitions, therefore, but rather sums of parameters for the two transitions. The same situation obtains with respect to potential surface information, contained in $\overline{W_{J^1A}}$ and $\overline{W_{J^2A}}$. Further separation of quantities for individual transitions requires additional theoretical input for the specific system being considered. As in the previous section, therefore, we find that moment analysis is easier than dispersion calculation, and can be carried out with fewer restrictive assumptions, but has at the same time less information content.

One further property of the moments in (130) and (136) should be illustrated. Consider, for example, the $\mathcal{A}$ terms in $\langle \Delta A \rangle_1^{\overline{\mathcal{E}}}$ and rewrite $\mathcal{A}_1^1 + \mathcal{A}_1^2 + \mathcal{A}_1'$, in the form

$$\mathcal{A}_1^1 + \mathcal{A}_1^2 + \mathcal{A}_1' = \frac{1}{d_A} \sum_{\alpha, i, \lambda_i, i', \lambda_{i'}} \Big\{ \Big[\langle A_\alpha | m_-^e | J^i_{\lambda_i} \rangle^\circ \langle J^{i'}_{\lambda_{i'}} | m_+^e | A_\alpha \rangle^\circ$$

$$- \langle A_\alpha | m_+^e | J^i_{\lambda_i} \rangle^\circ \langle J^{i'}_{\lambda_{i'}} | m_-^e | A_\alpha \rangle^\circ$$

$$\times \Big[\langle J^i_{\lambda_i} | L_z + 2S_z | J^{i'}_{\lambda_{i'}} \rangle^\circ - \langle A_\alpha | L_z + 2S_z | A_\alpha \rangle^\circ \delta_{ii'} \delta_{\lambda_i \lambda_{i'}} \Big] \Big\} \quad (139)$$

This expression is invariant under a unitary transformation on the $\psi^\circ_{J^i_{\lambda_i}}$ set

$$\psi^\circ_{J^i_{\lambda_i}} = \sum_j \phi_j^\circ U^{(5)}_{j, i\lambda_i} \quad (140)$$

Accordingly, in case that the $\psi^\circ_{J^i_{\lambda_i}}$ functions can be written in terms of the basis set ϕ_j°, the latter can be used in the calculation of $\mathcal{A}_1^1 + \mathcal{A}_1^2 + \mathcal{A}_1'$. For example, the diagonalization of $\langle J^i_{\lambda_i} | L_z + 2S_z | J^i_{\lambda_{i'}} \rangle^\circ$ can be avoided. If we start with a set $\psi^\circ_{J^i_{\mu_i}}$ that does not diagonalize $L_z + 2S_z$, then

$$\psi^\circ_{J^i_{\lambda_i}} = \sum_{i', \mu_{i'}} \psi^\circ_{J^{i'}\mu_{i'}} U^{(6)}_{i'\mu_{i'}; i\lambda_i} \quad (141)$$

and the $\psi^\circ_{J^i_{\mu_i}}$ functions can be equally well used. More importantly, consider the case in which the eigenfunctions $\psi^\circ_{J^i_{\lambda_i}}$ are approximated by an expansion in terms of a simpler basis of the form of (140). Calculation of $\psi^\circ_{J^i_{\lambda_i}}$, that is, of $U^{(5)}$, requires diagonalization of $\langle \phi_j^\circ | \mathcal{H}^\circ_{el} | \phi_{j'}^\circ \rangle$, which can range from trivial to impossible. Because of the invariance property of $\mathcal{A}_1^1+\mathcal{A}_1^2+\mathcal{A}_1'$ the ϕ_j° set can be used in its calculation and this diagonalization is not necessary.

For example, let us take the O_h system illustrated in Fig. 7. Without spin–orbit coupling, the ground state is ${}^1A_{1g}$ and the two excited states are ${}^3A_{1u}$ and ${}^1T_{1u}$. With spin–orbit coupling added these states become A_{1g}, $T_{1u}{}^1$ and $T_{1u}{}^2$. To first order we can write

$$
\begin{aligned}
|A_{1g}\rangle &= |{}^1A_{1g}\rangle \\
|T^i_{1u\lambda_i}\rangle &= \sum_{S,\Gamma,M_S,\gamma} |S\Gamma M_S\gamma\rangle\langle S\Gamma M_S\gamma|i\lambda_i\rangle \qquad (i=1,2)
\end{aligned}
\tag{142}
$$

where $|S\Gamma M_S\gamma\rangle$ span the ${}^1T_{1u}$ and ${}^3A_{1u}$ functions. The transformation determining the mixing of the ${}^1T_{1u}$ and ${}^3A_{1u}$ functions in the $T_{1u}{}^1$ and T_{1u}^2 states depends on the ${}^1T_{1u}-{}^3A_{1u}$ separation and the matrix of the spin-orbit interaction within the ${}^1T_{1u}$, ${}^3A_{1u}$ manifold. In this case, in evaluating $\mathcal{A}_1^1+\mathcal{A}_1^2+\mathcal{A}_1'$ the invariance property allows the use of the ${}^1T_{1u}$ and ${}^3A_{1u}$ excited state basis in place of the T_{1u}^1 and T_{1u}^2 eigenfunctions. Since the ground state is ${}^1A_{1g}$, and since $\langle {}^1A_{1g}|m_\pm^e|{}^3A_{1u}\rangle=0$, due to the spin independence of $m_\pm^e$, it also follows that the ${}^3A_{1u}$ states do not contribute, and the result obtained is identical to that assuming the transition to be just ${}^1A_{1g}\rightarrow{}^1T_{1u}$. As long as Eq. 142 remains a valid approximation, therefore, the $\mathcal{A}$ terms in $\langle\Delta A\rangle_1^{\mathcal{E}}$ are independent of the ${}^1T_{1u}-{}^3A_{1u}$ separation and spin–orbit interaction.

It is easy to show that this invariance property also holds for $\langle A_\pm^\circ\rangle_0$ and $\langle\Delta A\rangle_0$, but not for $\overline{\mathcal{E}}$ and the $\mathcal{B}$ and $\mathcal{C}$ terms of $\langle\Delta A\rangle_1^{\mathcal{E}}$, all of which depend on $\overline{W_{J'A}}$.

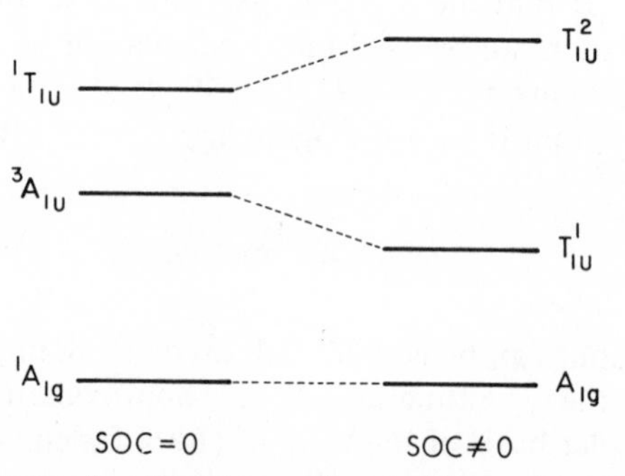

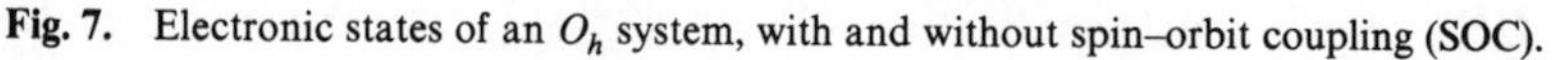

Fig. 7. Electronic states of an O_h system, with and without spin–orbit coupling (SOC).

C. A Jahn–Teller System

We return now to the situation of a single excited state and generalize the discussion of Section IV.A to allow for an excited state Jahn–Teller effect. As we indicated earlier, with the breakdown of the BO approximation consequent on a Jahn–Teller effect the calculation of the excited state vibronic levels—and, thence, of the dispersion of the ZFA and MCD—becomes very complex. Moment analysis, however, is able to avoid the explicit calculation of the excited state vibronic levels and their Zeeman effect and can be carried out with essentially no increase in difficulty over the case where the Jahn–Teller effect is absent and the BO approximation holds, as we shall now demonstrate. For simplicity we will restrict our system to have a nondegenerate ground state, when $\mathcal{C}$ terms are absent from the MCD. We will further ignore the intermixing of different electronic states by the magnetic field and the MCD $\mathcal{B}$ terms arising therefrom, leaving only the $\mathcal{A}$ terms resulting from the excited state degeneracy to be considered.

The zero-field eigenstates of the system using the generalized BO approximation for the excited states (Eq. A.20) are

$$\begin{aligned} |Aa\rangle &= \psi_A(\mathbf{r},\mathbf{R})\chi_a(\mathbf{R}) \\ |Jj^\circ\rangle &= \sum_{\lambda} \phi_{J_\lambda}(\mathbf{r},\mathbf{R})\chi_{J_\lambda j^\circ}(\mathbf{R}) \qquad (\lambda = 1 \text{ to } d_J) \end{aligned} \tag{143}$$

where $\phi_{J_\lambda} = \psi_{J_\lambda}$ at $\mathbf{R}_0$ and ϕ_{J_λ} is a slowly varying function of $\mathbf{R}$. As before, we expand the excited state vibrational functions in terms of the ground state set χ_a:

$$\chi_{J_\lambda j^\circ} = \sum_a \chi_a \langle a|\lambda j^\circ\rangle \tag{144}$$

when

$$\begin{aligned} |Jj^\circ\rangle &= \sum_{\lambda,a} |J_\lambda a\rangle U^{(7)}_{\lambda a;j^\circ} \\ |J_\lambda a\rangle &\equiv \phi_{J_\lambda}\chi_a;\; U^{(7)}_{\lambda a;j^\circ} = \langle a|\lambda j^\circ\rangle \end{aligned} \tag{145}$$

It follows from (145) that the $|J_\lambda a\rangle$ basis set can be used in ZFA moment calculation. The required matrix elements of $\mathcal{H}_0^\circ$, within the generalized BO approximation, and the transition moments within the FC approximation are

$$\begin{aligned} \langle J_\lambda a|\mathcal{H}_0^\circ|J_{\lambda'}a'\rangle &= \langle a|\langle\phi_{J_\lambda}|\mathcal{H}_{el}|\phi_{J_{\lambda'}}\rangle + T_n\delta_{\lambda\lambda'}|a'\rangle \\ \langle Aa|\mathcal{H}_0^\circ|Aa\rangle &= \langle a|W_A + T_n|a\rangle \\ \langle Aa|m_\pm|J_\lambda a'\rangle &= \langle A|m^e_\pm|J_\lambda\rangle^\circ\delta_{aa'} \end{aligned} \tag{146}$$

Note that use of the ϕ_{J_λ} set, rather than the ψ_{J_λ} eigenfunctions, not only simplifies the matrix of $\mathcal{H}_0^\circ$ (as discussed in Appendix I) but also permits the FC approximation to be invoked. The latter would not be the case with the ψ_{J_λ} functions that vary rapidly with $\mathbf{R}$. Then from (104) we obtain

$$
\begin{aligned}
\langle A_\pm^\circ \rangle_0 &= \gamma \mathcal{D}_0 cz \\
\langle A_\pm^\circ \rangle_1^{\overline{\mathcal{E}}} &= \gamma \left\{ \sum_{\lambda,\lambda'} \langle A|m_\pm^e|J_\lambda\rangle^\circ \langle J_{\lambda'}|m_\mp^e|A\rangle^\circ \overline{\Delta W_{\lambda\lambda'}} \right\} cz \\
\overline{\Delta W_{\lambda\lambda'}} &\equiv \left[\sum_a \frac{N_a}{N} \langle a|\langle \phi_{J_\lambda}|\mathcal{H}_{el}|\phi_{J_{\lambda'}}\rangle - W_A \delta_{\lambda\lambda'}|a\rangle - \overline{\mathcal{E}}\delta_{\lambda\lambda'} \right]
\end{aligned}
\tag{147}
$$

whence

$$
\begin{aligned}
\overline{\mathcal{E}} = \frac{1}{\mathcal{D}_0} \Bigg\{ &\sum_{\lambda,\lambda'} \langle A|m_\pm^e|J_\lambda\rangle^\circ \langle J_{\lambda'}|m_\mp^e|A\rangle^\circ \\
&\times \sum_a \frac{N_a}{N} \langle a|\langle \phi_{J_\lambda}|\mathcal{H}_{el}|\phi_{J_{\lambda'}}\rangle - W_A \delta_{\lambda\lambda'}|a\rangle \Bigg\}
\end{aligned}
\tag{148}
$$

In the presence of the magnetic field, the eigenstates become

$$
\begin{aligned}
|Aa\rangle' &= |Aa\rangle \\
|Jj\rangle &= \sum_{j^\circ} |Jj^\circ\rangle U_{j^\circ j}^{(8)}
\end{aligned}
\tag{149}
$$

if the mixings of the $|Aa\rangle$ and $|Jj^\circ\rangle$ manifolds with those of other electronic states are neglected. With (145), $|Jj\rangle$ further becomes

$$
\begin{aligned}
|Jj\rangle &= \sum_{\lambda,a} |J_\lambda a\rangle U_{\lambda a;j}^{(9)} \\
U^{(9)} &= U^{(7)} U^{(8)}
\end{aligned}
\tag{150}
$$

showing that the $|J_\lambda a\rangle$ basis can also be used for calculation of the MCD moments. The matrix elements of $\mathcal{H}_0$ and the transition moments required are

$$
\begin{aligned}
\langle J_\lambda a|\mathcal{H}_0|J_{\lambda'} a'\rangle = &\langle a|\langle \phi_{J_\lambda}|\mathcal{H}_{el}|\phi_{J_{\lambda'}}\rangle + T_n \delta_{\lambda\lambda'}|a'\rangle \\
&- \langle J_\lambda|\mu_z|J_\lambda\rangle^\circ H \delta_{\lambda\lambda'}\delta_{aa'}
\end{aligned}
$$

$$\langle Aa|\mathcal{H}_0|Aa\rangle = \langle a|W_A + T_n|a\rangle$$

$$\langle Aa|m_{\pm}|J_\lambda a'\rangle = \langle A|m^e_{\pm}|J_\lambda\rangle^\circ \delta_{aa'} \tag{151}$$

choosing $\langle J_\lambda|\mu_z|J_{\lambda'}\rangle^\circ$ to be diagonal. Substituting in (104) then gives

$$\langle \Delta A\rangle^\circ = 0$$

$$\langle \Delta A\rangle_1^{\bar{\mathcal{E}}} = \gamma \mathcal{A}_1(\beta H)cz \tag{152}$$

Equations 147 and 152 for $\langle A^\circ_{\pm}\rangle_0$, $\langle \Delta A\rangle_0$, and $\langle \Delta A\rangle_1^{\bar{\mathcal{E}}}$ are independent of the ground and excited state potential surfaces, and therefore replicate those obtained in Section IV.A (Eqs. 112 and 119) if $\mathcal{B}_0$ and $\mathcal{C}_0$ are put to zero. Equation 148 for $\bar{\mathcal{E}}$ generalizes Eq. 114.

The transformations $U^{(7)}$ and $U^{(8)}$ of (145) and (149), respectively, embody the solution of the excited state dynamic Jahn–Teller problem and the diagonalization of the Zeeman interaction within the resulting vibronic levels. Their explicit determination is unnecessary for moment analysis, which is carried out with an excited state basis set of the same complexity as that used in the system with zero Jahn–Teller effect.

So far we have examined only zeroth and first moments. In the rigid shift model $\langle \Delta A\rangle_0$ and $\langle \Delta A\rangle_1^{\bar{\mathcal{E}}}$ contain all information available from the MCD—the $\mathcal{A}_1$, $\mathcal{B}_0$, and $\mathcal{C}_0$ parameters—and there is no practical value in pursuing higher moments. In the present system it is also the case that $\mathcal{A}_1$ is obtained from $\langle \Delta A\rangle_1^{\bar{\mathcal{E}}}$ and, if this is the sole information sought, higher moments are redundant. However, it is no longer true that the higher moments do not contain further information relating to the ground and excited potential surfaces additional to that provided by the ZFA, and such moments can be used to study the excited state Jahn–Teller effect. To demonstrate this application we now continue the moment analysis, up to $n=2$ for the ZFA and $n=3$ for the MCD, and examine the information provided for a specific example.

Rewriting Eq. 102 to obtain expressions invariant to unitary transformations on the excited state manifold, we obtain

$$\langle A^\circ_{\pm}\rangle_2^{\bar{\mathcal{E}}} = \gamma'\Bigg\{\sum_{a,j,j',j''} \frac{N_a}{N}\langle a|m_{\pm}|j\rangle\Big[\langle j|\mathcal{H}_0^\circ|j'\rangle - \langle a|\mathcal{H}_0^\circ|a\rangle\delta_{jj'} - \bar{\mathcal{E}}\,\delta_{jj'}\Big]$$

$$\times\Big[\langle j'|\mathcal{H}_0^\circ|j''\rangle - \langle a|\mathcal{H}_0^\circ|a\rangle\delta_{j'j''} - \bar{\mathcal{E}}\,\delta_{j'j''}\Big]\langle j''|m_{\mp}|a\rangle\Bigg\}cz$$

$$\langle \Delta A \rangle_2^{\bar{\mathcal{E}}} = \gamma' \Bigg\{ \sum_{a,j,j',j''} \frac{N_a}{N} [\langle a|m_-|j\rangle\langle j''|m_+|a\rangle - \langle a|m_+|j\rangle\langle j''|m_-|a\rangle]$$

$$\times \Big[\langle j|\mathcal{H}_0|j'\rangle - \langle a|\mathcal{H}_0|a\rangle\delta_{jj'} - \bar{\mathcal{E}}\,\delta_{jj'}\Big]$$

$$\times \Big[\langle j'|\mathcal{H}_0|j''\rangle - \langle a|\mathcal{H}_0|a\rangle\delta_{j'j''} - \bar{\mathcal{E}}\,\delta_{j'j''}\Big]\Bigg\} cz$$

$$\langle \Delta A \rangle_3^{\bar{\mathcal{E}}} = \gamma \Bigg\{ \sum_{a,j,j',j'',j'''} \frac{N_a}{N} [\langle a|m_-|j\rangle\langle j'''|m_+|a\rangle - \langle a|m_+|j\rangle\langle j'''|m_-|a\rangle]$$

$$\times \Big[\langle j|\mathcal{H}_0|j'\rangle - \langle a|\mathcal{H}_0|a\rangle\delta_{jj'} - \bar{\mathcal{E}}\,\delta_{jj'}\Big]$$

$$\times \Big[\langle j'|\mathcal{H}_0|j''\rangle - \langle a|\mathcal{H}_0|a\rangle\delta_{j'j''} - \bar{\mathcal{E}}\,\delta_{j'j''}\Big]$$

$$\times \Big[\langle j''|\mathcal{H}_0|j'''\rangle - \langle a|\mathcal{H}_0|a\rangle\delta_{j''j'''} - \bar{\mathcal{E}}\,\delta_{j''j'''}\Big]\Bigg\} cz \tag{153}$$

where in $\langle A_\pm^\circ \rangle_2^{\bar{\mathcal{E}}}$ a and j are zero-field eigenstates of $\mathcal{H}_0^\circ$. As shown previously both ZFA and MCD moments can be evaluated using the $|J_\lambda a\rangle$ basis of (145). Proceeding as before, Eq. 153 then reduces to

$$\langle A_\pm^\circ \rangle_2^{\bar{\mathcal{E}}} = \gamma \mathfrak{D}_2^{\bar{\mathcal{E}}} cz$$

$$\langle \Delta A \rangle_2^{\bar{\mathcal{E}}} = \gamma \mathcal{Q}_2^{\bar{\mathcal{E}}}(\beta H) cz \tag{154}$$

$$\langle \Delta A \rangle_3^{\bar{\mathcal{E}}} = \gamma \mathcal{Q}_3^{\bar{\mathcal{E}}}(\beta H) cz$$

$$\mathfrak{D}_2^{\bar{\mathcal{E}}} = \sum_{\lambda,\lambda',\lambda''} \langle A|m_\pm^e|J_\lambda\rangle^\circ \langle J_{\lambda''}|m_\mp^e|A\rangle^\circ\, \overline{(\Delta W_{\lambda\lambda'})(\Delta W_{\lambda'\lambda''})}$$

$$\mathcal{Q}_2^{\bar{\mathcal{E}}} = \sum_{\lambda,\lambda',\lambda''} [\langle A|m_-^e|J_\lambda\rangle^\circ \langle J_{\lambda''}|m_+^e|A\rangle^\circ - \langle A|m_+^e|J_\lambda\rangle^\circ \langle J_{\lambda''}|m_-^e|A\rangle^\circ]$$

$$\times [\langle J_\lambda|L_z + 2S_z|J_{\lambda'}\rangle^\circ\, \overline{\Delta W_{\lambda'\lambda''}} + \langle J_{\lambda'}|L_z + 2S_z|J_{\lambda''}\rangle^\circ\, \overline{\Delta W_{\lambda\lambda'}}]$$

$$\mathcal{A}_3^{\bar{\mathcal{E}}} = \sum_{\lambda,\lambda',\lambda'',\lambda'''} \left[\langle A|m_-^e|J_\lambda\rangle^\circ\langle J_{\lambda'''}|m_+^e|A\rangle^\circ - \langle A|m_+^e|J_\lambda\rangle^\circ\langle J_{\lambda'''}|m_-^e|A\rangle^\circ\right]$$
$$\times\left[\langle J_\lambda|L_z+2S_z|J_{\lambda'}\rangle^\circ\,\overline{(\Delta W_{\lambda'\lambda''})(\Delta W_{\lambda''\lambda'''})}\right]$$
$$+\left[\langle J_{\lambda'}|L_z+2S_z|J_{\lambda''}\rangle^\circ\,\overline{(\Delta W_{\lambda\lambda'})(\Delta W_{\lambda''\lambda'''})}\right.$$
$$\left.+\langle J_{\lambda''}|L_z+2S_z|J_{\lambda'''}\rangle^\circ\,\overline{(\Delta W_{\lambda\lambda'})(\Delta W_{\lambda'\lambda''})}\right]$$

where we have not assumed $\langle J_\lambda|L_z+2S_z|J_{\lambda'}\rangle^\circ$ to be diagonalized.

To reduce Eq. 154 further we again consider the specific example of an $A_{1g}\rightarrow T_{1u}$ transition in O_h symmetry. The ground state potential surface is assumed to be given by (A.5) and is further approximated to be harmonic. The $\langle\phi_{T_{1u}\lambda}|\mathcal{H}_{el}|\phi_{T_{1u}\lambda'}\rangle$ matrix is given formally by (A.8); we further ignore terms higher than quadratic and assume the quadratic terms to be identical to those in the ground state. Then

$$\Delta W_{\lambda\lambda'} = W_{JA}^\circ\delta_{\lambda\lambda'} + \sum_i l_{\lambda\lambda'}^i q_i - \bar{\mathcal{E}}\,\delta_{\lambda\lambda'}$$

$$l_{\lambda\lambda'}^i = \langle\psi_{T_{1u}\lambda}^\circ|u_i|\psi_{T_{1u}\lambda'}^\circ\rangle \tag{155}$$

As a result of the ground state harmonic approximation

$$\langle a|q_i|a\rangle = 0 \tag{156}$$

when

$$\overline{\Delta W_{\lambda\lambda'}} = \left(W_{JA}^\circ - \bar{\mathcal{E}}\right)\delta_{\lambda\lambda'} \tag{157}$$

Substitution in (147) then gives

$$\langle A_\pm^\circ\rangle_1^{\bar{\mathcal{E}}} = \gamma\,\mathfrak{D}_0\left(W_{JA}^\circ - \bar{\mathcal{E}}\right)cz \tag{158}$$

whence

$$\bar{\mathcal{E}} = W_{JA}^\circ \tag{159}$$

and

$$\Delta W_{\lambda\lambda'} = \sum_i l_{\lambda\lambda'}^i q_i$$

$$\overline{\Delta W_{\lambda\lambda'}} = 0 \tag{160}$$

From (154) and (160), it follows immediately that

$$\mathcal{A}_2^{\bar{\mathcal{E}}} = \langle\Delta A\rangle_2^{\bar{\mathcal{E}}} = 0 \tag{161}$$

$\mathfrak{D}_2^{\bar{\mathcal{E}}}$ and $\mathcal{C}_3^{\bar{\mathcal{E}}}$ require evaluation of $\overline{(\Delta W_{\lambda\lambda'})(\Delta W_{\lambda''\lambda'''})}$; from (156) and (160)

$$\overline{(\Delta W_{\lambda\lambda'})(\Delta W_{\lambda''\lambda'''})} = \overline{\sum_{i,i'} l^i_{\lambda\lambda'} l^{i'}_{\lambda''\lambda'''} q_i q_{i'}}$$

$$= \sum_i l^i_{\lambda\lambda'} l^i_{\lambda''\lambda'''} \overline{q_i^2} \tag{162}$$

$$\overline{q_i^2} = \frac{\mathcal{E}_i}{2k^i{}_A} \coth\left(\frac{\mathcal{E}_i}{2kT}\right); \qquad \mathcal{E}_i = \hbar\left(k^i{}_A\right)^{1/2}$$

Now, from group theory, for $l^i_{\lambda\lambda'}$ to be nonzero q_i must belong to a_{1g}, e_g, t_{1g}, or t_{2g}. For simplicity we make the further assumption that only modes of a_{1g} and e_g symmetry contribute. Then, for the pth a_{1g} mode the nonzero $l^i_{\lambda\lambda'}$ elements are

$$l_{xx}{}^{a_{1g}p} = l_{yy}{}^{a_{1g}p} = l_{zz}{}^{a_{1g}p} = l_{ap} \tag{163}$$

and for the rth e_g mode whose components are $e_g\theta$ and $e_g\varepsilon$:

$$l_{xx}{}^{e_g\theta} = l_{yy}{}^{e_g\theta} = -\frac{1}{2} l_{er}$$

$$l_{zz}{}^{e_g\theta} = l_{er} \tag{164}$$

$$l_{xx}{}^{e_g\varepsilon r} = -l_{yy}{}^{e_g\varepsilon r} = \frac{\sqrt{3}}{2} l_{er}$$

Equations 162 through 164, together with the matrix elements of $m^e_\pm$ and $L_z + 2S_z$

$$\langle A_{1g}|m^e_\pm|T_{1u}x\rangle^\circ = \frac{m}{\sqrt{2}}$$

$$\langle A_{1g}|m^e_\pm|T_{1u}y\rangle^\circ = \pm\frac{im}{\sqrt{2}} \tag{165}$$

$$\langle T_{1u}x|L_z + 2S_z|T_{1u}y\rangle^\circ = -\langle T_{1u}y|L_z + 2S_z|T_{1u}x\rangle^\circ$$

$$= -ig$$

then lead to the following expressions for $\mathfrak{D}_2^{\bar{\mathcal{E}}}$ and $\mathcal{C}_3^{\bar{\mathcal{E}}}$:

$$\mathfrak{D}_2^{\bar{\mathcal{E}}} = m^2\left[\sum_p l_{ap}^2 \overline{q_{a_{1g}p}^2} + \sum_r l_{er}^2 \overline{q_{e_g r}^2}\right]$$

$$= \mathfrak{D}_0\left[\Delta_{a_{1g}}^2 + \Delta_{e_g}^2\right]$$

$$\mathcal{C}_3^{\bar{\mathcal{E}}} = 6gm^2\left[\sum_p l_{ap}^2 \overline{q_{a_{1g}p}^2} + \frac{1}{2}\sum_r l_{er}^2 \overline{q_{e_g r}^2}\right]$$

$$= 3\mathcal{C}_1\left[\Delta_{a_{1g}}^2 + \frac{1}{2}\Delta_{e_g}^2\right] \qquad (166)$$

Equations 154 and 166 can be reduced to

$$\frac{\langle A_{\pm}^{\circ}\rangle_2^{\bar{\mathcal{E}}}}{\langle A_{\pm}^{\circ}\rangle_0} = \Delta_{a_{1g}}^2 + \Delta_{e_g}^2$$

$$\frac{\langle \Delta A\rangle_3^{\bar{\mathcal{E}}}}{\langle \Delta A\rangle_1^{\bar{\mathcal{E}}}} = 3\left[\Delta_{a_{1g}}^2 + \frac{1}{2}\Delta_{e_g}^2\right] \qquad (167)$$

As is seen from (163) and (164), a_{1g} modes cause a displacement in the equilibrium geometry, but no splitting, of the T_{1u} excited state, while e_g modes cause a Jahn–Teller splitting. Quantitatively, with respect to the pth a_{1g} mode, the excited state minimum energy is moved to $q = -l_{ap}/k_{ap}$ and lowered from $W_{T_{1u}}^{\circ}$ by $\frac{1}{2}l_{ap}^2/k_{ap} \equiv \Delta W_{ap}$. The rth e_g mode causes a Jahn–Teller splitting of the T_{1u} potential surfaces into three displaced paraboloids, intersecting at $q_{e_g\theta,r} = q_{e_g\epsilon,r} = 0$ and energy $W_{T_{1u}}^{\circ}$, and with minima at $W_{T_{1u}}^{\circ} - \frac{1}{2}l_{er}^2/k_{er}$.[4] The energy decrease $\frac{1}{2}l_{er}^2/k_{er} \equiv \Delta W_{er}$ is often referred to as the Jahn–Teller energy. From (162) and (166) we can then write

$$\Delta_{a_{1g}}^2 = \sum_p (\Delta W_{ap})\mathcal{E}_{a_{1g}p} \coth\left(\frac{\mathcal{E}_{a_{1g}p}}{2kT}\right)$$

$$\Delta_{e_g}^2 = \sum_r (\Delta W_{er})\mathcal{E}_{e_g r} \coth\left(\frac{\mathcal{E}_{e_g r}}{2kT}\right) \qquad (168)$$

Equation 168 relates the contributions of the individual a_{1g} and e_g modes to the mean square band width to their displacement and splitting of the excited T_{1u} state potential surfaces.

Equations 166 through 168 show that in this example $\langle \Delta A \rangle_3^{\overline{\mathcal{E}}}$ enables the total contributions of all $\underline{a}_{1g}$ and of all e_g modes to the ZFA band shape to be separated, since $\langle \Delta A \rangle_3^{\overline{\mathcal{E}}} / \langle \Delta A \rangle_1^{\overline{\mathcal{E}}}$ is a different linear combination of $\Delta^2_{a_{1g}}$ and $\Delta^2_{e_g}$ than $\langle A^{\circ}_{\pm} \rangle_2^{\overline{\mathcal{E}}} / \langle A^{\circ}_{\pm} \rangle$. The MCD thus provides excited state potential surface information beyond that accessible from ZFA, as we set out to demonstrate.

V. VIBRATION-INDUCED TRANSITIONS

The discussion in Sections III and IV has been concerned exclusively with allowed electronic transitions, whose intensity is nonzero in the Franck–Condon (FC) approximation. In this section we turn to Franck–Condon forbidden electronic transitions where intensity is obtained through vibrational motion.

We discuss the system treated in Section III.A in the case that $\langle A_\alpha | m^e_{\pm} | J_\lambda \rangle^{\circ} = 0$. As seen from (22) in the FC approximation, $\langle A_\alpha a | m_{\pm} | J_\lambda j \rangle$ is then zero and the transition has zero absorption intensity. If $\langle A_\alpha | m^e_{\pm} | J_\lambda \rangle$ becomes nonzero as the nuclei are displaced away from the $\mathbf{R}_0$ geometry and the FC approximation is relaxed, however, $\langle A_\alpha a | m_{\pm} | J_\lambda j \rangle$ is no longer zero—that is, intensity is induced by vibrational motion.

The $\mathbf{R}$-dependence of $\langle A_\alpha | m^e_{\pm} | J_\lambda \rangle$ is determined by the variation with $\mathbf{R}$ of ψ_{A_α} and ψ_{J_λ}. Assuming these to be obtainable by perturbation theory and given by (A.7), $\langle A_\alpha | m^e_{\pm} | J_\lambda \rangle$ takes the form

$$\langle A_\alpha | m^e_{\pm} | J_\lambda \rangle = \langle A_\alpha | m^e_{\pm} | J_\lambda \rangle^{\circ} + \sum_i m^{\alpha\lambda}_{\pm_i} q_i + \cdots \tag{169}$$

where

$$\begin{aligned} m^{\alpha\lambda}_{\pm_i} &\equiv \left(\frac{\partial \langle A_\alpha | m^e_{\pm} | J_\lambda \rangle}{\partial q_i} \right)_{q_i = 0} \\ &= \sum_{K_\kappa \neq J} \frac{\langle A_\alpha | m^e_{\pm} | K_\kappa \rangle^{\circ} \langle K_\kappa | u_i | J_\lambda \rangle^{\circ}}{W^{\circ}_J - W^{\circ}_K} \\ &\quad + \sum_{K_\kappa \neq A} \frac{\langle A_\alpha | u_i | K_\kappa \rangle^{\circ} \langle K_\kappa | m^e_{\pm} | J_\lambda \rangle^{\circ}}{W^{\circ}_A - W^{\circ}_K} \end{aligned} \tag{170}$$

and q_i are the ground state normal coordinates, assumed henceforth to be real. When $\langle A_\alpha | m^e_\pm | J_\lambda \rangle^\circ = 0$, to first order

$$\langle A_\alpha | m^e_\pm | J_\lambda \rangle = \sum_i m^{\alpha\lambda}_{\pm_i} q_i \tag{171}$$

Equation (22) is then replaced by

$$\langle A_\alpha a | m_\pm | J_\lambda j \rangle = \sum_i m^{\alpha\lambda}_{\pm_i} \langle a | q_i | j \rangle \tag{172}$$

which, on substitution in (14), gives

$$\frac{A^\circ_\pm}{\mathcal{E}} = \gamma \left\{ \sum_{i,i'} \sum_{\alpha,\lambda} \frac{1}{d_A} m^{\alpha\lambda}_{\pm_i} m^{\lambda\alpha}_{\mp_{i'}} \sum_{a,j} \frac{N_a}{N} \langle a | q_i | j \rangle \langle j | q_{i'} | a \rangle \, \delta(\mathcal{E}_{ja} - \mathcal{E}) \right\} cz \tag{173}$$

In order to simplify our discussion, we now make the further assumption that, for each mode i that makes $m^{\alpha\lambda}_{\pm_i} \neq 0$, whenever $\langle a | q_i | j \rangle$ is nonzero, $\langle a | q_{i'} | j \rangle$ ($i' \neq i$) and $\langle a | j \rangle$ are simultaneously zero. This situation obtains, for example, when q_i is a separable and harmonic coordinate in both ground and excited states A and J; that is,

$$W_A = W^\circ_A + \frac{1}{2} k^i_A q_i^2 + w'_A$$

$$W_J = W^\circ_J + \frac{1}{2} k^i_A q_i^2 + w'_J \tag{174}$$

where w'_A and w'_J are independent of q_i. Then

$$\chi_a = \chi_{n^i_a}(q_i) \chi'_a$$

$$\chi_j = \chi_{n^i_j}(q_i) \chi'_j$$

$$\mathcal{E}_a = \mathcal{E}^i_a + \mathcal{E}'_a$$

$$\mathcal{E}_j = \mathcal{E}^i_j + \mathcal{E}'_j \tag{175}$$

$$\left. \begin{aligned} \mathcal{E}^i_a &= \left(n^i_a + \frac{1}{2} \right) \mathcal{E}_i \\ \mathcal{E}^i_j &= \left(n^i_j + \frac{1}{2} \right) \mathcal{E}_i \end{aligned} \right\} \qquad \mathcal{E}_i = \hbar \left(k^i_A \right)^{1/2}$$

whence

$$\langle a|q_i|j\rangle = \langle \chi_{n_a^i}|q_i|\chi_{n_j^i}\rangle\langle \chi_a'|\chi_j'\rangle$$

$$= \delta_{n_j^i, n_a^i \pm 1}\left[\binom{n_j^i}{a} \frac{\hbar}{2\sqrt{k_A^i}} \right]^{1/2} \langle \chi_a'|\chi_j'\rangle$$

$$\langle a|q_{i'}|j\rangle = \langle \chi_{n_a^i}|\chi_{n_j^i}\rangle\langle \chi_a'|q_{i'}|\chi_j'\rangle$$

$$= \delta_{n_a^i n_j^i}\langle \chi_a'|q_{i'}|\chi_j'\rangle \tag{176}$$

$$\langle a|j\rangle = \langle \chi_{n_a^i}|\chi_{n_j^i}\rangle\langle \chi_a'|\chi_j'\rangle$$

$$= \delta_{n_a^i, n_j^i}\langle \chi_a'|\chi_j'\rangle$$

These conditions may also be satisfied for other, such as group-theoretical, reasons. Whatever the origin, Eq. 173 under this assumption simplifies to

$$\frac{A_{\pm}^{\circ}}{\mathcal{E}} = \sum_i \left(\frac{A_{\pm}^{\circ}}{\mathcal{E}}\right)_i$$

$$\left(\frac{A_{\pm}^{\circ}}{\mathcal{E}}\right)_i = \gamma\left\{ \sum_{\alpha,\lambda} \frac{1}{d_A} |m_{\pm_i}^{\alpha\lambda}|^2 \sum_{a,j} \frac{N_a}{N} |\langle a|q_i|j\rangle|^2 \, \delta(\mathcal{E}_{ja} - \mathcal{E}) \right\} cz \tag{177}$$

and the contributions of the normal modes i to the ZFA are additive.

An important difference between vibration-induced and allowed transitions occurs in the T-dependence of the integrated ZFA intensity. As shown earlier (Eq. 24), $\int (A_{\pm}^{\circ}/\mathcal{E})\, d\mathcal{E}$ is independent of T for an allowed transition. In the case of (177), on the other hand,

$$\int \left(\frac{A_{\pm}^{\circ}}{\mathcal{E}}\right)_i d\mathcal{E} = \gamma\left\{ \sum_{\alpha,\lambda} \frac{1}{d_A} |m_{\pm_i}^{\alpha\lambda}|^2 \overline{q_i^2} \right\} cz$$

$$\overline{q_i^2} = \sum_{a,j} \frac{N_a}{N} |\langle a|q_i|j\rangle|^2 = \sum_a \frac{N_a}{N} \langle a|q_i^2|a\rangle \tag{178}$$

and since the thermal average of q_i^2, $\overline{q_i^2}$ increases with T, so does the integrated intensity. For example, if q_i is a harmonic coordinate in the ground state

$$\overline{q_i^2} = \frac{\hbar^2}{2\mathcal{E}_i} \coth\left(\frac{\mathcal{E}_i}{2kT}\right) \tag{179}$$

and

$$\int\left(\frac{A^{\circ}_{\pm}}{\mathcal{E}}\right)_i d\mathcal{E}=\gamma\left\{\sum_{\alpha,\lambda}\frac{1}{d_A}|m^{\alpha\lambda}_{\pm_i}|^2\frac{\hbar^2}{2\mathcal{E}_i}\coth\left(\frac{\mathcal{E}_i}{2kT}\right)\right\}cz \tag{180}$$

If we put Eq. 177 into the form of (26), therefore, the $\mathfrak{D}_0$ parameter is a function of both the active vibrational modes and T:

$$\left(\frac{A^{\circ}_{\pm}}{\mathcal{E}}\right)_i=\gamma\mathfrak{D}_0(i)f_i(\mathcal{E})cz$$

$$\mathfrak{D}_0(i)=\sum_{\alpha,\lambda}\frac{1}{d_A}|m^{\alpha\lambda}_{\pm_i}|^2\overline{q_i^2} \tag{181}$$

$$f_i(\mathcal{E})=\left(\overline{q_i^2}\right)^{-1}\sum_{a,j}\frac{N_a}{N}|\langle a|q_i|j\rangle|^2\delta(\mathcal{E}_{ja}-\mathcal{E})$$

$$\int f_i(\mathcal{E})d\mathcal{E}=1$$

Turning now to the MCD, we can take over unchanged equations 31, 34, and 38 for the vibronic state energies, wavefunctions, and populations in the presence of the magnetic field (again assuming "high" temperatures). Evaluation ofthe perturbed transition moment $\langle A_\alpha a|m_\pm|J_\lambda j\rangle'$, allowing for the change in transition mechanism and with our earlier assumptions leads to

$$\langle A_\alpha a|m_\pm|J_\lambda j\rangle'=\sum_i\left[m^{\alpha\lambda}_{\pm_i}+(m^{\alpha\lambda}_{\pm_i})'H\right]\langle a|q_i|j\rangle \tag{182}$$

in place of (35), where

$$(m^{\alpha\lambda}_{\pm_i})'=\left\{\sum_{K_\kappa\neq J}\frac{m^{\alpha\kappa}_{\pm_i}\langle K_\kappa|\mu_z|J_\lambda\rangle^{\circ}}{W^{\circ}_K-W^{\circ}_J}+\sum_{K_\kappa\neq A}\frac{\langle A_\alpha|\mu_z|K_\kappa\rangle^{\circ}m^{\kappa\lambda}_{\pm_i}}{W^{\circ}_K-W^{\circ}_A}\right\} \tag{183}$$

Substitution in (14) then gives

$$\frac{A_\pm}{\mathcal{E}}=\sum_i\left(\frac{A_\pm}{\mathcal{E}}\right)_i$$

$$\left(\frac{A_\pm}{\mathcal{E}}\right)_i=\gamma\left\{\sum_{\alpha,\lambda}\frac{1}{d_A}\left[1+\frac{\langle A_\alpha|\mu_z|A_\alpha\rangle^{\circ}H}{kT}\right]\left|m^{\alpha\lambda}_{\pm_i}+(m^{\alpha\lambda}_{\pm_i})'H\right|^2\overline{q_i^2}\,f'_{\alpha\lambda_i}(\mathcal{E})\right\}cz \tag{184}$$

where

$$f'_{\alpha\lambda_i}(\mathcal{E}) = \left(\overline{q_i^2}\right)^{-1}\left\{\sum_{a,j}\frac{N_a}{N}|\langle a|q_i|j\rangle|^2 \times \delta\left(\mathcal{E}_{ja} - [\langle J_\lambda|\mu_z|J_\lambda\rangle^\circ - \langle A_\alpha|\mu_z|A_\alpha\rangle^\circ]H - \mathcal{E}\right)\right\} \quad (185)$$

is identical in shape to $f_i(\mathcal{E})$ but shifted rigidly along the $\mathcal{E}$-axis by $-[\langle J_\lambda|\mu_z|J_\lambda\rangle^\circ - \langle A_\alpha|\mu_z|A_\alpha\rangle^\circ]H$. Again assuming the Zeeman shift to be much smaller than the band width of f_i and expanding $f'_{\alpha\lambda_i}$ in a Taylor expansion (as in Eq. 42), Eq. 184 becomes, to first order in H,

$$\left(\frac{A_\pm}{\mathcal{E}}\right)_i = \left(\frac{A_\pm^\circ}{\mathcal{E}}\right)_i + \gamma\left\{\sum_{\alpha,\lambda}\frac{1}{d_A}|m_{\pm_i}^{\alpha\lambda}|^2\,\overline{q_i^2} \times [\langle J_\lambda|\mu_z|J_\lambda\rangle^\circ - \langle A_\alpha|\mu_z|A_\alpha\rangle^\circ]\frac{\partial f_i}{\partial \mathcal{E}} + \sum_{\alpha,\lambda}\frac{2}{d_A}\operatorname{Re}\left[m_{\pm_i}^{\alpha\lambda}\left(m_{\pm_i}^{\alpha\lambda}\right)'^*\right]\overline{q_i^2}\, f_i + \sum_{\alpha,\lambda}\frac{1}{d_A}\frac{\langle A_\alpha|\mu_z|A_\alpha\rangle^\circ}{kT}|m_{\pm_i}^{\alpha\lambda}|^2\,\overline{q_i^2}\, f_i\right\}Hcz \quad (186)$$

whence

$$\frac{\Delta A}{\mathcal{E}} = \sum_i\left(\frac{\Delta A}{\mathcal{E}}\right)_i$$

$$\left(\frac{\Delta A}{\mathcal{E}}\right)_i = \gamma\left\{\mathcal{A}_1(i)\left(\frac{-\partial f_i}{\partial \mathcal{E}}\right) + \left[\mathcal{B}_0(i) + \frac{\mathcal{C}_0(i)}{kT}\right]f_i\right\}(\beta H)cz \quad (187)$$

where

$$\mathcal{A}_1(i) = \frac{1}{d_A}\sum_{\alpha,\lambda}\left[|m_{-_i}^{\alpha\lambda}|^2 - |m_{+_i}^{\alpha\lambda}|^2\right]\overline{q_i^2} \times [\langle J_\lambda|L_z + 2S_z|J_\lambda\rangle^\circ - \langle A_\alpha|L_z + 2S_z|A_\alpha\rangle^\circ]$$

$$\mathcal{B}_0(i) = -\frac{2}{d_A}\,\mathrm{Re}\sum_{\alpha,\lambda}\left\{\sum_{K_\kappa \neq J}\left[m_{-_i}^{\alpha\lambda} m_{+_i}^{\kappa\alpha} - m_{+_i}^{\alpha\lambda} m_{-_i}^{\kappa\alpha}\right]\frac{\langle J_\lambda|L_z+2S_z|K_\kappa\rangle^\circ}{W_K^\circ - W_J^\circ}\right.$$

$$\left. + \sum_{K_\kappa \neq A}\left[m_{-_i}^{\alpha\lambda} m_{+_i}^{\lambda\kappa} - m_{+_i}^{\alpha\lambda} m_{-_i}^{\lambda\kappa}\right]\frac{\langle K_\kappa|L_z+2S_z|A_\alpha\rangle^\circ}{W_K^\circ - W_A^\circ}\right\} \tag{188}$$

$$\mathcal{C}_0(i) = \frac{-1}{d_A}\sum_{\alpha,\lambda}\left[|m_{-_i}^{\alpha\lambda}|^2 - |m_{+_i}^{\alpha\lambda}|^2\right]\overline{q_i^2}\,\langle A_\alpha|L_z+2S_z|A_\alpha\rangle^\circ$$

The contribution of vibration i to the MCD is thus given by an equation of the rigid-shift form (Eq. 44), with parameters $\mathcal{A}_1$, $\mathcal{B}_0$, and $\mathcal{C}_0$ similar in form to those for an allowed transition but now dependent on the nature of the vibration. The total MCD is the sum of the contributions of all active intensity-inducing vibrations. When there is more than one such vibration, the total MCD cannot in general be represented by a single equation of the rigid-shift form. Exceptions occur in the special cases that (*a*) the vibrations are degenerate, when f_i is independent of i, or (*b*) all ratios $\mathcal{A}_1(i)/\mathcal{D}_0(i)$, $\mathcal{B}_0(i)/\mathcal{D}_0(i)$, and $\mathcal{C}_0(i)/\mathcal{D}_0(i)$ are independent of i. Case (*b*) occurs, for example, if all active vibrations are of the same symmetry and the parameter ratios are entirely determined by symmetry.

The existence of $\mathcal{A}$ and $\mathcal{C}$ terms in the MCD depends on the degeneracies of A and J exactly as in the case of allowed transitions. The magnitudes and signs of the parameters $\mathcal{A}_1(i)$ and $\mathcal{C}_0(i)$ further depend on the magnitudes and signs of the magnetic moments of A and J in an analogous fashion to $\mathcal{A}_1$ and $\mathcal{C}_0$ and their information content with respect to magnetic moments and electronic wavefunctions is the same. $\mathcal{A}_1(i)$ and $\mathcal{C}_0(i)$ also again depend on the symmetries and natures of A and J through the transition moments $m_{\pm_i}^{\alpha\lambda}$, but this dependence is more complex than in the case of an allowed transition, being a function of the vibration i and the intermediate electronic states through which intensity is derived. By the same token, however, the MCD can provide information about the intensity mechanism, beyond that obtainable from the ZFA alone.

To illustrate the foregoing discussion we will consider in more detail the specific example of an $A_{1g} \rightarrow T_{1g}$ transition in O_h symmetry allowed by either t_{1u} or t_{2u} vibrations. In this case there are no $\mathcal{C}$ terms, and we limit discussion to the $\mathcal{A}$ terms that arise from the excited state degeneracy. The Zeeman splitting of the T_{1g} state is shown in Fig. 8. Choosing t_{1u} and t_{2u} normal coordinates transforming as $T_{1u}x$, $T_{1u}y$, $T_{1u}z$, and $T_{2u}\xi$, $T_{2u}\eta$, $T_{2u}\zeta$ standard basis functions, Eq. 170 and group theory lead to the following

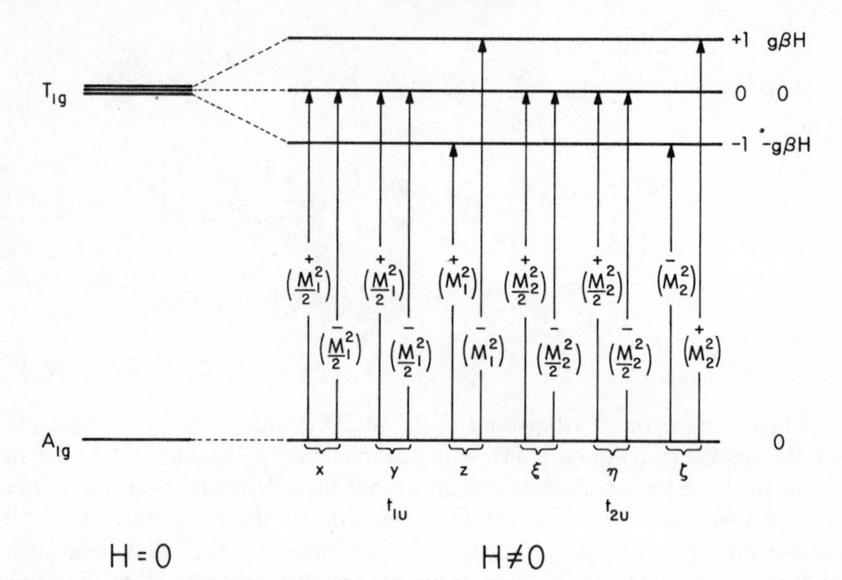

Fig. 8. The Zeeman effect in a vibration—induced $A_{1g}\rightarrow T_{1g}$ transition. The $T_{1g}+1$, 0, and -1 functions diagonalize $(L_z+2S_z)\,\beta H$, with expectation values shown at the right. Excitations for which $|m^{\lambda}_{\pm\mu}|^2$ is nonzero are indicated; the sign denotes the $m_{\pm}$ component involved; the magnitude in parenthesis is the value of $|m^{\lambda}_{\pm\mu}|^2$.

nonzero $m^{A_{1g},T_{1g}\lambda}_{\pm\, i\mu}\equiv m^{\lambda}_{\pm\,\mu}$ elements:

$$
\begin{array}{ll}
\quad t_{1u} & \quad t_{2u} \\
m^{0}_{\pm x}=\dfrac{\pm iM_1}{\sqrt{2}} & m^{0}_{\pm\xi}=\dfrac{\pm iM_2}{\sqrt{2}} \\
m^{0}_{\pm y}=\dfrac{-M_1}{\sqrt{2}} & m^{0}_{\pm\eta}=\dfrac{M_2}{\sqrt{2}} \\
m^{+1}_{\pm z}=iM_1 & m^{+1}_{\pm\zeta}=iM_2 \\
m^{-1}_{\pm z}=-iM_1 & m^{-1}_{\pm\zeta}=-iM_2
\end{array}
\tag{189}
$$

The corresponding $|m^{\lambda}_{\pm\mu}|^2$ quantities are shown in Fig. 8. Substituting in (181) and (188)

$$\mathcal{Q}_1(t_{1u}x)=\mathcal{Q}_1(t_{1u}y)=0$$

$$\mathcal{Q}_1(t_{1u}z)=2gM_1^2\,\overline{q^2_{t_{1u}}}$$

$$\mathcal{Q}_1(t_{2u}\xi)=\mathcal{Q}_1(t_{2u}\eta)=0$$

$$\mathcal{C}_1(t_{2u}\zeta) = -2gM_2^2\overline{q_{t_{2u}}^2}$$

$$\mathcal{D}_0(t_{1u}x) = \mathcal{D}_0(t_{1u}y) = \left(\frac{M_1^2}{2}\right)\overline{q_{t_{1u}}^2}$$

$$\mathcal{D}_0(t_{1u}z) = M_1^2\overline{q_{t_{1u}}^2} \tag{190}$$

$$\mathcal{D}_0(t_{2u}\xi) = \mathcal{D}_0(t_{2u}\eta) = \left(\frac{M_2^2}{2}\right)\overline{q_{t_{2u}}^2}$$

$$\mathcal{D}_0(t_{2u}\zeta) = M_2^2\overline{q_{t_{2u}}^2}$$

where we have put $\overline{q_{i\mu}^2} = \overline{q_i^2}$. Since only $m_{\pm\mu}^{\lambda}$ with $\lambda = +1$ or -1 contribute to $\mathcal{C}_1$, $\mathcal{C}_1$ is only nonzero for $t_{1u}z$ and $t_{2u}\zeta$ modes. Furthermore, since the selection rules are opposite for these modes $\mathcal{C}_1(t_{1u}z)$ is opposite in sign to $\mathcal{C}_1(t_{2u}\zeta)$. With $f_{i\mu} = f_i$, Eq. 190 then leads to

$$\frac{A_\pm^\circ}{\mathcal{E}} = \left(\frac{A_\pm^\circ}{\mathcal{E}}\right)_{t_{1u}} + \left(\frac{A_\pm^\circ}{\mathcal{E}}\right)_{t_{2u}}$$

$$\left(\frac{A_\pm^\circ}{\mathcal{E}}\right)_{t_{1u}} = \gamma\mathcal{D}_0(t_{1u})f_{t_{1u}}cz; \qquad \mathcal{D}_0(t_{1u}) = 2M_1^2\overline{q_{t_{1u}}^2}$$

$$\left(\frac{A_\pm^\circ}{\mathcal{E}}\right)_{t_{2u}} = \gamma\mathcal{D}_0(t_{2u})f_{t_{2u}}cz; \qquad \mathcal{D}_0(t_{2u}) = 2M_2^2\overline{q_{t_{2u}}^2} \tag{191}$$

$$\left(\frac{\Delta A}{\mathcal{E}}\right) = \left(\frac{\Delta A}{\mathcal{E}}\right)_{t_{1u}} + \left(\frac{\Delta A}{\mathcal{E}}\right)_{t_{2u}}$$

$$\left(\frac{\Delta A}{\mathcal{E}}\right)_{t_{1u}} = \gamma\mathcal{C}_1(t_{1u})\left(\frac{-\partial f_{t_{1u}}}{\partial\mathcal{E}}\right)(\beta H)cz; \qquad \mathcal{C}_1(t_{1u}) = 2gM_1^2\overline{q_{t_{1u}}^2}$$

$$\left(\frac{\Delta A}{\mathcal{E}}\right)_{t_{2u}} = \gamma\mathcal{C}_1(t_{2u})\left(\frac{-\partial f_{t_{2u}}}{\partial\mathcal{E}}\right)(\beta H)cz; \qquad \mathcal{C}_1(t_{2u}) = -2gM_2^2\overline{q_{t_{2u}}^2}$$

For the t_{1u} vibration, $\mathcal{C}_1/\mathcal{D}_0 = g$; for the t_{2u} vibration, $\mathcal{C}_1/\mathcal{D}_0 = -g$. Thus, if only one vibration is active, the MCD is of the simple rigid-shift form and $\mathcal{C}_1/\mathcal{D}_0$ is $+g$ or $-g$ depending on the symmetry of the vibration. If both vibrations are simultaneously active, the MCD is not of the simple rigid-shift form, but is the sum of two such effects with oppositely signed $\mathcal{C}$ terms. If the magnitude and sign of the T_{1g} excited state g-value are known, the MCD provides information concerning the vibration-induced intensity mechanism. Conversely, if the active vibrations are known, the T_{1g} g-value can be determined.

Returning to the general discussion, we emphasize, first, that Eqs. 181 and 187 for the ZFA and MCD rest on the assumption made initially concerning the simultaneous existence of $\langle a|q_i|j\rangle$, $\langle a|q_{i'}|j\rangle$, and $\langle a|j\rangle$ and are not valid for any arbitrary ground and excited state potential surfaces. In the most general case neither ZFA nor MCD are separable into single-mode contributions, and the MCD is not the sum of rigid-shift expressions. Second, even when Eqs. 181 and 187 are valid, the extraction of $\mathcal{A}_1(i)$, $\mathcal{B}_0(i)$, $\mathcal{C}_0(i)$, and $\mathcal{D}_0(i)$ parameters from experimental data cannot be carried out systematically and requires further theoretical input concerning the specific system to be practicable. Such input may consist in assumptions concerning the number and symmetries of active modes, relationships between parameters for different modes, or forms for the band shape functions of different modes. The analysis of the MCD of a vibration-induced transition is thus appreciably more complex than that for an allowed transition.

In addition, of course, the foregoing theoretical derivation has been based throughout on the assumption that the BO approximation is valid in both ground and excited states. As in the case of allowed transitions, the calculation of the dispersion of the MCD in the event that the BO approximation fails is very much more complex and moment analysis becomes a more practicable approach. This can be carried out straightforwardly after the manner illustrated in Section IV, but will not be further pursued here.

VI. CONCLUSION

We have attempted in this essay to exhibit the origins, the theoretical methodology, and the information content of MCD. Our approach has been illustrative, making use of simple model systems. By deliberate choice we have attempted neither a theoretical presentation of maximal generality and rigor nor a discussion of the MCD literature. These can be found elsewhere.[5] However, there has been lacking discussion of the theory of MCD intermediate between elementary rigid shift theory and general, algebraically complex, moment analysis. Our purpose here has been to bridge that gap. On the one hand, we have presented the simple rigid shift approach, but also tried to demonstrate explicitly its limitations. On the other hand, we have discussed the application of the method of moments, but only at a simple level. Since most real systems are complex, few of our models will be directly applicable to experimental data. However, it is our hope that our presentation will enable discussions of more complex systems to be easily understood, where existent, and to be easily pursued, where nonexistent.

APPENDIX I

In this appendix we summarize the ideas, methods, and results used in discussing electronic potential surfaces and nuclear motion thereon that are of relevance to this article. The basic problem is to solve the Schrödinger equation

$$\mathcal{H}_0^\circ \Psi_k(\mathbf{r},\mathbf{R}) = \mathcal{E}_k \Psi_k(\mathbf{r},\mathbf{R}) \tag{A.1}$$

where $\mathbf{r}$ and $\mathbf{R}$ denote electronic and nuclear coordinates, respectively, and

$$\mathcal{H}_0^\circ = \mathcal{H}_{el}(\mathbf{r},\mathbf{R}) + T_n(\mathbf{R}) \tag{A.2}$$

$\mathcal{H}_0^\circ$ is the zero-field Hamiltonian; $\mathcal{H}_{el}$ is the Hamiltonian for the system clamped at nuclear configuration $\mathbf{R}$; and T_n is the nuclear kinetic energy operator. The solutions to the electronic Schrödinger equation

$$\mathcal{H}_{el}\psi_K(\mathbf{r},\mathbf{R}) = W_K(\mathbf{R})\psi_K(\mathbf{r},\mathbf{R}) \tag{A.3}$$

are the electronic wavefunctions and potential surfaces.

We assume throughout our discussion that the ground state has a simple, single-welled potential with a minimum at $\mathbf{R}=\mathbf{R}_0$. At $\mathbf{R}_0$ Eq. 3 becomes

$$\mathcal{H}_{el}^\circ \psi_K^\circ(\mathbf{r}) = W_K^\circ \psi_K^\circ(\mathbf{r}) \tag{A.4}$$

We further assume that $\psi_K(\mathbf{r},\mathbf{R})$ and $W_K(\mathbf{R})$ can be developed by perturbation theory with (4) as the zeroth-order problem. Thus, we suppose that the ground state potential is of the form

$$W_A = W_A^\circ + \frac{1}{2}\sum_i k_A^i q_i^2 + \cdots \tag{A.5}$$

where q_i are mass-normalized normal coordinates and expand $\mathcal{H}_{el}$

$$\begin{aligned}\mathcal{H}_{el} &= \mathcal{H}_{el}^\circ + \sum_i (\partial \mathcal{H}_{el}/\partial q_i)_0 q_i \\ &\quad + \frac{1}{2}\sum_{i,i'}\left(\frac{\delta^2 \mathcal{H}_{el}}{\partial q_i \partial q_{i'}}\right) q_i q_{i'} + \cdots \\ &\equiv \mathcal{H}_{el}^\circ + \sum_i u_i q_i + \frac{1}{2}\sum_{i,i'} v_{ii'} q_i q_{i'} + \cdots \end{aligned} \tag{A.6}$$

Then, using perturbation theory, for a state K whose d_K-fold degeneracy is maintained at all q_i

$$\psi_{K_\kappa} = \psi^\circ_{K_\kappa} + \sum_{K' \neq K,\, i} \psi^\circ_{K'} \frac{\langle \psi^\circ_{K'} | u_i | \psi^\circ_{K_\kappa} \rangle}{W^\circ_K - W^\circ_{K'}} q_i + \cdots \qquad (\kappa = 1 \text{ to } d_K)$$

$$\begin{aligned} W_K = W^\circ_K &+ \sum_i \langle \psi^\circ_{K_\kappa} | u_i | \psi^\circ_{K_\kappa} \rangle q_i \\ &+ \frac{1}{2} \sum_{i,i'} \Big\{ \langle \psi^\circ_{K_\kappa} | v_{ii'} | \psi^\circ_{K_\kappa} \rangle \\ &+ 2 \sum_{K' \neq K} \langle \psi^\circ_{K_\kappa} | u_i | \psi^\circ_{K'} \rangle \langle \psi^\circ_{K'} | u_{i'} | \psi^\circ_{K_\kappa} \rangle / W^\circ_K - W^\circ_{K'} \Big\} q_i q_{i'} + \cdots \\ &\equiv W^\circ_K + \sum_i l^i_K q_i + \frac{1}{2} \sum_{i,i'} k^{ii'}_K q_i q_{i'} + \cdots \end{aligned} \qquad \text{(A.7)}$$

where κ runs over degenerate components of K. For an excited state K whose degeneracy is lifted at some q_i

$$\psi_{K_\kappa} = \sum_{\kappa'} c^\circ_{\kappa\kappa'} \left\{ \psi^\circ_{K_{\kappa'}} + \sum_{K' \neq K, i} \psi^\circ_{K'} \frac{\langle \psi^\circ_{K'} | u_i | \psi^\circ_{K_{\kappa'}} \rangle}{W^\circ_K - W^\circ_{K'}} q_i + \cdots \right\}$$

$$\equiv \sum_{\kappa} c^\circ_{\kappa\kappa'} \phi_{K_{\kappa'}} \qquad (\kappa = 1 \text{ to } d_K)$$

$$\sum_{\kappa''} c^\circ_{\kappa\kappa''} (W_{\kappa'\kappa''} - W_{K_\kappa} \delta_{\kappa'\kappa''}) = 0 \qquad (\kappa, \kappa', \kappa'' = 1 \text{ to } d_K)$$

$$\begin{aligned} W_{\kappa'\kappa''} = W^\circ_K \delta_{\kappa'\kappa''} &+ \sum_i \langle \psi^\circ_{K_{\kappa'}} | u_i | \psi^\circ_{K_{\kappa''}} \rangle q_i \\ &+ \frac{1}{2} \sum_{i,i'} \Bigg\{ \langle \psi^\circ_{K_{\kappa'}} | v_{ii'} | \psi^\circ_{K_{\kappa''}} \rangle \\ &+ 2 \sum_{K' \neq K} \frac{\langle \psi^\circ_{K_{\kappa'}} | u_i | \psi^\circ_{K'} \rangle \langle \psi^\circ_{K'} | u_{i'} | \psi^\circ_{K_{\kappa''}} \rangle}{W^\circ_K - W^\circ_{K'}} \Bigg\} q_i q_{i'} + \cdots \\ &\equiv W^\circ_K \delta_{\kappa'\kappa''} + \sum_i l^i_{\kappa'\kappa''} q_i + \frac{1}{2} \sum_{i,i'} k^{ii'}_{\kappa'\kappa''} q_i q_{i'} + \cdots \end{aligned} \qquad \text{(A.8)}$$

Equation 8 reduces to Eq. 7 when $l^i_{\kappa\kappa'}$ and $k^{ii'}_{\kappa\kappa'}$ are diagonal in, and independent of, κ.

For the ground state A Eq. 7 obtains with, further, $l^i_A = 0$ and $k^{ii'}_A = k^i_A \delta_{ii'}$ by choice of the coordinates q_i. For excited states K the quantities $l^i_{\kappa\kappa'}$ (or l^i_K) and $k^{ii'}_{\kappa\kappa'}$ (or $k^{ii'}_K$) determine the form of the potential surfaces. In the case of (7) and to second order in q, if $l^i_K = 0$ and $k^{ii'}_K = k^i_A \delta_{ii'}$ the ground and excited state potential surfaces, W_A and W_K are identical in equilibrium geometry and shape. If $l^i_K \neq 0$, but $k^{ii'}_K = k^i_A \delta_{ii'}$ the minimum in W_K is displaced from $\mathbf{R}_0$, but the shape (and therefore force constant) is the same as for the ground state; that is,

$$W_K = W^\circ_K + \sum_i \left(l^i_K q_i + \frac{1}{2} k^i_A q_i^2 \right)$$

$$= W^\circ_K - \frac{1}{2} \sum_i \frac{l^{i2}_K}{k^i_A} + \frac{1}{2} \sum_i k^i_A q_i'^2$$

$$q_i' = q_i + \frac{l^i_K}{k^i_A} \tag{A.9}$$

$$(W_K)_{\min} = W_K \left(q_i = -\frac{l^i_K}{k^i_A} \right) = W^\circ_K - \frac{1}{2} \sum_i \frac{l^{i2}_K}{k^i_A}$$

If $l^i_K \neq 0$ and $k^{ii'}_K = k^i_K \delta_{ii'} = (k^i_A + \Delta k^i_K)\delta_{ii'}$, Eq. 9 becomes

$$W_K = W^\circ_K - \frac{1}{2} \sum_i \frac{l^{i2}_K}{k^i_K} + \frac{1}{2} \sum_i (k^i_A + \Delta k^i_K) q_i'^2 \tag{A.10}$$

showing that the excited state force constant then differs by Δk^i_K from the ground state value k^i_A. In the general case where $k^{ii'}_K$ is not diagonal in i, the excited state potential surface differs in equilibrium configuration and shape from the ground state and its normal coordinates are related to the q_i set in a more complex manner than in (9) and (10).

In the case of (8) similar results obtain, excepting that different components of K have different potential surfaces. Thus, the $l^i_{\kappa\kappa'}$ terms not only displace the excited state equilibrium from $\mathbf{R}_0$, but also cause splittings of the degenerate K components when $l^i_{\kappa\kappa'} \neq l^i_K \delta_{\kappa\kappa'}$—the first order Jahn–Teller effect. Similarly, second-order Jahn–Teller effects occur when $k^{ii'}_{\kappa\kappa'} \neq k^{ii'}_K \delta_{\kappa\kappa'}$.

As a matter of nomenclature, terms contributing to a difference in ground and excited state potential surfaces first and second order in q, respectively, are often referred to as linear and quadratic electron–phonon interactions.

In the case of exactly (or non-) degenerate potential surfaces W_K the q-dependence of the electronic wavefunctions ψ_{K_κ} derives from intermixing of other electronic functions $\psi^\circ_{K'}$ with $\psi^\circ_{K_\kappa}$. If perturbation theory is valid, this q-dependent change is small. Also, it is inversely dependent on the separation $W^\circ_K - W^\circ_{K'}$. In the case of a state whose potential surfaces W_{K_κ} are split by the nuclear motion from $\mathbf{R}_0$, on the other hand, even without such off-diagonal mixing ψ_{K_κ} varies rapidly with q, through the zeroth-order coefficients $c^\circ_{\kappa\kappa'}$. That is, the correct zeroth-order wavefunctions vary according to the values of q_i. The mixing with other states $\psi^\circ_{K'}$ is still small and slowly q-dependent, however. For reasons made clear hereafter and in Section IV, it is convenient in this situation to also make use of the functions ϕ_{K_κ} defined in (8) and related to ψ_{K_κ} by a q-dependent unitary transformation. The ϕ_{K_κ} functions behave like ψ_{K_κ} functions of a non- or exactly-degenerate state, varying slowly with q_i due to mixing with other states K' and having a q-independent zeroth-order form $\psi^\circ_{K_\kappa}$. However, in contradistinction to the ψ_{K_κ} set, they do not diagonalize $\mathcal{H}_{el}$. From (8) it follows that

$$\langle \phi_{K_\kappa} | \mathcal{H}_{el} | \phi_{K_{\kappa'}} \rangle = W_{\kappa\kappa'} \tag{A.11}$$

that is, the $\langle \phi_{K_\kappa} | \mathcal{H}_{el} | \phi_{K_\kappa} \rangle$ matrix is just that diagonalized to obtain the potential surfaces W_{K_κ}.

We now turn to the dynamical problem of (1).[6] A general solution for Ψ_k can be written

$$\Psi_k = \sum_K \psi_K(\mathbf{r},\mathbf{R})\chi_{Kk}(\mathbf{R}) \tag{A.12}$$

which on substitution leads to the coupled equations

$$\sum_{K'} \left(W_K \delta_{KK'} + \langle K | T_n | K' \rangle - \mathcal{E}_k \delta_{KK'} \right) \chi_{K'k} = 0 \tag{A.13}$$

where

$$\langle K | T_n | K' \rangle \chi_{K'k} \equiv \int \psi_K(\mathbf{r},\mathbf{R}) T_n(\mathbf{R}) \psi_{K'}(\mathbf{r},\mathbf{R}) \chi_{K'k}(\mathbf{R})\, d\mathbf{r} \tag{A.14}$$

The Born–Oppenheimer (BO) approximation consists of assuming that T_n commutes with electronic functions:

$$\begin{aligned} T_n \psi_K &= \psi_K T_n \\ \langle K | T_n | K' \rangle &= \delta_{KK'} T_n \end{aligned} \tag{A.15}$$

when Eqs. 12 and 13 decouple to

$$\Psi_k = \psi_K \chi_{Kk}$$
$$[W_K + T_n]\chi_{Kk} = \mathcal{E}_k \chi_{Kk} \tag{A.16}$$

According to (16) the vibrational functions χ_{Kk} depend only on the W_K potential surface—the nuclear motion is on that surface alone. The BO approximation is justified when ψ_K is a slowly varying function of $\mathbf{R}$, and hence when K is either non- or exactly-degenerate.

In the case of degenerate states whose potential surfaces split and where ψ_K varies rapidly with $\mathbf{R}$ the simple BO approximation is not usable. However, we can obtain an equivalent level of approximation by using the slowly $\mathbf{R}$-dependent functions defined in (8). Thus, rewriting Eq. 12

$$\Psi_k = \sum_{K_\kappa} \phi_{K_\kappa}(\mathbf{r},\mathbf{R})\chi_{K_\kappa k}(\mathbf{R}) \tag{A.17}$$

Eq. 13 becomes

$$\sum_{K'_{\kappa'}} \{\langle \phi_{K_k}|\mathcal{H}_{el}|\phi_{K_{k'}}\rangle \delta_{KK'} + \langle K_\kappa|T_n|K'_{\kappa'}\rangle - \mathcal{E}_k \delta_{K_\kappa K'_{\kappa'}}\}\chi_{K'_{\kappa'}k} = 0 \tag{A.18}$$

Analogous to Eq. 15 we now put

$$T_n \phi_{K_\kappa} = \phi_{K_\kappa} T_n$$
$$\langle K_\kappa|T_n|K'_{\kappa'}\rangle = \delta_{K_\kappa K'_{\kappa'}} T_n \tag{A.19}$$

when Eqs. 17 and 18 reduce to

$$\Psi_k = \sum_{\kappa} \phi_{K_\kappa}\chi_{K_\kappa k}$$
$$\sum_{\kappa'} \{\langle \phi_{K_\kappa}|\mathcal{H}_{el}|\phi_{K_{\kappa'}}\rangle + T_n \delta_{\kappa\kappa'} - \mathcal{E}_k \delta_{\kappa\kappa'}\}\chi_{K_{\kappa'}k} = 0 \tag{A.20}$$

Equation 20 is the generalization of (16) for states exhibiting Jahn–Teller effects; we refer to it as the generalized BO approximation. The vibrational functions $\chi_{K_\kappa k}$ are now solutions of d_K coupled equations, coupled by the matrix element $\langle \phi_{K_\kappa}|\mathcal{H}_{el}|\phi_{K_{\kappa'}}\rangle$. Since by virtue of (11) the $\langle \phi_{K_\kappa}|\mathcal{H}_{el}|\phi_{K_{\kappa'}}\rangle$ matrix determines the potential surfaces of the K state, according to (20) the nuclear motion is determined by all d_K potential surfaces and the nuclei cannot be localized on any one surface.

The wavefunction of (20) can be written alternatively in terms of eigenfunctions of $\mathcal{H}_{el}$:

$$\Psi_k = \sum_{\kappa} \psi_{K_\kappa} \chi'_{K_\kappa k} \tag{A.21}$$

where

$$\sum_{\kappa} c^{\circ}_{\kappa\kappa'} \chi'_{K_\kappa k} = \chi_{K_{\kappa'} k} \tag{A.22}$$

An approximation often used in discussing the dynamical Jahn–Teller problem is the so-called "crude Jahn–Teller approximation," in which the vibronic states are written

$$\Psi_k = \sum_{\kappa} \psi^{\circ}_{K_\kappa} \chi_{K_\kappa k} \tag{A.23}$$

Equation 23 leads to coupled equations identical in form to (20) but with $\langle \phi_{K_\kappa} | \mathcal{H} |_{el} | \phi_{K_{\kappa'}} \rangle = \langle \psi^{\circ}_{K_\kappa} | \mathcal{H}_{el} | \psi^{\circ}_{K_{\kappa'}} \rangle$. The intermixing of $\mathbf{R}_0$ electronic states by the nuclear motion is thus ignored in this approximation both in the electronic wavefunctions and in the potential surfaces, the latter determining the vibrational functions. The quality of this approximation is the same as

$$\Psi_k = \psi^{\circ}_{K} \chi_{Kk} \tag{A.24}$$

for a nondegenerate BO function.

To summarize, the process of calculating the potential surfaces and vibronic wavefunctions using perturbation theory is as follows. Ground and excited potential surfaces are obtained from Eqs. 7 and 8 at the same time generating the electronic wavefunctions and the related set of slowly **R**-dependent functions. For states with a single potential surface, the vibrational functions are then obtained from (16). For degenerate states whose potential surfaces split the matrix used to determine the potential surfaces in (8) is again employed in (20) to render vibronic functions in the form of (20).

Appendix II

Equations 26 and 44 relate the ZFA and MCD to the parameters $\mathcal{A}_1$, $\mathcal{B}_0$, $\mathcal{C}_0$ and $\mathcal{D}_0$. With $\mathbf{m}^e$ in Debye units (D), when $\mathcal{A}_1$, $\mathcal{C}_0$ and $\mathcal{D}_0$ are in

D^2 and $\mathcal{B}_0$ is in $D^2\mathcal{E}^{-1}$,

$$\frac{A^\circ}{\mathcal{E}} = \frac{\epsilon^\circ}{\mathcal{E}} cz = 326.6\, \mathcal{D}_0\, fcz$$

$$\frac{\Delta A}{\mathcal{E}} = \frac{\Delta\epsilon}{\mathcal{E}} cz = 326.6\left[\mathcal{A}_1\left(-\frac{\partial f}{\partial \mathcal{E}}\right) + \left(\mathcal{B}_0 + \frac{\mathcal{C}_0}{kT}\right) f\right](\beta H)cz \qquad \text{(B.1)}$$

Equation B.1 is independent of the units of $\mathcal{E}$, as long as βH, kT, and W° are in identical units. If all energies are expressed in cm^{-1}, and H is in 10^4 gauss (= 1 tesla) the equation for ΔA can be written alternatively

$$\frac{\Delta A}{\mathcal{E}} = \frac{\Delta\epsilon}{\mathcal{E}} cz = 152.5\left[\mathcal{A}_1\left(-\frac{\partial f}{\partial \mathcal{E}}\right) + \left(\mathcal{B}_0 + \frac{\mathcal{C}_0}{kT}\right) f\right] Hcz \qquad \text{(B.2)}$$

The corresponding equations for the zeroth and first moments of the ZFA and MCD, Eq. 107, likewise become

$$\langle A^\circ \rangle_0 = \langle \epsilon^\circ \rangle cz = 326.6\, \mathcal{D}_0 cz$$

$$\langle \Delta A \rangle_0 = \langle \Delta\epsilon \rangle_0 cz = 326.6\left[\mathcal{B}_0 + \frac{\mathcal{C}_0}{kT}\right](\beta H)cz = 152.5\left[\mathcal{B}_0 + \frac{\mathcal{C}_0}{kT}\right] Hcz \qquad \text{(B.3)}$$

$$\langle \Delta A \rangle_1^{\bar{\mathcal{E}}} = \langle \Delta\epsilon \rangle_1^{\bar{\mathcal{E}}} cz = 326.6\, \mathcal{A}_1(\beta H)cz = 152.5\, \mathcal{A}_1 Hcz$$

where, in the equations for $\langle \Delta A \rangle_0$ and $\langle \Delta A \rangle_1^{\bar{\mathcal{E}}}$, the first form is independent of energy units, whereas the second form requires all energies to be in cm^{-1} and H in 10^4 gauss (T).

These equations have appeared in the MCD literature in a variety of forms. To convert to equations in terms of k° and Δk, where k is the absorption coefficient defined in Section II, put

$$k^\circ = \frac{\hbar c}{2\log_{10} e}\left(\frac{A^\circ}{\mathcal{E}}\right)\frac{1}{z} = \frac{\hbar c}{2\log_{10} e}\left(\frac{\epsilon^\circ}{\mathcal{E}}\right)c$$

$$\Delta k = \frac{\hbar c}{2\log_{10} e}\left(\frac{\Delta A}{\mathcal{E}}\right)\frac{1}{z} = \frac{\hbar c}{2\log_{10} e}\left(\frac{\Delta\epsilon}{\mathcal{E}}\right)c \qquad \text{(B.4)}$$

MCD is often expressed as molar ellipticity $[\theta]_M$, related to ΔA and $\Delta\epsilon$ through

$$[\theta]_M = \frac{4500}{\pi \log_{10} e}\Delta\epsilon = \frac{4500}{\pi \log_{10} e}\left(\frac{\Delta A}{cz}\right) \qquad \text{(B.5)}$$

where $4500/\pi \log_{10} e = 3298$. In terms of $[\theta]_M$, Eq. B.2 becomes

$$\frac{[\theta]_M}{\mathcal{E}} = 5.028 \times 10^5 \left[\mathcal{A}_1 \left(-\frac{\partial f}{\partial \mathcal{E}} \right) + \left(\mathcal{B}_0 + \frac{\mathcal{C}_0}{kT} \right) f \right] H \qquad (B.6)$$

The forms of $\mathcal{A}_1$, $\mathcal{B}_0$, $\mathcal{C}_0$, and $\mathcal{D}_0$, given in Eqs. 27 and 45, differ from those in previous publications of the author.[2] The relations between the new and old parameters are

$$\mathcal{A}_1(\text{new}) = -\frac{2}{3\beta} \mathcal{A}_1(\text{old})$$

$$\mathcal{B}_0(\text{new}) = -\frac{2}{3\beta} \mathcal{B}_0(\text{old}) \qquad (B.7)$$

$$\mathcal{C}_0(\text{new}) = -\frac{2}{3\beta} \mathcal{C}_0(\text{old})$$

$$\mathcal{D}_0(\text{new}) = \frac{1}{3} \mathcal{D}_0(\text{old})$$

The new parameters are preferable because: (1) the MCD due to $\mathcal{B}$ and $\mathcal{C}$ terms is of the same sign as $\mathcal{B}_0$ and $\mathcal{C}_0$; (2) the first moment of the MCD due to an $\mathcal{A}$ term is of the same sign as $\mathcal{A}_1$; (3) the parameters are related simply to a diagram showing circularly polarized transition probabilities and first-order Zeeman energies (e.g., Fig. 3 and 4).

References

1. A. D. Buckingham and P. J. Stephens, *Ann. Rev. Phys. Chem.*, **17**, 399 (1966).
2. P. J. Stephens, *Ann. Rev. Phys. Chem.*, **25**, 201 (1974).
3. In these and following examples, the definitions and nomenclature of O_h standard basis functions is that of J. S. Griffith, *The Theory of Transition Metal Ions*, Cambridge Univ. Press, Cambridge, 1961, Table A16.
4. W. Moffitt and W. Thorson, *Phys. Rev.*, **108**, 1251 (1957); R. Englman, *The Jahn–Teller Effect in Molecules and Crystals*, Wiley-Interscience, New York, 1972.
5. For a comprehensive review of the MCD literature, experimental and theoretical, through 1973, see Ref. 2.
6. For more expanded discussions of the Born–Oppenheimer approximation and the dynamic Jahn–Teller problem see: W. Moffitt and A. D. Liehr, *Phys. Rev.*, **106**, 1195 (1957); H. C. Longuet-Higgins, *Adv. Spectrosc.*, **2**, 429 (1961); W. D. Hobey and A. D. McLachlan, *J. Chem. Phys.*, **33**, 1695 (1960); A. D. McLachlan, *Mol. Phys.*, **4**, 417 (1961); R. Englman, *The Jahn–Teller Effect in Molecules and Crystals*, Wiley-Interscience, New York, 1972.

TIME-DEPENDENT PERTURBATION OF A TWO-STATE QUANTUM SYSTEM BY A SINUSOIDAL FIELD*

DAVID R. DION† AND JOSEPH O. HIRSCHFELDER

Theoretical Chemistry Institute,
University of Wisconsin,
Madison, Wisconsin 53706

CONTENTS

---

* Supported by the National Aeronautics and Space Administration Grant NGL-50-002-001 and the National Science Foundation Grant GP-28213.

† Present address: Department of Chemistry, State University of New York at Stony Brook, Stony Brook, New York 11794.

I. INTRODUCTION

This article is a study of the ways and means of solving the "two-level problem": What happens to a material system having only two nondegenerate energy levels when it is perturbed by an electromagnetic field that varies with time in a monochromatic sinusoidal fashion? This problem has many important applications to laser optics, nuclear magnetic resonance, and so on. From the mathematical standpoint, Einstein might say, "It is simple—just hard to do!" Many people have developed many different techniques for obtaining approximate solutions. We attempt to systematize, generalize, and improve the methods that they have used.

The two-level system exactly corresponds to a spin one-half particle in a sinusoidally oscillating monochromatic electric or magnetic field. However, it is also a good model for a many-level system in which only two nondegenerate quantum states strongly interact under the influence of the time-dependent field. Indeed, a thorough understanding of the two-level problem is valuable for the understanding of the time-dependent perturbation of any quantum mechanical system.

The two-level problem has a long history. The following are some of the highlights:

(1) I. Rabi[1] (1937)—"Space Quantization in a Gyrating Magnetic Field." Rabi considered the two-level system in a rotating field and thereby showed how to absolutely measure signs and magnitudes of magnetic moments.

(2) F. Bloch and A. Siegert[2] (1940)—"Magnetic Resonance for Nonrotating Fields." They show that when the field is linearly polarized rather than rotating, the resonance frequency for the two-level system depends on the field strength as well as the splitting between energy states.

(3) A. F. Stevenson[3] (1940)—"On the Theory of the Magnetic Resonance Method of Determining Nuclear Moments." He derives the Bloch–Siegert results in a more elegant way.

(4) S. H. Autler and C. H. Townes[4] (1955)—"Stark Effect in Rapidly Varying Fields." In developing the theory for resonant modulation in microwave spectroscopy, they find the wavefunction for the two-level system in terms of infinite continued fractions.

(5) J. -M. Winter[5] (1959)—"Theoretical and Experimental Study of Many-Photon Transitions between an Atom's Zeeman Levels." He studied the Bloch–Siegert shifts for one-photon and many-photon absorptions in a two-level system.

(6) J. Shirley (1963, unpublished thesis)[6]—"Interaction of a Quantum System with a Strong Oscillating Field." He reviewed the work done on the two-level system. He reformulated the problem as a static problem and was able to obtain solutions valid in some instances of large field strength. Most of his thesis work is in his (1965) article[25]: "Solution of the Schrödinger Equation with a Hamiltonian Periodic in Time."

(7) Recent work has been concerned with treating a specific theoretical or experimental problem by studying the two-level system's wavefunction. For example, in 1970 Series[31] ("Optical Pumping and Related Topics") considered extremely strong fields. In 1972 Silverman and Pipkin[7] ("Interaction of a Decaying Atom with a Linearly Polarized Oscillating

Field") considered the effect of letting the atom decay. Pegg[8] ("Semi-Classical Model of Magnetic Resonance in Intense RF Fields") reconsidered the Bloch–Siegert shift in 1973. This latter work was sparked by a disagreement between Shirley's results for the Bloch–Siegert shift and a fully quantized calculation by Chang and Stehle.[9,10] Stenholm[11] ("Quantum Theory of RF Resonances: The Semiclassical Limit") used a density matrix approach developed in his papers[12,13,14] to reconfirm Shirley's results and Pegg's results.

(8) Most recently, McGurk, Schmalz, and Flygare[15] ("A Density Matrix, Bloch Equation Description of Infrared and Microwave Transient Phenomena") studied the two-level system's density matrix to treat transient effects in microwave and infrared spectroscopy.

Most of the problems of spectroscopy and optics depend upon the effects of a time-dependent sinusoidally varying perturbation. Solutions to such time-dependent perturbation problems could be used to predict the spectroscopic probability of a transition from one quantum state to another as a function of field frequency and field strength. Such a solution would determine the Bloch–Siegert shift for one-photon or multi-photon absorptions. The same is true of induced (but not spontaneous) photon emissions. Furthermore, the behavior of light passing through any material system is understood by considering how the material system reacts to the incident radiation. The incident radiation drives the material system to "oscillate." The induced oscillating dipoles in turn radiate a field. Finding the driven motions of the system, therefore, determines the field radiated by the driven atoms. For example, if the incident field has angular frequency ω, the driven dipoles always have components varying as $\exp(in\omega t)$ for $n=1,2,3,\ldots$. The $n=2$ component is responsible for frequency doubling (or second harmonic generation) of light in a KDP crystal. Also, the field in the medium is a superposition of the incident field and the field produced by the induced oscillating dipoles. Determining the motion of a system in an oscillating field is therefore the major hurdle in finding the index of refraction as a function of the frequency and strength of the incident field. Such optical phenomena as optical activity, the Kerr effect, the Faraday effect, the Cotton–Moulton effect, the Pockels effect, and so on can be understood in a similar manner.

Although Dirac[16] showed how to quantize the field, the semiclassical theory of the interaction of matter and radiation explains many phenomena. For example, Wentzel[17] described the photoelectric effect, Klein and Nishina[18] correctly explained the scattering of radiation from a free electron and Klein[19] treated absorption and stimulated emission of radiation by an atom without quantizing the field. Bloembergen[20] has treated nonlinear optics in a completely semiclassical manner. Most linear optical effects can be treated similarly. However, the radiation field needs to be quantized in dealing with such problems as the Lamb shift and the spontaneous emission of radiation (see Scully and Sargent[21]). Haroche,[22] von Foerster,[23] and Stenholm[24] have recently applied quantum field theory to approximating solutions to the two-level problem.

Most of the other two-level problem research with which we are familiar is couched in terms of a semiclassical formulation: The material system is described by a time-dependent Schrödinger equation in which the effect of the radiation is represented by an effective Hamiltonian consistent with Maxwell's equations. The advantage of the two-level or two-state system is that the wavefunction can be expressed in terms of two coupled ordinary first-order differential equations. The general solution of these equations is a linear combination of two Floquet modes whose functional forms are known. Nevertheless, it is frequently difficult to obtain accurate solutions. No closed-form solution has ever been found. Thus many techniques have been proposed for approximating the solutions. The best procedure to use depends upon the strength and frequency of the radiative field and the properties of the material system.

With the availability of powerful analog and digital computers, one might very well consider the possibility of directly integrating the two coupled first-order differential equations. This possibility is discussed in Section I.D.

Autler and Townes[4] transformed the two-level problem into a matrix eigenvalue–eigenvector problem where each of the infinite dimensions of the matrix corresponds to the amplitude of a Fourier component. Then Shirley[6,25] converted the matrix eigenvalue–eigenvector problem into the analog of a time-independent (or static) Schrödinger equation. Most of the current research uses this approach. Of course, one has the choice of solving this "Schrödinger equation" in terms of either wavefunctions or density matrices. The present article is couched in terms of *wavefunctions*. Different mathematical techniques are required to solve directly for the density matrices. Feynman, Vernon, and Hellwarth[26] show how the equation of motion of the density matrix allows an analogy of the two-level system to the easily visualized problem of a constant length magnetic moment vector rotating in space under the influence of a sinusoidally oscillating classical magnetic field. McGurk, Schmaltz, and Flygare[15] have recently studied ways of approximately solving for the two-level system's density matrix. These studies were applied to transient phenomena in infrared and microwave spectroscopy. In Stroud and Jaynes'[27] novel semiclassical treatment of spontaneous emission and the Lamb shift, approximation techniques are used that also apply to finding solutions for the two-level system's density matrix. Stenholm[11,12,13,14] has used a density matrix approach in finding the Bloch–Siegert shift for a two-level system.

A. Formulation of the Problem

Before discussing the solutions of the model problem of a two-level material system in a sinusoidally oscillating field, it is necessary to write the equations that must be satisfied. We take the time-dependent Hamiltonian for the material system to be

$$H(\mathbf{r},t) = H^0(\mathbf{r}) + 2FV(\mathbf{r})\cos(\omega t) - \frac{i\hat{\gamma}}{2} \tag{1.1}$$

There are only two eigenstates for the unperturbed Hamiltonian $H^0(\mathbf{r})$ so that

$$H^0(\mathbf{r})\psi_j(\mathbf{r}) = W_j\psi_j(\mathbf{r}) \qquad (j=a,b) \tag{1.2}$$

For convenience, let $W_b > W_a$. In (1.1) F is the field strength parameter, and the field interacts with the system through the interaction operator $V(\mathbf{r})$. The operator $\hat{\gamma}$ introduces damping constants into the time-dependent wavefunctions. It is defined by

$$\hat{\gamma}\psi_j(\mathbf{r}) = \gamma_j\psi_j(\mathbf{r}) \qquad (j=a \text{ or } b)$$

where γ_a and γ_b are scalar constants. Phenomenologically $\hat{\gamma}$ takes into account transitions away from levels $\psi_a(\mathbf{r})$ and $\psi_b(\mathbf{r})$ that are not explicitly considered.* These "away transitions" could be, for instance, spontaneous relaxation of $\psi_a(\mathbf{r})$ or $\psi_b(\mathbf{r})$ into some lower state that is not explicitly considered in this treatment. Silverman and Pipkin[7] were the first to use $\hat{\gamma}$ in the two-level problem. Unfortunately, $\hat{\gamma}$ *does not* take into account spontaneous transitions from the state b to the state a. Indeed, we do not know how to construct a suitable semiclassical operator to be added to $H(\mathbf{r},t)$ corresponding to $b \to a$ spontaneous transitions.† The difficulty is that such an operator would be required to leave the total population in states a and b unchanged. This seems to be an inherent defect in the semiclassical treatment of the interaction of radiation and matter.

The time evolution of the two-level system is determined by the Schrödinger equation:

$$i\dot{\Psi}(\mathbf{r},t) = H(\mathbf{r},t)\Psi(\mathbf{r},t) \tag{1.3}$$

*See, for example, Weisskopf and Wigner[28] or Maitland and Dunn[29] (Chapter 3, Section 3).

†See von Foerster[23] who does this by using quantum field theory.

Throughout this article, we use atomic units so that $\hbar$ is set equal to unity; also, a dot over a function denotes differentiation with respect to either time or the reduced time. Since $\psi_a(\mathbf{r})$ and $\psi_b(\mathbf{r})$ span the Hilbert space of $H^0(\mathbf{r})$, the wavefunction for the two-level system has the form

$$\Psi(\mathbf{r},t)=\eta_a(t)\psi_a(\mathbf{r})+\eta_b(t)\psi_b(\mathbf{r}) \tag{1.4}$$

By substituting (1.4) into (1.3), multiplying by either $\psi_a(\mathbf{r})^*$ or $\psi_b(\mathbf{r})^*$, and integrating over $\mathbf{r}$, the following equations for $\eta_a(t)$ and $\eta_b(t)$ are obtained:

$$i\dot{\eta}_a(t)=W_a\eta_a(t)-i\frac{\gamma_a}{2}\eta_a(t)+2FV_{aa}\cos\omega t\eta_a(t)$$
$$+2FV_{ab}\cos\omega t\eta_b(t) \tag{1.5}$$

$$i\dot{\eta}_b(t)=W_b\eta_b(t)-i\frac{\gamma_b}{2}\eta_b(t)+2FV_{ab}\cos\omega t\eta_a(t)$$
$$+2FV_{bb}\cos\omega t\eta_b(t) \tag{1.6}$$

where

$$V_{ij}\equiv\langle\psi_i(\mathbf{r})|V(\mathbf{r})|\psi_j(\mathbf{r})\rangle$$

Here *all* of the V_{ij} are assumed to be *real* numbers.

The equations for $\eta_a(t)$ and $\eta_b(t)$ are considerably simplified by replacing $\eta_a(t)$ and $\eta_b(t)$ by the new variables $a(t)$ and $b(t)$ defined by

$$\eta_a(t)=a(t)\exp\left[-iW_at-\frac{\gamma_at}{2}-\frac{2iFV_{aa}}{\omega}\sin\omega t\right] \tag{1.7}$$

$$\eta_b(t)=b(t)\exp\left[-iW_at-\frac{\gamma_at}{2}-\frac{2iFV_{aa}}{\omega}\sin\omega t\right] \tag{1.8}$$

Let the reduced time be $\tau=\omega t$ and define the following reduced parameters:

$$\epsilon=\frac{(W_b-W_a)}{\omega},\qquad \alpha=\frac{FV_{ab}}{\omega}$$
$$\beta=\frac{F(V_{bb}-V_{aa})}{\omega},\qquad \delta=\frac{1}{2}\frac{(\gamma_b-\gamma_a)}{\omega} \tag{1.9}$$

In terms of the new independent variable τ, Eqs. 1.5 and 1.6 become

$$\dot{a}(\tau) = -2i\alpha\cos\tau b(\tau) \tag{1.10}$$

$$\dot{b}(\tau) = -i(\epsilon - i\delta + 2\beta\cos\tau)b(\tau) - 2i\alpha\cos\tau a(\tau) \tag{1.11}$$

The dots in (1.10) and (1.11) mean differentiation with respect to τ.

Equations 1.10 *and* 1.11 *are the working equations for the remainder of this article*. Note that they involve only four independent parameters α, β, ϵ, and δ; whereas, Eqs. 1.5 and 1.6 involve eight independent parameters.

In most of the previous articles on the two-level system, both β and δ have been neglected ($\beta = 0 = \delta$). The only previous use of δ with which we are familiar was Silverman and Pipkin[7] who took $\beta = 0$ and $\delta \neq 0$. The only usage of β that we have seen is Winter,[5] Ashby,[30] and Series,[31] all of whom took $\beta \neq 0$ and $\delta = 0$. However, Meadows[32] developed a numerical technique that is applicable to general systems of linear first-order homogeneous ordinary differential equations with periodic coefficients. The Meadows method is therefore more general than our treatment.

B. The Functional Form of the Exact Solutions

Equations 1.10 and 1.11 are the first-order coupled linear ordinary differential equations with periodic coefficients. Floquet[33] studied the solutions of the general nth order linear differential equation with periodic coefficients and Poincaré[34] investigated the practical construction of such solutions. Moulton[35,36] completely describes the solutions for n first-order coupled, linear, homogeneous differential equations with periodic coefficients. From these studies, the functional form of the solution to the two-level system is known. In Section II we show that the general solution to (1.10) and (1.11) is

$$a(\tau) = c_1\exp(-i\mu_1\tau)\phi_{a1}(\tau) + c_2\exp(-i\mu_2\tau)\phi_{a2}(\tau) \tag{1.12}$$

$$b(\tau) = c_1\exp(-i\mu_1\tau)\phi_{b1}(\tau) + c_2\exp(-i\mu_2\tau)\phi_{b2}(\tau) \tag{1.13}$$

where the general solution involves an arbitrary linear combination of two normal (or Floquet) modes. The jth Floquet mode in turn involves a characteristic constant (the "characteristic exponent") μ_j and functions $\phi_{ij}(\tau)$ where $i = a, b$. The $\phi_{ij}(\tau)$'s have the same periodicity as the periodic coefficients in (1.10) and (1.11):

$$\phi_{ij}(\tau + 2\pi) = \phi_{ij}(\tau); \qquad \begin{array}{l} i = a, b \\ j = 1, 2 \end{array}$$

The arbitrary constants c_1 and c_2 can be chosen so as to make the solution given by (1.12) and (1.13) obey any initial conditions.* Depending on the values of α, β, δ and ϵ, the value of μ_1 may or may not differ from μ_2 by an integer, or zero.

At first glance the solution written down in (1.12) and (1.13) bears little resemblance to the standard Rayleigh–Schrödinger time-dependent perturbation theory solution to (1.10) and (1.11). Using standard techniques [see Schiff[37] (Section 29) or any other elementary quantum mechanics text], secular terms (terms that are linear or not periodic in the time variable) occur. For example let the terms in (1.10) and (1.11) that depend on α and β be first-order perturbations. This is equivalent to taking all terms dependent on the field strength as perturbations. The standard time-dependent perturbation corresponds to expressing $a(\tau)$ and $b(\tau)$ as a double power series in α and β. If $\epsilon=1$ and $\delta=0$, the solution that satisfies the initial condition $a(0)=1$ and $b(0)=0$ has the double power series (correct through the terms that are linear in either α or β)

$$\begin{aligned} a(\tau) &= 1+\cdots \\ b(\tau) &= \frac{\alpha}{2}\left[\exp(-i\tau)-\exp(i\tau)\right]-i\alpha\tau\exp(-i\tau)+\cdots \end{aligned} \tag{1.14}$$

Note the secular term $-i\alpha\tau\exp(-i\tau)$ in the first-order correction to $b(\tau)$. It, as well as all other secular terms that appear in the higher order perturbation corrections, comes from not putting the characteristic exponents where they belong, that is, as arguments of an exponential function. To be specific, let each c_j in (1.12) and (1.13) be respectively replaced by K_j to denote that we are speaking about a solution satisfying specific initial conditions. The μ_j's, the $\phi_{ij}(\tau)$'s, and the K_j's are all expandable in a perturbation series:

$$\begin{aligned} \mu_j &= \mu_j^{(0)}+\mu_j^{(1)}+\cdots, && j=1,2 \\ \phi_{ij}(\tau) &= \phi_{ij}^{(0)}(\tau)+\phi_{ij}^{(1)}(\tau)+\cdots, && i=a,b;j=1,2 \\ K_j &= K_j^{(0)}+K_j^{(1)}+\cdots, && j=1,2 \end{aligned} \tag{1.15}$$

The exponentials may be expanded by

$$\exp x = 1+x+\frac{x^2}{2!}+\frac{x^3}{3!}+\cdots \tag{1.16}$$

*Throughout this article we use lower case "c" to indicate an arbitrary constant. Either capital "K" or capital "C" will be used to mean a nonarbitrary constant that is either known or must be determined.

Using the expansions (1.15) and (1.16) in (1.13) yields

$$\begin{aligned} b(\tau) = &\left(K_1^{(0)} + K_1^{(1)} + \cdots\right)\left(\phi_{b1}^{(0)}(\tau) + \phi_{b1}^{(1)}(\tau) + \cdots\right) \\ &\times \left(1 - i\left(\mu_1^{(0)} + \mu_1^{(1)} + \cdots\right)\tau + \cdots\right) \\ &+ \left(K_2^{(0)} + K_2^{(1)} + \cdots\right)\left(\phi_{b2}^{(0)}(\tau) + \phi_{b2}^{(1)}(\tau) + \cdots\right) \\ &\times \left(1 - i\left(\mu_2^{(0)} + \mu_2^{(1)} + \cdots\right)\tau + \cdots\right) \end{aligned} \tag{1.17}$$

Equation 1.17 is to be compared to (1.14) to note that the secular terms arise merely because the characteristic exponent has not been left where it belongs: as the argument of an exponential function. The expression for $a(\tau)$ that is analogous to (1.17) also contains secular terms for the same reason. Knowledge of the exact solution's functional form, therefore, allows us to obtain approximate solutions not having secular divergences. In fact, it should be reemphasized that the occurrence of secular terms is just an artifice of the standard perturbation treatment. Langhoff, Epstein, and Karplus[38] modified the standard time-dependent perturbation treatment for general many-state quantum mechanical systems so that secular terms no longer occur. As explained in Section V.D, their procedure reduces to our Floquet modes for two-state problems. It is important to note that their use of *adiabatic turn-on of the perturbation insures that the solution be a particular Floquet mode*.

The importance of the Floquet modes is that they are the steady-state solutions required for the determination of many properties such as the index of refraction. They are the solutions that remain after the transients have disappeared. It is helpful to think of the Floquet modes in terms of a classical damped forced harmonic oscillator. The Floquet modes are the solutions that have no components in the homogeneous (or zero driving force) solutions. Actually, the classical damped forced harmonic oscillator is a good model for optical properties. Indeed, Lorentz[89] and Drude[88] treat optical phenomena by assuming that atoms are composed of harmonically bound damped electrons. Both the classical and quantum mechanical versions of the Lorentz–Drude optics are given by Hirschfelder, Curtiss, and Bird[39] (see pp. 881 and 956).

C. Techniques Considered

In Table I each of seven different techniques for solving the two-level problem is described, and its range of applicability is given, along with the section number where it can be found in this article and references to related research. The Autler–Townes (AT)′ is the simplest and the easiest to comprehend, but it has the disadvantages that δ must be zero, and in order for the continued fractions to converge rapidly the field strength must be weak. The Meadows–Ashby (MA)′ procedure is a useful trick for determining the values of μ. The other five techniques involve different types of perturbation methods: The Sen Gupta (SG)′ uses nondegenerate Rayleigh–Schrödinger; the Shirley–Series (SSe)′ and the (DRS) involve degenerate Rayleigh–Schrödinger; the Salwen–Winter–Shirley (SWS)′ involves partitioning techniques; and the (SPT) is a singular perturbation expansion. All of these treatments are based upon the Floquet–Poincaré theorems that have determined the functional form of the exact solutions and provided a great deal of information regarding the properties of these solutions.

TABLE I.
Location, Description, and Range of
Applicability of Each of the Techniques

Section III—The Autler–Townes (AT)′ tridiagonal technique
Solutions are found for the values of μ and the amplitudes of the Fourier components in terms of continued fractions. The continued fractions converge rapidly when the field strength is weak. The formulation is very simple and is applicable even in the vicinity of the principal resonance, $\epsilon \approx 1$.
Applicability: $\beta = 0$; α, δ, and ϵ arbitrary but is only useful when α is less than unity.
Related Research: Autler–Townes.[4]
Section IV—The Meadows–Ashby (MA)′ method for evaluating μ
The values of μ are expressed in terms of the determinant of an infinite-dimensional unsymmetric tridiagonal matrix that is real if $\delta = 0$ and complex if $\delta \neq 0$. This matrix must be evaluated numerically.
Applicability: α, β, δ, and ϵ arbitrary.
Related Research: Meadows[32]; Ashby.[30]
Section V—The Sen Gupta (SG)′ technique using nondegenerate Rayleigh–Schrödinger perturbation theory
The Floquet ϕ's are expressed in terms of Fourier components. The amplitudes of these Fourier components and the values of μ are expanded in powers of the field strength. If $(\epsilon - i\delta)$ is equal to, or almost equal to, an integer n_ϵ, the Floquet modes and their values of μ can be determined by this technique through the $(n_\epsilon - 1)$st order if $\delta \neq 0$, or through the $(2n_\epsilon - 1)$st order if $\delta = 0$. If $(\epsilon - i\delta)$ does not have a value close to an integer, the power series can be determined up to a very high order.
Applicability: α and β much less than one. If the calculations are to be carried out to an order of the field strength greater than $n_\epsilon - 1$, then $(\epsilon - i\delta)$ should not be close to the integer n_ϵ.

TABLE I. (*Continued*)

Related Research: Sen Gupta[40,41,42]; Langhoff, Epstein, and Karplus[38]; Sambe[43]; Young, Deal, and Kestner[44]; Young and Deal[45]; Okuniewicz[46,47]; Hicks, Hess, and Cooper[48]; and Salzman.[49]

Section VI—The Salwen–Winter–Shirley (SWS)′ partitioning perturbation technique with $\delta=0$

The amplitudes of the Fourier components are expanded in powers of the field strength as in the (SG)′ method. However, the Certain–Hirschfelder partitioning procedure is used in order to treat the resonance regions where ϵ is close to an integer n_ϵ. For the principal resonance where $n_\epsilon=1$, the values of μ (accurate through the third-order in the field strength) are given by a quadratic equation; the amplitudes of the Fourier components (accurate through the first-order) are then known. Numerical results are presented in both tabular and graphical form for this case. If $\delta\neq 0$, the (SWS)′ method might still be used by replacing ϵ by $(\epsilon-i\delta)$ in the quadratic equation for μ (after the matrix elements have been determined) and in the equations for the $|\chi>$; however, there would be no guarantee as to the accuracy of the results.

Applicability: $\delta=0$; both α and β much less than one. For large values of ϵ, high orders of perturbation are required.

Related Research: Salwen[50]; Winter[5]; Shirley[6,25]; Chang and Stehle[9]; Bloch and Siegert[2]; Pegg and Series[51]; Stenholm[11,14,24]; Cohen-Tannoudji, DuPont-Roc, and Fabre[52]; Rabi[1]; Moulton, Buchanan, Pouck, Griffin, Longley, and MacMillan[35]; and Stevenson.[3]

Section VII—The Shirley–Series (SSe)′ degenerate Rayleigh–Schrödinger expansion in powers of $(\epsilon-i\delta)$

New functions $F(\tau)$ and $G(\tau)$ are defined in terms of $a(\tau)$ and $b(\tau)$. The values of μ and the amplitudes of the Fourier expansion coefficients of $F(\tau)\exp(iu\tau)$ and $G(\tau)\exp(iu\tau)$ are expanded in powers of $(\epsilon-i\delta)$ by using a degenerate Rayleigh–Schrödinger procedure. This technique is useful for considering problems involving high frequency fields.

Applicability: Both ϵ and δ much less than one; α or β much greater than one.

Related Research: Shirley[6]; Series.[31]

Section VIII—The degenerate Rayleigh–Schrödinger (DRS) technique for considering high-frequency fields

This is a perturbation technique for considering those high-frequency field problems in the parameter range (δ and ϵ are both smaller than either α or β) where the (SSe)′ method does not apply. The (DRS) formulation uses the same Floquet effective Hamiltonian as was used in the (SG)′ and (SWS)′ procedures. However, the (DRS)′ makes a different resolution into the zeroth- and first-order Hamiltonians.

Applicability: α, β, δ, and ϵ are *all* much smaller than unity; both δ and ϵ are much smaller than either α or β.

Section IX–The $(1/\epsilon)$ singular perturbation expansion (SPT) technique

The (SPT) technique is useful in considering problems where the energy gap between the two quantum states is large compared to $\hbar\omega$ so that ϵ is large compared to both unity and the parameters α, β, and δ. In this case the differential equations for the ratios $b(\tau)/a(\tau)$ and

TABLE I. (*Continued*)

$a(\tau)/b(\tau)$ are "stiff" and correspond to singular perturbations. The series expansions of these ratios in powers of $(1/\epsilon)$ are atypical "outer" solutions. The "outer" solution for one of these ratios corresponds to the "inner" solution for the other ratio. Thus, the two normal Floquet modes are determined from the two power series.

Applicability: ϵ much larger than 1, α, β, and δ.

One of the best ways to determine μ involves the direct numerical integration of (1.10) and (1.11) to obtain the "fundamental" solutions at the time $\tau = 2\pi$. The "fundamental" solutions are defined as those particular solutions (indicated by a caret) that correspond to the starting conditions: $\hat{a}_1(0) = 1$, $\hat{b}_1(0) = 0$ and $\hat{a}_2(0) = 0$, $\hat{b}_2(0) = 1$. The values of μ corresponding to the two linearly independent Floquet modes are then obtained by solving Eqs. 2.7 and 2.8. Since this procedure is so simple, there was no need to devote a whole section to it! This formalism is derived in connection with the proof of Theorem 1 in Section II.A.

The (DRS) and the (SPT) techniques are novel. In all of the other techniques, we have modified the treatments given by the named authors. The modifications are indicated by a prime, as for example (AT)′. In the case of the (AT)′ and (MA)′ our contribution is the trivial generalization to include nonvanishing values of δ. In the (SG)′, (SWS)′, and (SSe)′, the procedures that we use are quite different from the original authors. Wherever possible, we have labeled techniques by the names of a few of the authors who have used somewhat similar procedures. We feel that it is important to give credit to the pioneers in this research area. However, there are a great many other people who have made important contributions to these techniques; we have discussed the work of some of them in the last part of some of our sections. We hope that we have not offended those people whose research we have inadvertently neglected to include.

D. Numerical Procedures

Numerical procedures are not discussed in this article. However, digital, analog, and hybrid computers can be extremely useful in obtaining numerical solutions to specific two-level problems. Let us consider how these machines might be used.

1. *Analog Computers*

There are a number of excellent analog computers on the market such as the Applied Dynamics Company's 256 and the Electronic Associates, Inc. TR680. The present generation of analog computers can calculate the solutions of suitable equations (or coupled equations) to within an error of between 0.1 and 1%. One advantage of the analog computers is that they provide very convenient graphical output of the solutions. Analog computers would be ideal for directly integrating Eqs. 1.10 and 1.11 to obtain numerical solutions to the two-level problem. However, in order to obtain solutions for the Floquet normal modes, it is necessary to use the "fundamental" solution formalism (derived in connection with the proof of Theorem 1 in Section II.A) and solve Eqs. 2.7 and 2.8. See Section III.B for numerical values of the μ obtained by this procedure, but using a digital computer.

2. *Digital Computers*

Of course, the digital computers can also integrate Eqs. 1.10 and 1.11 directly to obtain the "fundamental" solutions at $\tau = 2\pi$ that are then used in accordance with (2.7) and (2.8) to obtain values of μ with whatever precision is required. This procedure was used to determine the values of μ given in Section III.B.

Digital computers can also carry out iterative algebraic manipulations, such as determining the solutions to our (SG)′ Rayleigh–Schrödinger perturbation equations to arbitrarily high orders. There are examples in the literature of perturbed simple harmonic oscillators where computer solutions have been obtained through the 150th order.

There are a number of other ways in which digital computers might be used in connection with the two-level problem. For example, they might be used to evaluate the eigenvalues (the characteristic constants μ) and the eigenvectors (the Fourier components of the ϕ's) of the infinite-dimensional matrices: M [given in (2.36)], M' [given in (3.6)], and $\underline{\Delta}$ [given in (4.8)]. All three of these matrices are real if $\delta = 0$ and complex non-Hermitian if δ does not vanish. M is a symmetric quintidiagonal band matrix; M' is a symmetric tridiagonal matrix; and $\underline{\Delta}$ is an unsymmetrical quintidiagonal band matrix. In practice, the eigenvalues and eigenvectors would be calculated with the infinite matrix truncated so that it had a large but finite dimensionality. The convergence of the eigenvalues and eigenvectors with the dimensionality can be determined by repeating the calculations with matrices of large dimensionality.

In the (MA)′ method of Section IV it is necessary to evaluate the determinant of either $\underline{\Delta}_1(0)$ [given in (4.2)] or $\underline{\Delta}_2(0)$ [given in (4.3)]. Both of these matrices are unsymmetric quintidiagonal infinite-dimensional, and if $\delta = 0$, they are real; but if δ does not vanish they are complex non-Hermitian.

There are a large number of excellent books on the algebra and algorithms used in the computation of eigenvalues, eigenvectors, and determinants, for example, Wilkinson[53] and Conté and de Boor.[54] However, the best book for our purposes is Wilkinson and Reinsch,[55] *Handbook for Automatic Computation*. Wilkinson and Reinsch[55] contains a thorough discussion of the best procedures to use together with the nature of the matrix and the properties that are to be computed. For each of these procedures, ALGOL programs are given. These procedures have also been translated into FORTRAN IV that are called either EISPACK or IMSL. The EISPACK programs are currently available in single precision and very shortly they will also be available in double precision. For information regarding EISPACK, write to: Dr. Barton S. Garbow, Applied Mathematics Division, Argonne National Laboratory, 9700 Cass Avenue, Argonne, Illinois 60439. For information regarding IMSL, write to: International Mathematics Software Library, 7500 Bellaire Boulevard, Houston, Texas 77036. We list the currently available EISPACK program sequences for doing the sorts of computations which might be useful in connection with the two-level problem. Here the EISPACK codes are followed, in parenthesis, by a reference to Wilkinson and Reinsch[55] such as (II/4, p. 241) meaning Volume II, contribution 4, page 241.

(1) Currently there is no EISPACK program for calculating the determinant of a band matrix. However, (I/6, p. 70) gives a procedure and ALGOL program for putting a general band matrix into upper triangular form. The value of the determinant is then the product of the diagonal elements of the upper triangular form. This procedure should be easy to translate into FORTRAN.

(2) To determine all of the eigenvalues and eigenvectors of a *real symmetric tridiagonal matrix* use IMTQL2 (II/4, p. 241). This uses the QL algorithm.

(3) To determine all of the eigenvalues and eigenvectors of a *real symmetric band matrix* use BANDR (II/8, p. 273) followed by IMTQL2 (II/4, p. 241). The BANDR uses orthogonal transformations to reduce the symmetric band matrix to a symmetric tridiagonal matrix.

(4) To determine all of the eigenvalues and eigenvectors of a real nonsymmetric matrix requires a whole sequence of routines: BALANC (II/11, p. 315)→ELMHES (II/13, p. 339)→ELTRAN (II/15, p. 372) →HQR2 (II/15, p. 372)→BALBAK (II/11, p. 315). Here BALANC balances the matrix by using diagonal similarity transformations to reduce the norm of the matrix. ELMHES reduces the real general matrix to upper Hessenberg form (zeros up to the stripe just below the diagonal) by real stabilized elementary similarity transformations. ELTRAN accumulates the transformations in ELMHES. HQR2 uses the QR algorithm to determine the eigenvalues and eigenvectors of an upper Hessenberg matrix.

Finally, BALBAK back-transforms the eigenvectors, which had been transformed by BALANC, to correspond to the original form of the matrix.

(5) There do not seem to be any EISPACK programs that take advantage of the large number of zeros in a banded non-Hermitian complex matrix. Thus, the available programs apply to a general complex matrix. To determine all of the eigenvalues and eigenfunctions for a *general complex matrix* requires the sequence: CBAL (II/11, p. 315) →COMHES (II/13, p. 339)→COMLR2 (II/15, p. 372)→CBABK2 (II/11, pp. 315). Here CBAL balances a complex general matrix in much the same manner as BALANC does for a real matrix. COMHES reduces the general complex matrix to upper Hessenberg form much like ELMHES. COMLR2 uses a modified LR method to compute the complex eigenvalues and eigenvectors of the complex upper Hessenberg matrix. Finally, CBABK2 back-transforms the eigenvectors (which had been transformed by CBAL) to correspond to the original form of the matrix.

3. *Hybrid Computers*

The hybrid computers are analog computers linked to digital computers. Thus, they can take advantage of the capabilities of both. The hybrid computers can solve large systems of coupled differential equations. Thus, they should be most useful in treating the many-level problem where the time-dependent field perturbs a large number of interacting quantum states.

II. THE MATHEMATICAL MACHINERY

In order to derive the seven methods of solution that are considered in Sections III through IX, it is necessary to develop three types of mathematical machinery. First, we need a set of Floquet theorems that determine important properties of the exact solutions. Then we need to set the stage for the nonlinear quotient techniques, the matrix methods, and the "static" formulations.

A. The Floquet Theorems

Floquet[33] and Poincaré[34] studied the properties of sets of coupled homogeneous first order ordinary differential equations with periodic coefficients.* We are interested in two such coupled equations involving

*These n coupled equations are equivalent to an nth order homogeneous differential equation with periodic coefficients. For detailed discussion of the properties of the coupled equations, see Moulton[36] (Chapter 17) or Dion[56] (Appendix A). For less detailed discussions see Ince[57] (Section 15.7), Margenau and Murphy[58] (p. 80), or Brillouin.[59,60]

two dependent variables,

$$\begin{aligned} \dot{a}(\tau) &= -i\theta_{11}(\tau)a(\tau) - i\theta_{12}(\tau)b(\tau) \\ \dot{b}(\tau) &= -i\theta_{21}(\tau)a(\tau) - i\theta_{22}(\tau)b(\tau) \end{aligned} \tag{2.1}$$

where the $\theta_{ij}(\tau)$'s are continuous finite functions that are periodic with periodicity 2π

$$\theta_{ij}(\tau+2\pi)=\theta_{ij}(\tau) \qquad (i,j=1,2)$$

Clearly, Eqs. 2.1 are the same as Eqs. 1.10 and 1.11 where

$$\begin{aligned} &\theta_{11}(\tau)=0 \\ &\theta_{12}(\tau)=\theta_{21}(\tau)=2\alpha\cos\tau \\ &\theta_{22}(\tau)=\epsilon+2\beta\cos\tau-i\delta \end{aligned} \tag{2.2}$$

Floquet and Poincaré derived nine theorems that define the functional form of the exact solutions to these equations and also provide a set of conditions that the exact solutions must satisfy.

Theorem 1. If in (2.1) the time-dependent functions $\theta_{ij}(\tau)$ are periodic with period 2π, then the solutions to Eqs. 2.1 are arbitrary linear combinations of two normal (or Floquet) modes each of which involves both a characteristic constant (the "characteristic exponent") μ and a function $\phi(\tau)$. The $\phi(\tau)$ has the same periodicity as the $\theta_{ij}(\tau)$

$$\phi(\tau+2\pi)=\phi(\tau)$$

There are three and only three possible forms* that the general solutions to (2.1) can take.

Form I. Here $\mu_1 \neq \mu_2 + n$ where n is any integer or zero.
In this instance, one particular solution is

$$a_1(\tau)=\exp(-i\mu_1\tau)\phi_{a1}(\tau); \qquad b_1(\tau)=\exp(-i\mu_1\tau)\phi_{b1}(\tau) \tag{2.3a}$$

The second linearly independent solution is

$$a_2(\tau)=\exp(-i\mu_2\tau)\phi_{a2}(\tau); \qquad b_2(\tau)=\exp(-i\mu_2\tau)\phi_{b2}(\tau) \tag{2.3b}$$

When $\mu_1=\mu_2+n$, two forms are possible.

*Note that most statements of Floquet's theorem appearing in the literature are incomplete; that is, they only allow for the first two of the three forms. However, Moulton[36] and Dion[56] give a very complete statement.

Form II. Here $\mu_1=\mu_2+n$ where n is any integer or zero. The two linearly independent Floquet modes are

$$\begin{aligned} a_1(\tau)&=\exp(-i\mu_1\tau)\phi_{a1}(\tau); \qquad b_1(\tau)=\exp(-i\mu_1\tau)\phi_{b1}(\tau) \\ a_2(\tau)&=\exp(-i\mu_1\tau)\phi_{a2}(\tau); \qquad b_2(\tau)=\exp(-i\mu_1\tau)\phi_{b2}(\tau) \end{aligned} \tag{2.4}$$

Form III. Here again, $\mu_1=\mu_2+n$ where n is any integer or zero. The linearly independent Floquet modes for Form III are

$$a_1(\tau)=\exp(-i\mu_1\tau)\phi_{a1}(\tau); \qquad b_1(\tau)=\exp(-i\mu_1\tau)\phi_{b1}(\tau) \tag{2.5}$$

$$\begin{aligned} a_2(\tau)&=\exp(-i\mu_1\tau)[\tau\phi_{a1}(\tau)+\phi_{a2}(\tau)]; \\ b_2(\tau)&=\exp(-i\mu_1\tau)[\tau\phi_{b1}(\tau)+\phi_{b2}(\tau)] \end{aligned} \tag{2.6}$$

Proof of Theorem 1. The periodicity of the $\theta_{ij}(\tau)$'s and the relation

$$\frac{d}{d(\tau+2\pi)}=\frac{d}{d\tau}$$

require that if $a(\tau), b(\tau)$ is any solution to (2.1), then so is $a(\tau+2\pi), b(\tau+2\pi)$. If there exists a solution of the Floquet form

$$a(\tau)=\exp(-i\mu\tau)\phi_a(\tau) \quad \text{and} \quad b(\tau)=\exp(-i\mu\tau)\phi_b(\tau)$$

where μ is constant and the ϕ's are periodic, then $a(\tau+2\pi)$ and $b(\tau+2\pi)$ also form a solution. The only effect of changing variables is to multiply the original solution by the constant $\exp(-2i\pi\mu)$.

There are always two "fundamental" solutions to (2.1) that satisfy

$$\begin{aligned} \hat{a}_1(0)&=1; \qquad \hat{b}_1(0)=0 \\ \hat{a}_2(0)&=0; \qquad \hat{b}_2(0)=1 \end{aligned} \tag{2.7}$$

They are linearly independent and their determinant $(\hat{a}_1(\tau)\hat{b}_2(\tau)-\hat{a}_2(\tau)\hat{b}_1(\tau))$ never vanishes for finite values of τ. Therefore, *all* solutions to (2.1) (satisfying arbitrary boundary conditions) can be expressed as a linear combination of the "fundamental solutions." Thus,

$$\begin{aligned} \exp(-i\mu\tau)\phi_a(\tau)&=C_1\hat{a}_1(\tau)+C_2\hat{a}_2(\tau) \\ \exp(-i\mu\tau)\phi_b(\tau)&=C_1\hat{b}_1(\tau)+C_2\hat{b}_2(\tau) \end{aligned}$$

If the ϕ's are to be periodic, then by (2.7) the constants C_1 and C_2 must satisfy

$$\begin{pmatrix} \hat{a}_1(2\pi)-s & \hat{a}_2(2\pi) \\ \hat{b}_1(2\pi) & \hat{b}_2(2\pi)-s \end{pmatrix}\begin{pmatrix} C_1 \\ C_2 \end{pmatrix}=0 \tag{2.8a}$$

where $s=\exp(-2i\pi\mu)$. A nontrivial solution exists only when

$$s^2-s(\hat{a}_1(2\pi)+\hat{b}_2(2\pi))+\hat{a}_1(2\pi)\hat{b}_2(2\pi)-\hat{a}_2(2\pi)\hat{b}_1(2\pi)=0 \tag{2.8b}$$

By Theorem 3 of this section, the factor in (2.8b) proportional to s^0 must always be finite and nonzero. The factor linear in s must, therefore, always be finite, although it could vanish. Thus, s is always finite and nonzero. Consider these two possibilities.

1. s is nondegenerate. In this case, there are two linearly independent solutions for C_1 and C_2 corresponding to two linearly independent solutions of the Floquet form; that is, the Form I solutions of (2.3a) and (2.3b). The two values of s correspond to two values of the characteristic exponent μ for which $(\mu_1-\mu_2)$ is never an integer or zero.

2. s is doubly degenerate. Because of the degeneracy, values of μ satisfying $s=\exp(-2i\pi\mu)$ can at most differ by a real integer. When a degeneracy occurs, there are usually two linearly independent solutions for C_1 and C_2. This case corresponds to the Form II solutions.

The unlikely (but possible) occurrence of only one solution to (2.8) corresponds to the Form III solutions. The two coupled first-order equations for $a(\tau), b(\tau)$ are equivalent to an ordinary second-order linear homogeneous equation for $a(\tau)$. Since one solution for $a(\tau)$ is known, then by the method of Wronskians [see Coddington[61] (Chapter 3, Theorem 9)] another linearly independent solution may easily be generated. When only one solution for C_1 and C_2 exists, this new solution will be of the same form as Eqs. 2.6.

Discussion of Theorem 1. The appearance of the secular terms $\tau \exp(-i\mu\tau)\phi(\tau)$ in Form III solutions has an analogy in a more familiar problem: the problem of two linear, homogeneous, coupled, first-order differential equations with constant coefficients. Here, when the characteristic roots are doubly degenerate, the general solution contains secular terms. (See Coddington[61] or any other elementary differential equations text.)

Note that if the characteristic constants μ_1 and μ_2 were replaced by $\mu_1 + n'$ and $\mu_2 + n'$ where n' is any integer or zero, the Floquet solutions would remain unchanged with the exception that the original $\phi_1(\tau)$ and $\phi_2(\tau)$ would be replaced by $\exp(+in'\tau)\phi_1(\tau)$ and $\exp(+in'\tau)\phi_2(\tau)$ that are still periodic functions with period 2π. Thus, the definition of μ_1 is not unique and *μ_1 may be replaced by μ_1 plus an integer.*

Simple examples of Forms I, II, and III solutions are given in the appendix.

Theorem 2. Let $a_1(\tau), b_1(\tau)$, and $a_2(\tau), b_2(\tau)$ be *any* two solutions of (2.1). Define $D(\tau)$ by

$$D(\tau) = a_1(\tau)b_2(\tau) - a_2(\tau)b_1(\tau) \tag{2.9}$$

then,

$$D(\tau) = D(\tau_0)\exp\left[-i\int_{\tau_0}^{\tau}(\theta_{11}(\tau') + \theta_{22}(\tau'))\,d\tau'\right] \tag{2.10}$$

Discussion of Theorem 2. It is clear from (2.10) that as long as the $\theta_{ij}(\tau)$'s are continuous and bounded for all values of τ, $D(\tau)$ can be zero for finite values of τ only if $D(\tau_0) = 0$. Furthermore, if $D(\tau_0) = 0$, then $D(\tau)$ is zero for all times.

The function $D(\tau)$ plays an important role in the study of (2.1). Evaluation of $D(\tau)$ tells whether or not any two particular solutions to (2.1) form a linearly independent pair of solutions.

Proof of Theorem 2. The proof of this theorem is very simple and does not depend on the periodicity of the $\theta_{ij}(\tau)$'s. Differentiate $D(\tau)$ and use (2.1) to find

$$\dot{D}(\tau) = -i[\theta_{11}(\tau) + \theta_{22}(\tau)]D(\tau)$$

This is a differential equation for $D(\tau)$, which is easily solved to give Eq. 2.10.

Applied to (1.10) and (1.11), Theorem 2 gives

$$D(\tau) = D(\tau_0)\exp[-i(\epsilon - i\delta)(\tau - \tau_0) - 2i\beta(\sin\tau - \sin\tau_0)] \tag{2.11}$$

Theorem 3. The solutions $a_1(\tau), b_1(\tau)$ and $a_2(\tau), b_2(\tau)$ form a linearly independent set of solutions to (2.1) if and only if $D(\tau)$ never vanishes for any finite value of τ.

Proof of Theorem 3. The proof of this theorem is simple. Two solutions are said to be linearly independent if any particular solution obeying arbitrary initial conditions may be expressed as a linear combination of the two solutions *and* neither can be expressed as a constant times the other:

$$a(\tau)=c_1a_1(\tau)+c_2a_2(\tau)$$

$$b(\tau)=c_1b_1(\tau)+c_2b_2(\tau)$$

where c_1 and c_2 are constants. If $a(\tau_0)$ and $b(\tau_0)$ are to be arbitrarily chosen numbers, then the constants c_1 and c_2 are both nonvanishing if and only if

$$D(\tau_0)=a_1(\tau_0)b_2(\tau_0)-b_1(\tau_0)a_2(\tau_0)\neq 0 \tag{2.12}$$

for arbitrary finite τ_0.

Discussion of Theorem 3. It follows from Theorems 2 and 3 that in order to test the linear independence of any two solutions, it is sufficient to evaluate $D(\tau)$ at any finite value of τ. If $D(\tau)$ does not vanish, then the two tested solutions are linearly independent.

Two interesting theorems concerning the characteristic exponents associated with (1.10) and (1.11) are immediately derivable from Theorem 2.

Theorem 4. If the solutions to (1.10) and (1.11) are Form I solutions, then the characteristic exponents are related by

$$\mu_1+\mu_2=\epsilon-i\delta \tag{2.13}$$

Theorem 5. If the solutions to (1.10) and (1.11) are either Form II or Form III solutions, then the characteristic exponent μ_1 is explicitly given by

$$\mu_1=\frac{1}{2}(\epsilon-i\delta) \tag{2.14}$$

Proof of Theorems 4 and 5. To prove Theorems 4 and 5, set $\tau=2\pi$ and $\tau_0=0$ in (2.11) to obtain

$$D(2\pi)=D(0)\exp[-i(\epsilon-i\delta)2\pi] \tag{2.15}$$

The proof consists of a comparison of (2.15) with $D(2\pi)$ as determined by direct substitution of the Theorem 1 expressions for the Form I, II, and III Floquet modes into (2.9). Note that the $\phi(2\pi)=\phi(0)$.

Discussion of Theorems 4 and 5. Theorem 5 gives an exact explicit expression for the characteristic exponent for either Form II or III solutions. Theorem 4 [or (2.13)] is useful for Form I solutions since it expresses one of the characteristic exponents in terms of the other. Thus, only one of the two characteristic exponents needs to be computed directly.

Once an *arbitrary* particular solution to (2.1) has been found, Theorem 6 tells how to write down another linearly independent solution.

Theorem 6. If $\theta_{11}(\tau)$ vanishes,* $\theta_{12}(\tau)=\theta_{21}(\tau)$ and $\theta_{22}(\tau)=g(\tau)-i\delta$ where $\theta_{12}(\tau)$ and $g(\tau)$ are real, continuous, bounded, and periodic functions with period 2π and δ is a real parameter; then if $a_1(\tau)$ and $b_1(\tau)$ are continuous functions that form an arbitrary (nontrivial) particular solution to (2.1), it follows that another *linearly independent* solution to (2.1) is given by

$$\begin{aligned} a_2(\tau) &= -\,\underline{b}_1(\tau)^*\exp\left[-\int_0^{\tau}\left[ig(\tau')+\delta\right]d\tau'\right] \\ b_2(\tau) &= +\,\underline{a}_1(\tau)^*\exp\left[-\int_0^{\tau}\left[ig(\tau')+\delta\right]d\tau'\right] \end{aligned} \tag{2.16}$$

where the "barred" functions, $\underline{a}_1(\tau)$ and $\underline{b}_1(\tau)$, are generated by replacing δ by $-\delta$ wherever it appears in $a_1(\tau)$ and $b_1(\tau)$, respectively. The asterisk in (2.16) denotes the complex conjugate.

For the case of (1.10) and (1.11) where $g(\tau)=\varepsilon+2\beta\cos\tau$, Eq. 2.16 becomes

$$\begin{aligned} a_2(\tau) &= -\,\underline{b}_1(\tau)^*\exp\left[-i\varepsilon\tau-\delta\tau-2i\beta\sin\tau\right] \\ b_2(\tau) &= +\,\underline{a}_1(\tau)^*\exp\left[-i\varepsilon\tau-\delta\tau-2i\beta\sin\tau\right] \end{aligned} \tag{2.17}$$

Proof of Theorem 6. First of all, it is easy to substitute $a_2(\tau)$ and $b_2(\tau)$ into (2.1) and check that the solution given by (2.16) does indeed satisfy (2.1). Then, substituting $a_2(0)$ and $b_2(0)$ into (2.9),

$$D(0)=a_1(0)\,\underline{a}_1(0)^*+b_1(0)\,\underline{b}_1(0)^* \tag{2.18}$$

However, since $a_1(\tau), b_1(\tau)$ is an *arbitrary* particular solution of (2.1), we can regard $a_1(0)$ and $b_1(0)$ as being given constants. Then for small values of τ, if we express $a_1(\tau)$ and $b_1(\tau)$ as power series in τ and use our rules for

* If $\theta_{11}(\tau)$ does not vanish, (2.1) can be transformed into an equation of the same form with the exception that the new $\theta_{11}(\tau)=0$, by replacing $a(\tau)$ by $a'(\tau)=a(\tau)\exp[+i\int\theta_{11}(\tau)d\tau]$ and $b(\tau)$ by $b'(\tau)=b(\tau)\exp[+i\int\theta_{11}(\tau)d\tau]$.

forming $\underline{a}_1(\tau)$ and $\underline{b}_1(\tau)$ to obtain

$$\underline{a}_1(\tau)=\underline{a}_1(0)+\underline{a}_1^{(1)}\tau+\cdots \qquad \text{and} \qquad \underline{b}_1(\tau)=\underline{b}_1(0)+\underline{b}_1^{(1)}\tau+\cdots \tag{2.19}$$

it is clear that $\underline{a}_1(0)=a_1(0)$ and $\underline{b}_1(0)=b_1(0)$. Thus,

$$D(0)=|a_1(0)|^2+|b_1(0)|^2 \tag{2.20}$$

Since $a_1(\tau)$, $b_1(\tau)$ is not a trivial solution, both $a_1(0)$ and $b_1(0)$ cannot be simultaneously zero. We have therefore shown that $D(0)$ does not vanish. According to Theorems 2 and 3 this proves that $a_2(\tau)$, $b_2(\tau)$ is linearly independent of $a_1(\tau)$, $b_1(\tau)$.

Theorem 7. Solutions of (2.1) can never be Form III provided that $\theta_{11}(\tau)$ vanishes, $\theta_{12}(\tau)=\theta_{21}(\tau)$, $\theta_{22}=g(\tau)-i\delta$, and the functions $\theta_{12}(\tau)$ and $g(\tau)$ are real, bounded, continuous, and periodic with period 2π. Thus Form III can never occur for solutions of (1.10) and (1.11).

Proof of Theorem 7. According to Theorem 1, there is *always* one particular solution to (2.1) having the form

$$a_1(\tau)=\exp(-i\mu_1\tau)\phi_{a1}(\tau) \qquad \text{and} \qquad b_1(\tau)=\exp(-i\mu_1\tau)\phi_{b1}(\tau) \tag{2.21}$$

According to Theorem 6 another linearly independent solution to (2.1) is given by

$$\begin{aligned} a_2(\tau)&=-\exp(+i\,\underline{\mu}_1^*\tau)\,\underline{\phi}_{b1}(\tau)^*\exp\left[-\int_0^\tau[ig(\tau')+\delta]\,d\tau'\right] \\ b_2(\tau)&=+\exp(+i\,\underline{\mu}_1^*\tau)\,\underline{\phi}_{a1}(\tau)^*\exp\left[-\int_0^\tau[ig(\tau')+\delta]\,d\tau'\right] \end{aligned} \tag{2.22}$$

Here again the "barred" quantities are related to the "unbarred" by replacing δ wherever it appears in the "unbarred" functions by $-\delta$. Since $\phi_{a1}(\tau)$ and $\phi_{b1}(\tau)$ are periodic, it follows that $\underline{\phi}_{a1}(\tau)$ and $\underline{\phi}_{b1}(\tau)$ must also be periodic. However, in order for $a_2(\tau)$, $b_2(\tau)$ to be a Form III solution linearly independent of $a_1(\tau)$, $b_1(\tau)$ as given by (2.21), it would be necessary for $\underline{\phi}_{a1}(\tau)$ and $\underline{\phi}_{b1}(\tau)$ to contain secular terms linear in τ.

Theorem 8. For solutions to (1.10) and (1.11), $\underline{\mu}_1^*=\mu_1$ and if $\delta=0$, then both μ_1 and μ_2 must be real numbers.

Proof of Theorem 8. Comparing Eq. 2.17 or 2.22 with Eq. 2.3*b*, it follows that for Form I solutions

$$\mu_2 = -\ \underline{\mu}_1^* + \epsilon - i\delta \tag{2.23}$$

But from Theorem 4 or Eq. 2.13, $\mu_2 = -\mu_1 + \epsilon - i\delta$. Thus $\mu_1 = \underline{\mu}_1^*$. If $\delta = 0$, then by the definition of the "barring" process, $\mu_1 = \mu_1^*$ and therefore μ_1 is real. Also, if $\delta = 0$, by (2.13) if μ_1 is real then μ_2 is also real. Now comparing Eq. 2.17 or 2.22 with Eq. 2.4, it follows that for Form II solutions

$$\mu_1 = -\ \underline{\mu}_1^* + \epsilon - i\delta \tag{2.24}$$

But from Theorem 5 or Eq. 2.14, $\mu_1 = \frac{1}{2}[\epsilon - i\delta]$. Thus, again, $\underline{\mu}_1^* = \mu_1$. Of course, from (2.14) it is clear that if $\delta = 0$, then $\mu_1 = \epsilon/2$, a real number.

B. The Nonlinear Quotient Formulation

The problem of finding linearly independent solutions to the linear, homogeneous, first-order coupled system of equations (1.10) and (1.11) is equivalent to finding two particular solutions of a nonlinear equation. Consider the quotient

$$\Phi_1(\tau) = \frac{b_1(\tau)}{a_1(\tau)} \tag{2.25}$$

Simply differentiating $\Phi_1(\tau)$ with respect to τ and utilizing (1.10) and (1.11) gives the equation that $\Phi_1(\tau)$ obeys

$$\dot{\Phi}_1(\tau) = -2i\alpha\cos\tau - i[\epsilon - i\delta + 2\beta\cos\tau]\Phi_1(\tau) + 2i\alpha\cos\tau[\Phi_1(\tau)]^2 \tag{2.26}$$

Once a solution to (2.26) is found, $a_1(\tau)$ is recovered by substituting

$$b_1(\tau) = a_1(\tau)\Phi_1(\tau)$$

into (1.10) to obtain

$$a_1(\tau) = c_1 \exp\left[-i\int 2\alpha\cos\tau\,\Phi_1(\tau)\,d\tau\right] \tag{2.27}$$

where c_1 is an arbitrary constant of integration corresponding to the normalization of the wavefunction: $b_1(\tau)$ is, of course, recovered by using Eq. 2.25.

Similarly, if $\Phi_2(\tau)$ is defined by

$$\Phi_2(\tau) = \frac{a_2(\tau)}{b_2(\tau)} \tag{2.28}$$

$\Phi_2(\tau)$ obeys

$$\dot{\Phi}_2(\tau) = -2i\alpha\cos\tau + i[\epsilon - i\delta + 2\beta\cos\tau]\Phi_2(\tau) + 2i\alpha\cos\tau[\Phi_2(\tau)]^2 \quad (2.29)$$

Once a solution to (2.29) is found, the solutions to (1.10) and (1.11) are given in terms of $\Phi_2(\tau)$

$$\begin{aligned} a_2(\tau) &= \Phi_2(\tau)b_2(\tau) \\ b_2(\tau) &= c_2\exp\left[-i(\epsilon - i\delta)\tau - 2i\beta\sin\tau - i\int 2\alpha\cos\tau\Phi_2(\tau)\,d\tau\right] \end{aligned} \quad (2.30)$$

where c_2 is again an arbitrary constant of integration.

Equations 2.26 and 2.29 are generalized Riccati equations.* Since both equations are first order, the specification of $\Phi_j(\tau)$ requires *one* arbitrary constant of integration. Therefore, there is a particular solution for each and every (in general, complex) value of $\Phi_j(0)$. An important result concerning the particular solutions for $\Phi_j(\tau)$ is the following theorem.

Theorem 9. There are always at least two particular solutions for (2.26) *and* for (2.29) that are periodic (or Floquet) solutions:

$$\Phi_j(\tau + 2\pi) = \Phi_j(\tau)$$

Proof of Theorem 9. The proof of this theorem is simple. It is sufficient to only consider $\Phi_1(\tau)$ since

$$\Phi_2(\tau) = [\Phi_1(\tau)]^{-1}$$

From Floquet's theorem, the general solution to (2.26) is†

$$\Phi_1(\tau) = \frac{b(\tau)}{a(\tau)} = \frac{c_1\exp(-i\mu_1\tau)\phi_{b1}(\tau) + c_2\exp(-i\mu_2\tau)\phi_{b2}(\tau)}{c_1\exp(-i\mu_1\tau)\phi_{a1}(\tau) + c_2\exp(-i\mu_2\tau)\phi_{a2}(\tau)} \quad (2.31)$$

where the c_j's and μ_j's are constants and the ϕ_{ij}'s have period 2π. There are *always* two periodic solutions given by (2.31), and these correspond to

$$c_1 = 1, \qquad c_2 = 0; \qquad \Phi_1(\tau) = \frac{\phi_{b1}(\tau)}{\phi_{a1}(\tau)}$$

$$c_2 = 0, \qquad c_2 = 1; \qquad \Phi_1(\tau) = \frac{\phi_{b2}(\tau)}{\phi_{a2}(\tau)}$$

*See Ince[57] (Section 2.15).

†Recall that by Theorem 7 Form III solutions can never occur for (1.10) and (1.11).

These particular solutions are called the Floquet particular solutions for $\Phi_1(\tau)$ since each corresponds to a ratio of Floquet solutions to (1.10) and (1.11). Note that when α, β, δ, and ϵ are such that Form II solutions occur, *all* solutions for $\Phi_1(\tau)$ are periodic with period 2π.

C. The Fourier Expansion Coefficients and the Matrix Formulation

Following Autler and Townes,[4] we can express the solutions to the two-level problem in terms of a matrix **M** whose eigenvalues are the values of the characteristic constant and whose eigenvectors are the coefficients of the Fourier components of $\phi_a(\tau)$ and $\phi_b(\tau)$. If $\beta=0$, then **M** is an infinite tridiagonal matrix whose eigenvalues and eigenvectors can be expressed by recursion relations. This is the result that Autler and Townes obtained. With modern high-speed computing machines, there are a number of numerical techniques that are more efficient than making direct evaluations of the continued fractions. These methods are discussed in Section I.D. If β does not vanish, then **M** is an infinite quintidiagonal matrix that is more difficult to diagonalize.

According to Floquet's theorem, (1.10) and (1.11) have solutions of the form

$$a(\tau)=\exp(-i\mu\tau)\phi_a(\tau); \qquad b(\tau)=\exp(-i\mu\tau)\phi_b(\tau) \tag{2.32}$$

Since the $\phi_j(\tau)$'s are periodic with period 2π, by Fourier's theorem they can be expressed by

$$\phi_a(\tau)=\sum_{j=-\infty}^{\infty} A_j \exp(ij\tau); \qquad \phi_b(\tau)=\sum_{j=-\infty}^{\infty} B_j \exp(ij\tau) \tag{2.33}$$

where the A_j's and B_j's are the Fourier expansion coefficients. Substitute Eqs. 2.33 and 2.32 into (1.10) and (1.11) and group the terms multiplying each and every $\exp(ij\tau)(j=\infty$ to $-\infty)$. Since the two resulting equations are valid for *all* values of τ, the coefficient of each and every $\exp(ij\tau)$ in the two equations must vanish. Equating these coefficients to zero leads to the two sets of equations,

$$(j-\mu)A_j+\alpha(B_{j+1}+B_{j-1})=0 \tag{2.34}$$

$$(j+\epsilon-i\delta-\mu)B_j+\alpha(A_{j+1}+A_{j-1})+\beta(B_{j+1}+B_{j-1})=0 \tag{2.35}$$

The matrix representation of these equations is

$$(\mathbf{M}-\mu\mathbf{I})\mathbf{C}=0 \tag{2.36}$$

Here μ is the scalar characteristic exponent of expressions (2.32), $\mathbf{I}$ is the infinite unit matrix, $\mathbf{M}$ is an infinite square matrix, and $\mathbf{C}$ is an infinite column vector.

The elements of $\mathbf{C}$ are of two kinds

$$(\mathbf{C})_{A,n} = A_n; \qquad (\mathbf{C})_{B,n} = B_n \tag{2.37}$$

and these elements are ordered in the following manner:

$$\ldots, A_{n+1}, B_{n+1}, A_n, B_n, A_{n-1}, B_{n-1}, \ldots \tag{2.38}$$

The rows and columns of $\mathbf{M}$ are ordered in the same manner and all matrix elements of $\mathbf{M}$ vanish except for

$$\begin{aligned}(\mathbf{M})_{A,j;A,j} &= j \\ (\mathbf{M})_{B,j;B,j} &= j + \varepsilon - i\delta \\ (\mathbf{M})_{A,j;B,j\pm1} &= \alpha = (\mathbf{M})_{B,j;A,j\pm1} \\ (\mathbf{M})_{B,j;B,j\pm1} &= \beta\end{aligned} \tag{2.39}$$

In order that there be a solution to (2.36), it is necessary that

$$\det|\mathbf{M} - \mu\mathbf{I}| = 0 \tag{2.40}$$

Note that this is an infinite secular equation that determines μ. Once μ is found, the Fourier expansion coefficients may be computed, and thus the problem is solved. This is explained in Section III where the (AT)′ technique is discussed.

When $\delta = 0$, simplifications arise in the computation of the Fourier expansion coefficients. The first simplification is that the Fourier expansion coefficients can be chosen to be real. This follows from the fact that when $\delta = 0$, both the matrix $\mathbf{M}$ in (2.36) and the characteristic exponent μ are real. The second simplification is that when one solution for $\mathbf{C}$ in (2.36) is known, a second solution may immediately be written down in terms of the first:

$$A_{j2} = -\sum_{k=-\infty}^{\infty} (B_{k1}) J_{-j-k}(2\beta)$$

$$B_{j2} = \sum_{k=-\infty}^{\infty} (A_{k1}) J_{-j-k}(2\beta)$$

where the subscripts 1 and 2 on the expansion coefficients refer respectively to the first (the presumed known) and second Floquet solutions.

$J_q(2\beta)$ is the integer order Bessel function of order q and argument (2β).

If β is quite small, then $J_q(2\beta) \approx (-1)^q \beta^{|q|}$. Thus, very few terms must be retained in the numerical evaluation of the sums. When $\beta = 0$, only one term in the sums remains because $J_q(0) = \delta_{q0}$.

To demonstrate these relations, use Eq. 2.17 of Theorem 6 to write the second solution in terms of the first. Since by Theorem 4, $\mu_1 + \mu_2 = \varepsilon$; since by Theorem 8, both μ_1 and μ_2 are real; and since the expansion coefficients are real, we have

$$\sum_{j=-\infty}^{\infty} A_{j2} \exp(ij\tau) = -\sum_{j=-\infty}^{\infty} B_{j1} \exp(-ij\tau) \exp(-2i\beta \sin \tau)$$

$$\sum_{j=-\infty}^{\infty} B_{j2} \exp(ij\tau) = \sum_{j=-\infty}^{\infty} A_{j1} \exp(-ij\tau) \exp(-2i\beta \sin \tau)$$

Our proposition is then confirmed by using (see Olver[62])

$$\exp(-2i\beta \sin \tau) = \sum_{q=-\infty}^{\infty} J_q(2\beta) \exp(-iq\tau)$$

and matching the coefficients of $\exp(ij\tau)$

D. Reduction of the Dynamical Problem to an Equivalent Static Problem

Shirley[6,25] restated Autler and Townes'[4] matrix eigenvalue–eigenvector formulation (Eq. 2.36) as a quantum-mechanical stationary state problem in which μ corresponds to the system's energy, the Fourier expansion coefficients correspond to an orthonormal complete basis set and the matrix M corresponds to the matrix representation of a Hamiltonian operator H_F (the F stands for Floquet) in terms of the complete basis. If δ does not vanish, M and H_F are non-Hermitian.

To be specific, define a basis $|k,n\rangle$ such that the index k can be either A or B. Let the indices m and n range in integer steps from $-\infty$ to $+\infty$. Also let the basis be orthonormal, that is,

$$\langle k,n | l,m \rangle = \delta_{k,l} \delta_{n,m} \tag{2.41}$$

where the deltas are Krönecker deltas and are not to be confused with the parameter δ that has no subscripts on it. The Floquet Hamiltonian H_F is defined by the equations:

$$\begin{aligned} H_F|A,j\rangle &= j|A,j\rangle + \alpha\left[|B,j+1\rangle + |B,j-1\rangle\right] \\ H_F|B,j\rangle &= (j+\varepsilon - i\delta)|B,j\rangle + \alpha\left[|A,j+1\rangle + |A,j-1\rangle\right] \\ &\quad + \beta\left[|B,j+1\rangle + |B,j-1\rangle\right] \end{aligned} \tag{2.42}$$

Seeking the eigenvalues μ and eigenvectors $|\mu\rangle$ of the following Schrödinger-type equation:

$$H_F|\mu\rangle = \mu|\mu\rangle \tag{2.43}$$

is equivalent to solving (2.36). To demonstrate this assertion, note that the function $|\mu\rangle$ is expandable as a linear combination of the complete set of basis functions $|A,j\rangle$ and $|B,j\rangle$ so that

$$|\mu\rangle = \sum_{j=-\infty}^{\infty} (\langle A,j|\mu\rangle|A,j\rangle + \langle B,j|\mu\rangle|B,j\rangle) \tag{2.44}$$

where the $\langle A,j|\mu\rangle$'s and $\langle B,j|\mu\rangle$'s are expansion coefficients. When Eq. 2.44 is substituted into (2.43) and after the result is multiplied by each and every bra $\langle k',n'|$, we are led to (2.36), which is the matrix representation of the following infinite set of equations:

$$\sum_{k,n} \langle k',n'|H_F|k,n\rangle\langle k,n|\mu\rangle - \mu\langle k',n'|\mu\rangle = 0$$

$$k' = A \text{ or } B; \qquad n' = -\infty,\ldots,\infty \tag{2.45}$$

Here, the $\langle k',n'|H_F|k,n\rangle$ correspond to the elements of $\mathbf{M}$ and the $\langle k,n|\mu\rangle$ correspond to the elements of the column vector $\mathbf{C}$:

$$\langle A,j|\mu\rangle = A_j; \qquad \langle B,j|\mu\rangle = B_j \tag{2.46}$$

After an eigenvalue μ has been found and after the corresponding expansion coefficients in (2.44) have been found, the time-dependent functions $a(\tau)$ and $b(\tau)$ are recovered by

$$\begin{aligned} a(\tau) &= \exp(-i\mu\tau) \sum_{j=-\infty}^{\infty} \langle A,j|\mu\rangle \exp(ij\tau) \\ b(\tau) &= \exp(-i\mu\tau) \sum_{j=-\infty}^{\infty} \langle B,j|\mu\rangle \exp(ij\tau) \end{aligned} \tag{2.47}$$

This static formulation forms the basis for the (SG)′, (SWS)′, (SSe)′, and (DRS) techniques that are discussed in Sections V through VIII.

III. THE AUTLER–TOWNES (AT)′ TRIDIAGONAL MATRIX TECHNIQUE

The Autler and Townes[4] formulation* of a two-level problem in terms of a matrix eigenvalue–eigenvector problem (see Section II.C) leads to some of the simplest and most direct methods of solution of the two-level problem. If $\beta=0$, then $\mathbf{M}$ becomes a symmetric tridiagonal matrix (real if $\delta=0$ and complex non-Hermitian if δ does not vanish). The (AT)′ technique corresponds to expressing the eigenvalues, which are the μ's, and the eigenvectors, which are the Fourier components of the ϕ's, in terms of continued fractions. By truncating the continued fractions after a few quotients, practical formulas are obtained for the μ's and for the amplitudes of the Fourier components, which are accurate through the first few orders of the field strength. These formulas even give good approximations to the behavior near the principal resonance ($\varepsilon\approx 1$). Near the principal resonance, somewhat better formulas are given by the (SWS)′ technique in Section VI. Autler and Townes[4] assumed that δ, as well as β, vanished. Our modification is to permit nonvanishing values of δ.

When β does not vanish, then $\mathbf{M}$ is a quintidiagonal matrix of infinite dimensionality. The only practical methods of approximating its eigenvalues and eigenvectors are either numerical (see Section I.D) or perturbation procedures. Sections V through IX are devoted to different types of perturbation treatments.

A. The (AT)′ Continued Fraction Formulation when $\beta=0$

The formulation of the (AT)′ technique starts with $\beta=0$ in Eqs. 2.34 and 2.35

$$(j-\mu)A_j+\alpha(B_{j-1}+B_{j+1})=0 \tag{3.1}$$

and

$$(j+(\varepsilon-i\delta)-\mu)B_j+\alpha(A_{j-1}+A_{j+1})=0 \tag{3.2}$$

It is clear from (3.1) and (3.2) that the A_j's with even values of j are coupled to B_j's with odd j's; also, A_j's with odd j's are coupled to B_j's with

*The notation in Autler and Townes[4] is somewhat different from ours. Equating our symbol (on the left) to the Autler–Townes corresponding symbol on the right: $\eta_a=\Gamma_a$, $\eta_b=\Gamma_b$, $FV_{ab}=\beta_{ab}=\beta_{ba}$, $\mu=-\lambda\omega$, $\alpha=\beta_{ab}/\omega$, $\varepsilon=\omega_{ab}/\omega$, $-\mu/\alpha=L_a$, $(\varepsilon-\mu)/\alpha=L_b$, $A_j=A_{-j}$, $B_j=B_{-j}$.

even j's. The first Floquet mode has the A_j's with even j's and B_j's with odd j's; the second Floquet mode has the A_j's with odd j's and B_j's with even j's.

If we know a solution for the first mode: $\mu_1 y$ and the values of all the A_{j1}'s and B_{j1}'s, then it is easy to show that a solution for the second linearly independent Floquet mode is given by

$$\mu_2 = (\varepsilon - i\delta) - \mu_1, \qquad A_{j2} = -B_{(-j)1}, \qquad \text{and} \quad B_{j2} = A_{(-j)1} \tag{3.3}$$

Substituting Eq. 3.3 into (3.1) and (3.2) has the effect of interchanging Eqs. 3.1 and 3.2 but otherwise leaving them unchanged. Note that $\mu_2 = (\varepsilon - i\delta) - \mu_1$ is in accord with Theorem 4 of Section II. Thus, we can concentrate our attention on a determination of a solution for the first mode.

For the first mode we restrict our consideration to A_j's with even j's and B_j's with odd j's. Thus, instead of considering the array given by (2.38), our elements are ordered in a simple sequence according to the value of j,

$$\ldots, B_{-3}, A_{-2}, B_{-1}, A_0, B_1, A_2, B_3, \ldots \tag{3.4}$$

The matrix M is then compacted to conform to this array. According to (2.39) with $\beta = 0$, the only nonvanishing elements of M are then

$$\begin{aligned} M_{jj} &= j && \text{when } j \text{ is even} \\ M_{jj} &= j + (\varepsilon - i\delta) && \text{when } j \text{ is odd} \\ M_{j,j+1} &= \alpha = M_{j+1,j} \end{aligned} \tag{3.5}$$

This matrix is then tridiagonal with the elements α lying along the two stripes adjacent to the diagonal. In Section III.B a good computational procedure for determining the eigenvalues and eigenvectors of M is discussed.

In order to express the matrix problem in the form of continued fractions, let us define

$$\begin{aligned} M_j' &= \frac{(j-\mu)}{\alpha} && \text{when } j \text{ is even} \\ M_j' &= \frac{[j + (\varepsilon - i\delta) - \mu]}{\alpha} && \text{when } j \text{ is odd} \\ x_j &= \frac{B_{j-1}}{A_j} \quad \text{and} && y_j = \frac{B_{j+1}}{A_j} \end{aligned} \tag{3.6}$$

Then dividing Eq. 3.1 by A_j gives

$$M_j' + x_j + y_j = 0 \tag{3.7}$$

Next in (3.2) replacing each j by $j+1$ and dividing by B_{j+1} gives

$$M_{j+1}' + \left(\frac{1}{y_j}\right) + \left(\frac{1}{x_{j+2}}\right) = 0 \tag{3.8}$$

Equation 3.8 can be rewritten in either of two forms

$$x_j = \frac{-1}{\left[M_{j-1}' + (1/y_{j-2})\right]} \tag{3.9}$$

or

$$y_j = \frac{-1}{\left[M_{j+1}' + (1/x_{j+2})\right]} \tag{3.10}$$

Using Eq. 3.7 to eliminate the y_{j-2} in (3.9) and the x_{j+2} in (3.10), we obtain

$$x_j = \frac{-1}{\left[M_{j-1}' - 1/(M_{j-2}' + x_{j-2})\right]} \tag{3.11}$$

and

$$y_j = \frac{-1}{\left[M_{j+1}' - 1/(M_{j+2}' + y_{j+2})\right]} \tag{3.12}$$

Equation 3.11 with the j's decreased by two can be used to replace the x_{j-2} with an expression involving x_{j-4}, and this process can be iterated to obtain a continued fraction. In a similar manner, Eq. 3.12 can be iterated to obtain a continued fraction solution for y_j. Thus,

$$\frac{B_{j-1}}{A_j} = x_j = -1/\left[M_{j-1}' - 1/\left[M_{j-2}' - 1/\left[M_{j-3}' - 1/[\cdots]\right]\right]\right] \tag{3.13}$$

and

$$\frac{B_{j+1}}{A_j} = y_j = -1/\left[M_{j+1}' - 1/\left[M_{j+2}' - 1/\left[M_{j+3}' - 1/[\cdots]\right]\right]\right] \tag{3.14}$$

Taking $j=0$ and using (3.7), (3.13), and (3.14), we obtain a continued fraction expression for μ,

$$\frac{\mu}{\alpha}=-\cfrac{1}{M_1'-\cfrac{1}{M_2'-\cfrac{1}{M_3'-\cfrac{1}{M_4'-\cdots}}}}-\cfrac{1}{M_{-1}'-\cfrac{1}{M_{-2}'-\cfrac{1}{M_{-3}'-\cfrac{1}{M_{-4}'-\cdots}}}} \tag{3.15}$$

Once a value of μ has been determined by solving Eq. 3.15, then each of the x_j and y_j can be determined by (3.13) and (3.14). Since Eqs. 1.10 and 1.11 are linear, the set of Fourier coefficients can be multiplied by an arbitrary constant. Therefore, we can take $A_0=1$. Then, from (3.13) and (3.14)

$$\begin{gathered}B_{-1}=x_0, \qquad B_1=y_0\\ A_j=\frac{B_{j-1}}{x_j}, \qquad A_{-j}=\frac{B_{-(j-1)}}{y_{-j}}\\ B_{j+1}=A_jy_j, \qquad B_{-(j+1)}=A_{-j}x_{-j}\end{gathered} \tag{3.16}$$

Using these relations, all of the A_j's and B_j's can be determined.

Thus,* Eqs. 3.13 through 3.16 together with (3.3) completely determine two linearly independent Floquet modes and, hence, a general solution to the two-level problem for arbitrary boundary conditions. To obtain useful approximate solutions to (3.15), the continued fractions can be truncated after a finite number of quotients. The larger the number of quotients retained, the greater the accuracy. For large values of j or $-j$, the denominators are dominated by terms of the form j/α, which is a necessary condition for the fractions to converge. The smaller $\alpha=FV_{ab}/\omega$ is, the more rapid the convergence is, so it is necessary to retain fewer quotients for high frequencies or weak fields. No proof of the convergence of the continued fractions has been obtained. However, for the case where ω approaches zero the continued fractions converge to the correct steady-state solution, and this is the situation where one would expect the slowest convergence.

*This paragraph is taken almost directly from Autler and Townes.[4]

Let us consider a few special cases.

1. *The Limit of Zero Frequency*

Since $\alpha = FV_{ab}/\omega$, the value of α becomes infinite as ω approaches zero and j/α becomes negligible for finite values of j. Therefore,

$$\begin{aligned} M_j' &= \frac{-\mu}{\alpha}, & &\text{for } j \text{ even} \\ M_j' &= \frac{[-\mu+(\varepsilon-i\delta)]}{\alpha}, & &\text{for } j \text{ odd} \end{aligned} \tag{3.17}$$

Equation 3.15 then becomes

$$\frac{-\mu}{(2\alpha)} = \frac{1}{[-\mu+(\varepsilon-i\delta)]/\alpha - 1/[-(\mu/\alpha)+(\mu/2\alpha)]} \tag{3.18}$$

Here the $\mu/(2\alpha)$ in the denominator follows from the repetitive pattern of the continued fraction. Equation 3.18 can be rewritten in the form of the quadratic equation

$$\frac{\mu^2}{\alpha^2} - \left(\frac{\varepsilon-i\delta}{\alpha}\right)\left(\frac{\mu}{\alpha}\right) - 4 = 0 \tag{3.19}$$

which has the solutions

$$\frac{\mu}{\alpha} = \left(\frac{\varepsilon-i\delta}{2\alpha}\right) \pm \left[\left(\frac{\varepsilon-i\delta}{2\alpha}\right)^2 + 4\right]^{1/2} \tag{3.20}$$

For $\delta = 0$ this is the same result that is obtained by direct diagonalization of the time-independent Schrödinger equation.

2. *The Simplest Weak Field Approximation (Far from Resonance)*

The simplest weak field approximation, if the frequency is far from resonance ($\varepsilon \ll 1$), corresponds to retaining only the first quotient in the continued fractions. Thus, from (3.6)

$$-\frac{\mu}{\alpha} = \frac{1}{M_1'} + \frac{1}{M_{-1}'} = \frac{2\alpha[\varepsilon-i\delta-\mu]}{[\varepsilon-i\delta-\mu]^2-1} \tag{3.21}$$

Equation 3.21 has one solution accurate through terms in the square of α

or the field strength

$$\mu = \frac{2\alpha^2(\varepsilon - i\delta)}{\left[1-(\varepsilon - i\delta)^2\right]} \tag{3.22}$$

To the same approximation

$$x_0 = y_{-2} = \frac{-1}{M'_{-1}} = \frac{-\alpha}{\left[(\varepsilon - i\delta)-1\right]} = B_{-1}$$

$$y_0 = x_2 = \frac{-1}{M'_1} = \frac{-\alpha}{\left[(\varepsilon - i\delta)+1\right]} = B_1 \tag{3.23}$$

and

$$A_2 = A_0 = A_{-2} = 1$$

3. *The Second-Order Approximation Including Resonance*

In order to obtain μ/α, x_j, and y_j accurate through terms of the order of α^2 and $(\alpha/\varepsilon)^2$, it is necessary to consider the quotients $M'_{j\pm 1}$, $M'_{j\pm 2}$, and parts of the $M'_{j\pm 3}$. This approximation is sufficiently accurate to give a good description of the solution near the principal resonance ($\varepsilon \approx 1$). The results that Autler and Townes[4] obtained can be trivially modified to include δ by replacing each ε by $\varepsilon' = \varepsilon - i\delta$. In our notation, their solutions are

$$\frac{\mu}{\alpha} = -\rho\left[1 - \left(\frac{\alpha}{2\varepsilon'}\right)^2\left(\frac{2\rho^2-1}{\rho^2+1}\right)\right] \tag{3.24}$$

where

$$\rho = -\frac{\varepsilon'-1}{2\alpha} \pm \left[\left(\frac{\varepsilon'-1}{2\alpha}\right)^2 + 1 + \frac{\varepsilon'-1}{\varepsilon'+1}\right]^{1/2} \tag{3.25}$$

$+$for $\varepsilon' > 1$, $-$for $\varepsilon' < 1$. Here $1 > |\rho| > (2)^{-1/2}$ and therefore, the factor $\frac{1}{2} > (2\rho^2 - 1/\rho^2 + 1) > 0$. In order to simplify the expressions for the amplitudes, it is also convenient to define the parameter

$$K = \frac{\varepsilon'-1}{2} \pm \left[\left(\frac{\varepsilon'-1}{2}\right)^2 + \frac{2\alpha^2\varepsilon'}{\varepsilon'+1}\right]^{1/2} \tag{3.26}$$

Then

$$\frac{B_{-1}}{A_0}=-\frac{\alpha}{K}\left[1-\frac{\alpha^2}{2K}+\frac{\alpha^2}{4\varepsilon'^2}\left(\frac{\alpha^2}{K^2}-\frac{\alpha\rho}{K}\left\{\frac{2-\rho^2}{1+\rho^2}\right\}\right)\right]$$

$$\frac{B_1}{A_0}=-\left[\frac{\alpha}{2\varepsilon'}+\left(\frac{\alpha}{2\varepsilon'}\right)^2\left(\frac{\varepsilon'-1}{\alpha}-\rho\right)\right]=\frac{-A_{-1}}{B_0}$$

$$\frac{A_{-2}}{A_0}=\left(\frac{B_{-1}}{A_0}\right)\left[\frac{\alpha}{2\varepsilon'}+\left(\frac{\alpha}{2\varepsilon'}\right)^2\left(\frac{2(\varepsilon'-1)}{\alpha}+\rho\right)\right]=\frac{B_2}{B_0} \tag{3.27}$$

$$\frac{B_{-3}}{A_0}=\left(\frac{B_{-1}}{A_0}\right)\left(\frac{\alpha}{2\varepsilon'}\right)^2=-\left(\frac{A_{-2}}{A_0}\right)\left(\frac{\alpha}{\varepsilon'-3}\right)=\frac{-A_2}{B_0}$$

$$\frac{A_2}{A_0}=\left(\frac{\alpha}{2\varepsilon'}\right)^2=\frac{B_{-2}}{B_0}$$

If the system is in the first Floquet mode, then the principal resonance occurs when (see Bloch and Siegert[2])

$$\frac{|B_{-1}/A_0|^2}{1+|B_{-1}/A_0|^2}=\text{Maximum} \tag{3.28}$$

If $\delta=0$, this maximum occurs when

$$\frac{1}{\varepsilon}=1+\left(\frac{\alpha}{\varepsilon}\right)^2 \tag{3.29}$$

Thus, there is a small shift in the frequency of the resonance as the field becomes stronger. Near the principal resonance, somewhat better formulas are given by the (SWS)′ technique in Section VI.

Autler and Townes[4] give a great deal more interesting details and they have an excellent discussion of the difference between the exact and the approximate solutions.

B. Numerical Calculations of μ

Autler and Townes define the first Floquet mode as the one for which $\mu_1\to 0$ in the limit as the field strength approaches zero. Thus, the weak field (or small α) dependence of μ_1 is given by (3.22). However, if one considers the higher order approximations to μ, such as is given by (3.24) (see also Eq. 6.11), we find that the μ's of the two Floquet modes may be labeled μ_+ and μ_- where the plus and minus correspond to the sign of the square root in (3.25). If $\varepsilon' < 1$, then $\mu_1 = \mu_+$; if $\varepsilon' = 1, \mu$ is not determined; and if $\varepsilon' > 1$, $\mu_1 = \mu_-$. Indeed, if one takes $\delta = 0$ and holds α/ε equal to a constant, then μ_1 as a function of ε has a discontinuity at $\varepsilon = 1$. However, μ_+ is a smooth continuous function of ε. Therefore we decided to tabulate the values of μ_+.

Tables II and III give the exact values of μ_+ as a function of ε for various values of α/ε, β/ε, and δ/ε. Here $\alpha/\varepsilon = 0.2$ corresponds to a weak field; $\alpha/\varepsilon = 0.5$ corresponds to a strong field. Small values of ε correspond to a high frequency. Two different procedures were used for performing the calculations with a digital computer:

(1) For $\beta = 0 = \delta$, $\alpha/\varepsilon < 0.4$, and $\varepsilon < 1.5$, an initial value of μ_+ was assumed. This value was used to evaluate the continued fractions on the right-hand side of (3.15). The right-hand side of (3.15) is then an improved value of μ_+ divided by α. This procedure is iterated until the difference between successive values of μ_+ lies within the desired precision.

(2) For other values of the parameters, a digital computer was used to directly integrate Eqs. 1.10 and 1.11 to obtain the fundamental solutions at $\tau = 2\pi$; then Eqs. 2.7 and 2.8 were used to determine the values of μ_+. The values that were obtained by using the continued fractions agreed with the values obtained from the "fundamental" solutions.

When μ_+ is known, then $\mu_- = \varepsilon' - \mu_+$ is also known. The other infinite solutions to (3.15) correspond to adding or subtracting an integer from either μ_+ or μ_-.

TABLE II. Exact Values of μ_+ for $\beta = 0 = \delta$ as a function of ε for a Range of Values of α/ε[a]

				μ_+(exact)			
α/ε	$\varepsilon=0.6$	$\varepsilon=0.8$	$\varepsilon=0.9$	$\varepsilon=1.0$	$\varepsilon=1.1$	$\varepsilon=1.2$	$\varepsilon=1.4$
0.1	0.0066	0.0252	0.0508	0.0999	0.1731	0.2602	0.4506
0.2	0.0253	0.0805	0.1314	0.1990	0.2794	0.3684	0.5607
0.3	0.0530	0.1455	0.2148	0.2964	0.3869	0.4836	0.6880
0.4	0.0867	0.2130	0.2971	0.3908	0.4913	0.5964	0.8141
0.5	0.1241	0.2800	0.3765	0.4805	0.5893	0.7006	0.9214
0.6	0.1633	0.3448	0.4510	0.5623	0.6752	0.7858	0.9745
0.7	0.2029	0.4055	0.5180	0.6314	0.7394	0.8335	1.0613

[a] All of these values were calculated by the "fundamental" solution technique of (2.7) and (2.8). These values of μ_+ for $0.1 \leqslant \alpha/\varepsilon < 0.4$ agreed with the solution of (3.15). For larger values of α/ε the (AT)′ continued fractions converged too slowly for our computing machine calculations.

TABLE III. The Effect of β and of δ on the Exact Values of μ_+ as a Function of ε[a]

α/ε	β/ε	δ/ε	μ_+(exact) $\varepsilon=0.6$	$\varepsilon=0.8$	$\varepsilon=0.9$	$\varepsilon=1.0$	$\varepsilon=1.1$	$\varepsilon=1.2$	$\varepsilon=1.4$
0.2	0.2	0.0	0.0250	0.0789	0.1286	0.1944	0.2725	0.3583	0.5387
0.2	0.5	0.0	0.0235	0.0712	0.1144	0.1718	0.2397	0.3130	0.4572
0.5	0.2	0.0	0.1225	0.2719	0.3578	0.4308	0.4635	0.4665	0.4536
0.5	0.5	0.0	0.1144	0.2365	0.2938	0.3346	0.3556	0.3615	0.3551
0.2	0.0	0.1	0.0247 $-0.0048i$	0.0771 $-0.0210i$	0.1262 $-0.0368i$	0.1929 $-0.0552i$	0.2737 $-0.0727i$	0.3635 $-0.0883i$	0.5570 $-0.1147i$
0.2	0.0	0.2	0.0231 $-0.0092i$	0.0674 $-0.0397i$	0.1095 $-0.0719i$	0.1735 $-0.1115i$	0.2562 $-0.1484i$	0.3487 $-0.1798i$	0.5462 $-0.2320i$
0.5	0.0	0.1	0.1231 $-0.0180i$	0.2784 $-0.0401i$	0.3746 $-0.0520i$	0.4786 $-0.0637i$	0.5876 $-0.0749i$	0.6990 $-0.0858i$	0.9216 $-0.1052i$
0.5	0.0	0.2	0.1203 $-0.0357i$	0.2733 $-0.0801i$	0.3693 $-0.1040i$	0.4733 $-0.1275i$	0.5824 $-0.1500i$	0.6942 $-0.1718i$	0.9169 $-0.2110i$

[a]The values of μ_+ were calculated by the "fundamental" solution technique, (2.7) and (2.8).

IV. THE MEADOWS AND ASHBY (MA)′ METHOD FOR EVALUATING μ

Meadows[32] and Ashby[30] developed a convenient method for determining the values of μ that is applicable for all values of α, β, and ε. Our contribution is to extend the treatment so that it applies to all values of δ.

The present method was given by Meadows in 1962. Meadows considered a *general* system of N first-order linear homogeneous differential equations with periodic coefficients. He made no requirements such as stipulating that the matrix of coefficients be Hermitian and so on. As a numerical example he applies his technique to the Mathieu equation.

Ashby independently formulated this technique in 1968, and he is the first author to apply it to the problem of a two-state quantum system in an oscillating classical field. Ashby considered the case of nonvanishing β but did not allow the "energy" to have an imaginary component (i.e., he required $\delta=0$).

First, let us state the basic theorem that may be used even in the vicinity of resonances where ε is approximately equal to an odd integer. Next we state the modification of the theorem that is applicable in the vicinity of resonances where ε is approximately equal to an even integer. Then we present the proof of these formulas.

A. The Basic Theorem

$$\sin^2\left[\pi\left(\mu-\tfrac{1}{2}(\varepsilon-i\delta)\right)\right]=\sin^2\left[\frac{\pi}{2}(\varepsilon-i\delta)\right]\det(\underline{\Delta}_1(0)) \qquad (4.1)$$

where $\underline{\Delta}_1(0)$ is the limit as $u=\mu-\frac{1}{2}(\varepsilon-i\delta)$ approaches zero of the matrix $\underline{\Delta}_1(u)$. Here $\underline{\Delta}_1(u)$ is an unsymmetric quintidiagonal matrix with rows and columns corresponding to the Eq. 2.38 array. Its only nonvanishing elements are

$$
\begin{aligned}
(\underline{\Delta}_1(u))_{A,j;\,A,j}&=(\underline{\Delta}_1(u))_{B,j;\,B,j}=1\\
(\underline{\Delta}_1(u))_{A,j;\,B,j\pm1}&=\frac{\alpha}{j-\frac{1}{2}(\varepsilon-i\delta)-u}\\
(\underline{\Delta}_1(u))_{B,j;\,B,j\pm1}&=\frac{\beta}{j+\frac{1}{2}(\varepsilon-i\delta)-u}\\
(\underline{\Delta}_1(u))_{B,j;\,A,j\pm1}&=\frac{\alpha}{j+\frac{1}{2}(\varepsilon-i\delta)-u}
\end{aligned}
\tag{4.2}
$$

The determinant of $\underline{\Delta}_1(0)$ can be determined by standard methods such as those given in Wilkinson.[53] In practice the determinant is truncated at some high, but finite, order. Successively higher order truncations can be used to check the convergence. If $\delta=0$ so that all the elements of the determinant are real, the available program packages are somewhat simpler. Unfortunately, we do not know any available program for evaluating a determinant that takes advantage of the large number of zeros in a banded matrix.

The only difficulty that might arise is that if the value of ε is close to $2n$ where n is an integer or zero, then the values of $(\underline{\Delta}_1(0))_{A,n;\;B,n\pm1}$, $(\underline{\Delta}_1(0))_{B,-n;\;B,-n\pm1}$, and $(\underline{\Delta}_1(0))_{B,-n;\;A,-n\pm1}$ become very large. Indeed, if $\varepsilon=2n$ and if $\delta=0$, these elements would become indeterminate. Thus, special consideration must be given to this type of resonance.

B. The Resonance Modification When ε is Almost Equal to $2n$

When ε is almost equal to $2n$ (where n is an integer), the nondiagonal elements in the A,n and the $B,-n$ rows of $\underline{\Delta}_1(0)$ become very large. This difficulty can be overcome by multiplying these two rows of $\underline{\Delta}_1(0)$ by $\sin[(\pi/2)(\varepsilon-i\delta)]$ to form the new matrix $\underline{\Delta}_2(0)$; the other rows of $\underline{\Delta}_2(0)$ remain the same as the corresponding rows of $\underline{\Delta}_1(0)$. If $\varepsilon-i\delta=2n+z$, then the elements of $\underline{\Delta}_2(0)$ that differ from those of $\underline{\Delta}_1(0)$ are

$$
\begin{aligned}
(\underline{\Delta}_2(0))_{A,n;\,A,n}&=(\underline{\Delta}_2(0))_{B,-n;\,B,-n}=(-1)^n\sin\left(\frac{\pi z}{2}\right)\\
(\underline{\Delta}_2(0))_{B,-n;\,A,-n\pm1}&=-(\underline{\Delta}_2(0))_{A,n;\,B,n\pm1}=2\alpha(-1)^n\left(\frac{1}{z}\right)\sin\left(\frac{\pi z}{2}\right)\\
(\underline{\Delta}_2(0))_{B,-n;\,B,-n\pm1}&=2\beta(-1)^n\left(\frac{1}{z}\right)\sin\left(\frac{\pi z}{2}\right)
\end{aligned}
\tag{4.3}
$$

To evaluate these elements we note that

$$\left(\frac{1}{z}\right)\sin\left(\frac{\pi z}{2}\right)=\left(\frac{\pi}{2}\right)\left[1-\left(\frac{\pi z}{2}\right)^2/3!+\left(\frac{\pi z}{2}\right)^4/5!-\cdots\right] \tag{4.4}$$

Since the determinant of $\underline{\Delta}_2(0)$ differs from the determinant of $\underline{\Delta}_1(0)$ by the factor $\sin^2[(\pi/2)(\varepsilon-i\delta)]$, (4.1) becomes

$$\sin^2\left[\pi\left(\mu-\frac{1}{2}(\varepsilon-i\delta)\right)\right]=\det(\underline{\Delta}_2(0)) \tag{4.5}$$

C. The Proof of the Basic Theorem

It is convenient to prove the basic Meadows and Ashby theorem by paralleling the development in Ross'[63] thesis. Define the "shifted" characteristic exponent u by

$$\mu=u+\frac{1}{2}(\varepsilon-i\delta) \tag{4.6}$$

where μ is the usual characteristic exponent. Write Eqs. 2.34 and 2.35 in terms of u and divide the first of them by $[j-\frac{1}{2}(\varepsilon-i\delta)]$ and divide the second by $[j+\frac{1}{2}(\varepsilon-i\delta)]$. The result is the compact matrix equation

$$\underline{\Delta}(u)\mathbf{C}=0 \tag{4.7}$$

where $\mathbf{C}$ is exactly the same infinite column vector of Fourier coefficients that appears in (2.36). The rows and columns of the u-dependent infinite square matrix $\underline{\Delta}(u)$ are ordered according to expression (2.38). All of its elements vanish except for the following:

$$\begin{aligned}
(\underline{\Delta}(u))_{A,j;\,A,j}&=\frac{\left[j-\frac{1}{2}(\varepsilon-i\delta)-u\right]}{j-\frac{1}{2}(\varepsilon-i\delta)}\\
(\underline{\Delta}(u))_{B,j;\,B,j}&=\frac{\left[j+\frac{1}{2}(\varepsilon-i\delta)-u\right]}{j+\frac{1}{2}(\varepsilon-i\delta)}\\
(\underline{\Delta}(u))_{A,j;\,B,j\pm1}&=\frac{\alpha}{j-\frac{1}{2}(\varepsilon-i\delta)}\\
(\underline{\Delta}(u))_{B,j;\,B,j\pm1}&=\frac{\beta}{j+\frac{1}{2}(\varepsilon-i\delta)}\\
(\underline{\Delta}(u))_{B,j;\,A,j\pm1}&=\frac{\alpha}{j+\frac{1}{2}(\varepsilon-i\delta)}
\end{aligned} \tag{4.8}$$

Now define the matrix $\underline{\Delta}_1(u)$ as the matrix obtained by dividing every row of $\underline{\Delta}(u)$ by its diagonal element. $\underline{\Delta}_1(u)$, therefore, has all its diagonal elements equal to unity and all its elements vanish except for those given in (4.2).

By the definition of $\underline{\Delta}_1(u)$ in terms of $\underline{\Delta}(u)$, the determinants of these two matrices must be related by

$$\det(\underline{\Delta}(u))=\left[\prod_{j=-\infty}^{\infty}\left(\frac{j-\frac{1}{2}(\varepsilon-i\delta)-u}{j-\frac{1}{2}(\varepsilon-i\delta)}\right)\left(\frac{j+\frac{1}{2}(\varepsilon-i\delta)-u}{j+\frac{1}{2}(\varepsilon-i\delta)}\right)\right]\det(\underline{\Delta}_1(u)) \tag{4.9}$$

This relation can be written in a more convenient form if we make use of the trigonometric identity (see Zucker,[64] Eq. 4.3.89),

$$\sin\pi z=\pi z\prod_{k=1}^{\infty}\left[1-\frac{z^2}{k^2}\right]=\pi z\prod_{\substack{j=-\infty\\ j\neq 0}}^{\infty}\left[\frac{j+z}{j}\right] \tag{4.10}$$

If we let $z_1=-\frac{1}{2}(\varepsilon-i\delta)-u$, $z_2=\frac{1}{2}(\varepsilon-i\delta)-u$, and $z_3=-\frac{1}{2}(\varepsilon-i\delta)=-z_4$, then Eq. 4.9 becomes

$$\det(\underline{\Delta}(u))=\left[\frac{\sin(\pi z_1)\sin(\pi z_2)}{\sin(\pi z_3)\sin(\pi z_4)}\right]\det(\underline{\Delta}_1(u)) \tag{4.11}$$

Equation 4.11 can easily be transformed into

$$\det(\underline{\Delta}(u))=\left[1-\frac{\sin^2(\pi u)}{\sin^2[(\pi/2)(\varepsilon-i\delta)]}\right]\det(\underline{\Delta}_1(u)) \tag{4.12}$$

It is necessary to establish four properties of $\det(\underline{\Delta}_1(u))$.

(1) $\det(\underline{\Delta}_1(u))$ is periodic:

$$\det(\underline{\Delta}_1(u))=\det(\underline{\Delta}_1(u+n))$$

where n is any positive or negative integer. This follows immediately from the fact that the infinite matrix $\underline{\Delta}_1(u)$ is equivalent to the infinite matrix $\underline{\Delta}_1(u+n)$.

(2) The limit of $\det(\underline{\Delta}_1(u))$ as the imaginary part of u goes to $\pm\infty$ is unity. This follows from the fact that when the imaginary part of u tends to

$\pm\infty$, the off-diagonal elements of $\underline{\Delta}_1(u)$ tend to zero. $\underline{\Delta}_1(u)$, therefore, becomes the infinite unit matrix in this limit and its determinant tends to unity.

(3) $\det(\underline{\Delta}_1(u))$ has simple poles at $u=q\pm\frac{1}{2}(\varepsilon-i\delta)$ where q is any positive or negative integer including zero. That the poles are simple is demonstrated by noting that

$$\left(q\pm\tfrac{1}{2}(\varepsilon-i\delta)-u\right)\det\left(\underline{\Delta}_1(u)\right) \tag{4.13}$$

has no poles at $u=q\pm\frac{1}{2}(\varepsilon-i\delta)$. Since multiplication of one row of a determinant by a scalar is equivalent to multiplication of the determinant itself by the same scalar, expression (4.13) is the determinant of the matrix obtained by multiplying the row in $\underline{\Delta}_1(u)$ containing the denominator $(q\pm\frac{1}{2}(\varepsilon-i\delta)-u)$ by the quantity $(q\pm\frac{1}{2}(\varepsilon-i\delta)-u)$. The determinant of the matrix obtained in this manner has no poles at $u=q\pm\frac{1}{2}(\varepsilon-i\delta)$.

(4) $\det(\underline{\Delta}_1(u))$ is absolutely convergent except at its poles.

The proof is based on a theorem by St. Bobr[65] that states that an infinite determinant having elements of the form $(\delta_{ij}+a_{ij})$ absolutely converges if the infinite sums

$$\sum_{i=-\infty}^{\infty} |a_{ii}| \tag{4.14}$$

and

$$\sum_{i=-\infty}^{\infty}\left[\sum_{\substack{j=-\infty\\ j\neq i}}^{\infty} |a_{ij}|^{m/m-1}\right]^{m-1} \tag{4.15}$$

converge, where δ_{ij} is the Krönecker delta and m is in the interval $1<m\leqslant 2$. The sum in (4.14) converges since all the a_{ii} in the present example are zero.

Let $m=2$ and rewrite Eq. 4.15 explicitly in terms of $\underline{\Delta}_1(u)$ and its double indices:

$$\sum_{A,p}\sum_{\substack{l,q\\ l,q\neq A,p}} |(\underline{\Delta}_1(u))_{A,p;\,l,q}|^2+\sum_{B,p}\sum_{\substack{l,q\\ l,q\neq B,p}} |(\underline{\Delta}_1(u))_{B,p;\,l,q}|^2 \tag{4.16}$$

where p and q range from $-\infty$ to ∞ and $l=A$ or B. Consider the first double sum in (4.16). For a given (A,p) only two terms are nonvanishing:

those for which $l,q=B,p\pm 1$. The double sum reduces to

$$2\alpha^2 \sum_{A,p} \left(\frac{1}{p-\frac{1}{2}(\varepsilon - i\delta) - u} \right) \left(\frac{1}{p-\frac{1}{2}(\varepsilon + i\delta) - u^*} \right) \tag{4.17}$$

where u^* is the usual complex conjugate of u. According to Ashby[30] (see bottom of p. 609) the series

$$\sum_{p=-\infty}^{\infty} \left[(p+z_1)(p+z_2) \right]^{-1} \tag{4.18}$$

converges absolutely for arbitrary complex z_i as long as $(p+z_i)$ never equals zero. The series (4.17) thus converges everywhere except at the poles of $\underline{\Delta}_1(u)$. Similar considerations are applied to the second of the double summations in (4.16) to show that $\det(\underline{\Delta}_1(u))$ is absolutely convergent except at its poles.

Now define the function $f(u)$:

$$f(u) = \det(\underline{\Delta}_1(u)) - \sum_{q=-\infty}^{\infty} \frac{K_A}{\left[q + \frac{1}{2}(\varepsilon - i\delta) - u \right]}$$

$$- \sum_{q=-\infty}^{\infty} \frac{K_B}{\left[q - \frac{1}{2}(\varepsilon - i\delta) - u \right]} \tag{4.19}$$

where K_A and K_B are *constants* chosen so that $f(u)$ has no poles at $u = \pm\frac{1}{2}(\varepsilon - i\delta)$. These constants therefore correspond to the residues at the simple poles $u=\pm\frac{1}{2}(\varepsilon - i\delta)$. $f(u)$ is periodic since $\det(\underline{\Delta}_1(u))$ as well as both of the subtracted summations are periodic. Because of this periodicity, $f(u)$ has *no* poles in the complex u-plane, and it is therefore everywhere analytic.

Using the identity [see Jolley,[66] Eq. (450*a*)]

$$\sum_{q=-\infty}^{\infty} \frac{1}{q+z} = \pi \cot \pi z \tag{4.20}$$

Equation 4.19 is rewritten as

$$f(u) = \det(\underline{\Delta}_1(u)) - K_A \pi \cot\left[\pi\left(\tfrac{1}{2}(\varepsilon - i\delta) - u \right) \right]$$

$$- K_B \pi \cot\left[-\pi\left(\tfrac{1}{2}(\varepsilon - i\delta) + u \right) \right] \tag{4.21}$$

Since all three terms on the right-hand side of (4.21) are bounded except at the poles $u = q \pm \frac{1}{2}(\varepsilon - i\delta)$ and $f(u)$ is bounded at the poles, $f(u)$ is bounded everywhere. According to Liouville's theorem,* if $f(u)$ is analytic and if $|f(u)|$ is bounded everywhere in the complex u-plane, then $f(u)$ must be a constant f.

Now take the limits of (4.21) as the imaginary part of u goes first to $+\infty$ and then to $-\infty$ in order to obtain equations for f, K_A, and K_B:

$$\begin{aligned} f &= 1 + i\pi(K_A + K_B) \\ f &= 1 - i\pi(K_A + K_B) \end{aligned} \tag{4.22}$$

These equations require $f = 1$ and $K_A = -K_B$. Using these values together with some usual trigonometric identities, Eq. 4.21 becomes

$$\det(\underline{\Delta}_1(u)) = 1 + \frac{2\pi K_A \sin[(\pi/2)(\varepsilon - i\delta)] \cos[(\pi/2)(\varepsilon - i\delta)]}{\sin^2[(\pi/2)(\varepsilon - i\delta)] - \sin^2[\pi u]} \tag{4.23}$$

Equation 4.23 is valid for all values of u. In particular, set $u = 0$ to find K_A:

$$K_A = \frac{\tan[(\pi/2)(\varepsilon - i\delta)]}{2\pi}[\det(\underline{\Delta}_1(0)) - 1] \tag{4.24}$$

The important result is obtained by substituting Eqs. 4.23 and 4.24 into (4.12),

$$\det(\underline{\Delta}(u)) = \det(\underline{\Delta}_1(0)) - \frac{\sin^2(\pi u)}{\sin^2((\pi/2)(\varepsilon - i\delta))} \tag{4.25}$$

In the original Floquet problem, those values of u that satisfy $\det(\underline{\Delta}(u)) = 0$ are required. These characteristic exponents therefore satisfy

$$\sin^2(\pi u) = \sin^2\left[\frac{\pi}{2}(\varepsilon - i\delta)\right] \det(\underline{\Delta}_1(0)) \tag{4.26}$$

Replacing u by $(\mu - \frac{1}{2}(\varepsilon - i\delta))$ in (4.26) gives Eq. 4.1: The basic theorem for finding the characteristic exponent in terms of the parameters α, β, δ, and ε.

*See Churchill[67] (p. 125).

V. THE SEN GUPTA (SG)′ TECHNIQUE USING NONDEGENERATE RAYLEIGH–SCHRÖDINGER PERTURBATION THEORY

Sen Gupta;[40,41,42] Langhoff, Epstein, and Karplus;[38] and Sambe[43] in effect used nondegenerate Rayleigh–Schrödinger perturbation theory to expand the Floquet ϕ_a's, ϕ_b's, and μ's in powers of the field strength. These authors considered a general many-level problem and, although they used different techniques, they obtained the same results when applied to a two-level system. Our (SG)′ procedure differs from these previous treatments in two respects. First, we include the parameter δ that makes the problem considerably more difficult; and second, we use as our starting point, Shirley's[6,25] static formulation that reduces the two-level problem to the solution of a time-independent Schrödinger equation

$$H_F|\mu\rangle = \mu|\mu\rangle \tag{2.43}$$

where H_F is the effective (or Floquet) Hamiltonian, μ plays the role of the energy, and $|\mu\rangle$, the vector describing the Fourier components, plays the role of the wavefunction. We split H_F into the sum of a zeroth-order Hamiltonian that is independent of the radiation field and a first-order Hamiltonian that is proportional to the field intensity. Then we get exact solutions for each order of the resulting Rayleigh–Schrödinger perturbation equations even though H_F is non-Hermitian when $\delta \neq 0$.

A. Description of the (SG)′ Technique

Let H_F (defined by Eq. 2.42) be broken up into a zeroth-order part $H_F^{(0)}$ and a perturbation $H_F^{(1)}$ according to

$$H_F^{(0)}|A,j\rangle = j|A,j\rangle; \qquad H_F^{(0)}|B,j\rangle = (j+\varepsilon - i\delta)|B,j\rangle \tag{5.1}$$

and

$$\begin{aligned} H_F^{(1)}|A,j\rangle &= \alpha\left[|B,j+1\rangle + |B,j-1\rangle\right] \\ H_F^{(1)}|B,j\rangle &= \beta\left[|B,j+1\rangle + |B,j-1\rangle\right] + \alpha\left[|A,j+1\rangle + |A,j-1\rangle\right] \end{aligned} \tag{5.2}$$

Since $\alpha = FV_{ab}/\omega$ and $\beta = F(V_{bb} - V_{aa})/\omega$, the perturbation potential $H_F^{(1)}$ is proportional to the field intensity F.

The Rayleigh–Schrödinger perturbation treatment corresponds to expanding H_F, μ, and $|\mu\rangle$ in powers of an "ordering" parameter λ whose value is unity. Thus,

$$H_F = H_F^{(0)} + \lambda H_F^{(1)} \tag{5.3}$$

Furthermore, we assume that

$$\mu = \sum_{n=0}^{\infty} \lambda^n \mu^{(n)} \quad \text{and} \quad |\mu\rangle = \sum_{n=0}^{\infty} \lambda^n |\mu^{(n)}\rangle \tag{5.4}$$

where the nth order terms $\mu^{(n)}$ and $|\mu^{(n)}\rangle$ are proportional to F^n. The Rayleigh–Schrödinger perturbation equations are obtained by substituting Eqs. 5.3 and 5.4 into (2.43) and requiring that the coefficient of each power of λ vanish (see Hirschfelder and Certain[68])

$$\begin{aligned}(H_F^{(0)} - \mu^{(0)})|\mu^{(n)}\rangle + (H_F^{(1)} - \mu^{(1)})|\mu^{(n-1)}\rangle \\ = \mu^{(2)}|\mu^{(n-2)}\rangle + \cdots + \mu^{(n)}|\mu^{(0)}\rangle\end{aligned} \tag{5.5}$$

If $n=1$, the terms on the right-hand side vanish. It is convenient to use intermediate normalization so that

$$\langle \mu^{(0)}|\mu^{(n)}\rangle = \delta_{0,n} \tag{5.6}$$

We expand $|\mu^{(0)}\rangle$ and all of the higher order $|\mu^{(n)}\rangle$ in our basis set, which s complete and orthonormal,

$$\langle A,j|B,j'\rangle = 0 \qquad \text{and} \qquad \langle A,j|A,j'\rangle = \delta_{j,j'} = \langle B,j|B,j'\rangle$$

Let us define the two sets of resolvent operators

$$\begin{aligned} R_{A,k} &= \frac{-|B,k\rangle\langle B,k|}{(\varepsilon - i\delta)} + \sum_{\substack{p \\ p\neq k}} \left[\frac{|A,p\rangle\langle A,p|}{k-p} + \frac{|B,p\rangle\langle B,p|}{(k-p)-(\varepsilon-i\delta)} \right] \\ R_{B,k} &= + \frac{|A,k\rangle\langle A,k|}{(\varepsilon - i\delta)} + \sum_{\substack{p \\ p\neq k}} \left[\frac{|A,p\rangle\langle A,p|}{(k-p)+(\varepsilon-i\delta)} + \frac{|B,p\rangle\langle B,p|}{k-p} \right] \end{aligned} \tag{5.7}$$

If $\delta=0$ *and* $\varepsilon = n_\varepsilon$ (where n_ε is an integer), there is a degeneracy between the states $|A,j\rangle$ and $|B,j-n_\varepsilon\rangle$. Such degeneracies are treated in Section VI. Throughout the present section we assume that the states are nondegenerate. Then, if $|\mu_{A,k}^{(0)}\rangle = |A,k\rangle$, it is easy to prove by direct substitution into

(5.5) that

$$\begin{aligned} |\mu_{A,k}^{(n)}\rangle &= R_{A,k}\left[H_F^{(1)}|\mu_{A,k}^{(n-1)}\rangle - \sum_{s=1}^{n-1} \mu_{A,k}^{(s)}|\mu_{A,k}^{(n-s)}\rangle \right] \\ \mu_{A,k}^{(n)} &= \langle A,k|H_F^{(1)}|\mu_{A,k}^{(n-1)}\rangle \end{aligned} \tag{5.8}$$

Or, if $|\mu_{B,k}^{(0)}\rangle = |B,k\rangle$,

$$\begin{aligned} |\mu_{B,k}^{(n)}\rangle &= R_{B,k}\left[H_F^{(1)}|\mu_{B,k}^{(n-1)}\rangle - \sum_{s=1}^{n-1} \mu_{B,k}^{(s)}|\mu_{B,k}^{(n-s)}\rangle \right] \\ \mu_{B,k}^{(n)} &= \langle B,k|H_F^{(1)}|\mu_{B,k}^{(n-1)}\rangle \end{aligned} \tag{5.9}$$

Equations 5.5 and 5.7 through 5.9 are valid irrespective of whether H_F is Hermitian. If δ is not zero, then H_F is *not* Hermitian and many of the other relations which are commonly used in Rayleigh–Schrödinger perturbation theory are *not* correct. For example, if $\delta = 0$,

$$\begin{aligned} \mu^{(2n)} &= \langle \mu^{(n-1)}|H_F^{(1)}|\mu^{(n)}\rangle - \sum_{s=1}^{n-1}\sum_{r=1}^{n} \mu^{(2n-s-r)}\langle \mu^{(s)}|\mu^{(r)}\rangle \\ \mu^{(2n+1)} &= \langle \mu^{(n)}|H_F^{(1)}|\mu^{(n)}\rangle - \sum_{s=1}^{n}\sum_{r=1}^{n} \mu^{(2n+1-s-r)}\langle \mu^{(s)}|\mu^{(r)}\rangle \end{aligned} \tag{5.10}$$

However, if $\delta \neq 0$, Eq. 5.10 may not be valid.

We start the solution of a nondegenerate problem by picking the zeroth-order energy. According to the zeroth-order equation, $\mu^{(0)}$ can either be equal to some integer k or else $k + \varepsilon - i\delta$. However, the final solutions for $a(\tau)$ and $b(\tau)$ are independent of the choice of k since adding k to $\mu^{(0)}$ would only result in multiplying the ϕ's by $\exp(ik\tau)$. Thus, we need to consider the cases of $\mu^{(0)} = 0$ and $\mu^{(0)} = \varepsilon - i\delta$. These two choices for $\mu^{(0)}$ generate the two linearly independent Floquet modes.

For a particular value of $\mu^{(0)}$, the function $|\mu^{(0)}\rangle$ is determined by the zeroth-order equation and the normalization condition. Then the first-order equation together with the normalization condition determines both $\mu^{(1)}$ and $|\mu^{(1)}\rangle$. Thus, we can proceed in this step-by-step process to determine the solutions accurate through any desired order of the perturbation. The first few terms in the Floquet modes are found to be as follows.

1. First Floquet Mode

$|\mu_1^{(0)}\rangle = |A,0\rangle$ and $\mu_1^{(0)} = 0.$

$$\mu_1^{(1)} = 0 = \mu_1^{(3)} \quad \text{and} \quad \mu_1^{(2)} = \frac{2\alpha^2(\varepsilon - i\delta)}{1-(\varepsilon - i\delta)^2}$$

$$|\mu_1^{(1)}\rangle = -\alpha\left[\frac{|B,1\rangle}{1+\varepsilon - i\delta} - \frac{|B,-1\rangle}{1-\varepsilon + i\delta}\right] \tag{5.11}$$

$$|\mu_1^{(2)}\rangle = \frac{\alpha^2}{2}\left[\frac{|A,2\rangle}{1+\varepsilon - i\delta} + \frac{|A,-2\rangle}{1-\varepsilon + i\delta}\right]$$

$$+\alpha\beta\left[\frac{|B,-2\rangle}{(1-\varepsilon+i\delta)(2-\varepsilon+i\delta)} - \frac{2|B,0\rangle}{1-(\varepsilon-i\delta)^2} + \frac{|B,2\rangle}{(1+\varepsilon-i\delta)(2+\varepsilon-i\delta)}\right]$$

The time-dependent solutions $a_1(\tau)$ and $b_1(\tau)$ are then given by (2.47), which provides the link between the static and dynamic formulations. Thus, through the second-order in the field strength,

$$a_1(\tau) = \exp(-i\mu_1\tau)\left[1 + \frac{\alpha^2}{2}\left(\frac{\exp(2i\tau)}{1+\varepsilon - i\delta} + \frac{\exp(-2i\tau)}{1-\varepsilon+i\delta}\right) + \cdots\right]$$

and

$$b_1(\tau) = \exp(-i\mu_1\tau)\phi_{b1}(\tau) \tag{5.12}$$

where

$$\mu_1 = \frac{2\alpha^2(\varepsilon - i\delta)}{1-(\varepsilon-i\delta)^2} + \cdots$$

$$\phi_{b1}(\tau) = \alpha\left[\frac{\exp(-i\tau)}{1-\varepsilon+i\delta} - \frac{\exp(i\tau)}{1+\varepsilon-i\delta}\right]$$

$$+\alpha\beta\left[\frac{\exp(2i\tau)}{(1+\varepsilon-i\delta)(2+\varepsilon-i\delta)} - \frac{2}{1-(\varepsilon-i\delta)^2}\right.$$

$$\left.+\frac{\exp(-2i\tau)}{(1-\varepsilon+i\delta)(2-\varepsilon+i\delta)}\right] + \cdots \tag{5.13}$$

2. *Second Floquet Mode*

$|\mu_2^{(0)}\rangle = |B,0\rangle$ and $\mu_2^{(0)} = \varepsilon - i\delta.$

$$\mu_2^{(1)} = 0 = \mu_2^{(3)} \quad \text{and} \quad \mu_2^{(2)} = -\frac{2\alpha^2(\varepsilon - i\delta)}{1-(\varepsilon - i\delta)^2}$$

$$|\mu_2^{(1)}\rangle = \alpha\left[\frac{|A,-1\rangle}{1+\varepsilon-i\delta} - \frac{|A,1\rangle}{1-\varepsilon+i\delta}\right] + \beta[|B,-1\rangle - |B,1\rangle] \quad (5.14)$$

$$|\mu_2^{(2)}\rangle = \frac{\alpha^2}{2}\left[\frac{|B,-2\rangle}{1+\varepsilon-i\delta} + \frac{|B,2\rangle}{1-\varepsilon+i\delta}\right]$$

$$+\alpha\beta\left[\frac{|A,-2\rangle}{2+\varepsilon-i\delta} + \frac{|A,2\rangle}{2-\varepsilon+i\delta}\right] + \frac{\beta^2}{2}[|B,-2\rangle + |B,2\rangle]$$

The time-dependent solutions are then determined by (2.47)

$$a_2(\tau) = \exp(-i\mu_2\tau)\left[\alpha\left[\frac{\exp(-i\tau)}{1+\varepsilon-i\delta} - \frac{\exp(i\tau)}{1-\varepsilon+i\delta}\right]\right.$$

$$\left. + \alpha\beta\left[\frac{\exp(-2i\tau)}{2+\varepsilon-i\delta} + \frac{\exp(2i\tau)}{2-\varepsilon+i\delta}\right] + \cdots\right]$$

$$b_2(\tau) = \exp(-i\mu_2\tau)\left[1 + \beta[\exp(-i\tau) - \exp(i\tau)]\right.$$

$$+ \frac{\beta^2}{2}[\exp(-2i\tau) + \exp(2i\tau)]$$

$$\left. + \frac{\alpha^2}{2}\left[\frac{\exp(-2i\tau)}{1+\varepsilon-i\delta} + \frac{\exp(2i\tau)}{1-\varepsilon+i\delta}\right] + \cdots\right] \quad (5.15)$$

where

$$\mu_2 = (\varepsilon - i\delta) - \frac{2\alpha^2(\varepsilon - i\delta)}{1-(\varepsilon-i\delta)^2} + \cdots \quad (5.16)$$

B. Convergence of the (SG)′ Technique

Each of the coefficients of the $|A,j\rangle$ or $|B,j\rangle$ in $|\mu^{(m)}\rangle$ can be expressed as the sum of terms each of which is the product of m factors of the form*

$$G=\frac{\langle k,n|H_F^{(1)}|k',n'\rangle}{\mu_{k,n}^{(0)}-\mu_{k_0,0}^{(0)}} \tag{5.17}$$

Here the k's can be either A or B, $k_0,0$ corresponds to the zeroth-order state, and the denominator of G can never vanish. The only nonvanishing numerators of G (or matrix components of $H_F^{(1)}$) are either $\alpha[\delta_{n,n'+1}+\delta_{n,n'-1}]$ or $\beta[\delta_{n,n'+1}+\delta_{n,n'-1}]$. Because of this special form of the matrix elements of $H_F^{(1)}$, the values of j that correspond to nonvanishing coefficients of the $|A,j\rangle$ or $|B,j\rangle$ range from $-m$ to m. Similarly, the value of n in (5.17) must range between $-m$ and m. The denominators of G have three possible values $n, n+(\varepsilon-i\delta)$, and $n-(\varepsilon-i\delta)$ where $n\neq 0$. Thus, there are only five possible expressions for the absolute value of G,

$$|G|=0, \qquad \frac{\alpha}{|n|}, \qquad \frac{\beta}{|n|}, \qquad \frac{\alpha}{\left[(n+\varepsilon)^2+\delta^2\right]^{1/2}},$$

and

$$\frac{\beta}{\left[(n+\varepsilon)^2+\delta^2\right]^{1/2}} \tag{5.18}$$

Also, note that the denominator of G can never vanish. In order for the (SG)′ technique to converge rapidly, all of the possible values of $|G|$ should be much less than one for $-N_F<n<N_F$ where the "ϕ-parts" of the Floquet solutions are only required to be correct through the N_Fth order in the field strength. The requirements that the (SG)′ technique converges rapidly are therefore:

$$\alpha,\beta\ll 1$$

and if $N_F<n_\varepsilon$ [where n_ε is the integer that minimizes $(\varepsilon-n_\varepsilon)^2$], then

$$\alpha,\beta\ll\left[(\varepsilon-N_F)^2+\delta^2\right]^{1/2} \tag{5.19}$$

or, if $N_F>n_\varepsilon$, then

$$\alpha,\beta\ll\left[(\varepsilon-n_\varepsilon)^2+\delta^2\right]^{1/2}$$

*See Hirschfelder, Byers Brown, and Epstein.[69]

The expansion of the Floquet μ's and the ϕ's is generally *asymptotically* convergent. However, in this sense the solutions are frequently rapidly convergent and convenient. When $\delta=0$, all of the standard Rayleigh–Schrödinger formalism applies. This can be useful in examining the convergence of the series. Also, if $\delta=0$, the Wigner $(2N_F+1)$-rule applies: "If the ϕ-parts of the Floquet solutions are known through the N_F-th order, the value of μ correct through the $(2N_F+1)$-th order can be ascertained." (See Hirschfelder et al.[69])

Whenever ε *is almost or exactly equal to an integer, resonance occurs*, and some of the denominators in the perturbation expansion nearly vanish. This occurs when the energy difference between the two states $(W_b - W_a)$ is equal to $\hbar\omega$. Special techniques such as the (SSa)′ are required to treat either resonances or near-resonances.

C. Discussion of Sen Gupta's and Sambe's Treatments

The derivations used by Sen Gupta[40,41,42] and by Sambe[43] are quite different from our (SG)′ but the results are the same when applied to the two-level system. They solve the time-dependent equations without first making a Fourier series expansion to obtain the static analog. They assume that $a(\tau)$ and $b(\tau)$ have the Floquet form without any secular terms. Substituting Eqs. 2.32 into (1.10) and (1.11) gives the equations for $\phi_a(\tau)$ and $\phi_b(\tau)$,

$$\begin{aligned}\dot{\phi}_a(\tau) &= i\mu\phi_a(\tau) - 2i\lambda\alpha\cos\tau\,\phi_b(\tau)\\ \dot{\phi}_b(\tau) &= i\mu\phi_b(\tau) - i(\varepsilon - i\delta)\phi_b(\tau) - 2i\lambda\beta\cos\tau\,\phi_b(\tau) - 2i\lambda\alpha\cos\tau\,\phi_a(\tau)\end{aligned} \tag{5.20}$$

where the ordering parameter λ appears multiplying those terms that are proportional to the field strength. Perturbation equations are obtained by expanding μ, $\phi_a(\tau)$, and $\phi_b(\tau)$ in power series in λ and requiring that the coefficient of each power of λ vanishes. Starting with the zeroth-order, the nth order equation is then solved for $\mu^{(n)}$ by requiring that both $\phi_a^{(n)}(\tau)$ and $\phi_b^{(n)}(\tau)$ be periodic with period of 2π. The two solutions to the zeroth-order equation are

$$\mu_1^{(0)}=0, \qquad \phi_{a1}^{(0)}(\tau)=c_1, \qquad \text{and} \qquad \phi_{b1}^{(0)}(\tau)=0$$

and

$$\mu_2^{(0)}=\varepsilon - i\delta, \qquad \phi_{a2}^{(0)}(\tau)=0, \qquad \text{and} \qquad \phi_{b2}^{(0)}(\tau)=c_2$$

where the constants c_1 and c_2 are arbitrary. These two solutions are the zeroth-order of the two Floquet modes. Details of Sen Gupta's treatment are given in Chapter 8 of Dion's[56] thesis.

D. Discussion of the Langhoff–Epstein–Karplus Treatment

Langhoff, Epstein, and Karplus[38] and Epstein[70] have developed a "steady-state time-dependent perturbation theory" for general many-level systems. When applied to our two-level problem it is very similar to making a field strength expansion of the equations for the quotients $b(\tau)/a(\tau)$ and $a(\tau)/b(\tau)$. The interesting difference between their approach and ours is that they establish their steady state by invoking an adiabatic turn-on of the radiation field, whereas we obtain a steady state by requiring that the solution be a Floquet mode.

First, let us explain the technique which we would use in expanding the $b(\tau)/a(\tau)$ and $a(\tau)/b(\tau)$ quotients in powers of the field strength to obtain results identical with the (SG)′. In this procedure, no Fourier expansions are made.* Instead, we start with (2.26) and (2.29). In these equations we replace every α and β by $\lambda\alpha$ and $\lambda\beta$, respectively; then we expand $\Phi_1(\tau)$ and $\Phi_2(\tau)$ in power series in the ordering parameter λ. By requiring that the coefficient of each power of λ vanish, we obtain two sets of perturbation equations. These equations can be solved successively for the $\Phi^{(n)}(\tau)$ functions with the constraint that each of the $\Phi^{(n)}(\tau)$ be periodic. The approximations to the two Floquet modes $a_1(\tau)$, $b_1(\tau)$ and $a_2(\tau)$, $b_2(\tau)$ are then recovered by using Eqs. 2.25, 2.27, and 2.30. Details of this technique are given in Dion's[56] thesis (Chapter 8).

In contrast, Langhoff, Epstein, and Karplus have obtained their steady-state solutions by using an adiabatic turn-on of the time-dependent perturbation. If the time-dependent Hamiltonian that they wish to consider is $H^{(0)}(\mathbf{r})+H^{(1)}(\mathbf{r},t)$, where $H^{(1)}(\mathbf{r},t)$ is periodic in the time, then the adiabatic turn-on procedure consists of:

(1) Replace $H^{(1)}(\mathbf{r},t)$ by $f(t)H^{(1)}(\mathbf{r},t)$ where $f(t)$ is called the "switching function." The switching function is required to vanish when $t=-\infty$ and very gradually approach unity at $t=0$. It suffices to use†

$$\begin{aligned} f(t)&=\exp(gt) \qquad \text{when } t<0 \\ f(t)&=1 \qquad \text{when } t>0 \end{aligned} \tag{5.21}$$

where g is a positive real number. At the end of the calculation, the limit is taken as g approaches zero.

* Bloch and Siegert[2] approached the problem of finding the shift of the two-level system's main resonance frequency as a function of field strength by considering the quotient equations. Their method of solution is, however, quite different from the method of solution about to be discussed. (See Section VI.)

† Note that $f(t)$ as given by (5.21) has a discontinuous first derivative at $t=0$. However, this does not affect the results. Musher[71] recommends using another form for the switching function that does not have this discontinuity.

(2) At time $t=-\infty$, the system is in a pure eigenstate of the unperturbed Hamiltonian. For our two-state system this would mean that the two steady-state solutions would be obtained by taking

$$a_1(-\infty)=c_1, b_1(-\infty)=0 \quad \text{and} \quad a_2(-\infty)=0, b_2(-\infty)=c_2$$

where c_1 and c_2 are constants. Details of the Langhoff–Epstein–Karplus technique as applied to our two-level problem are given in Chapter 8 of Dion's[56] thesis.

The adiabatic turn-on procedure was first developed by Born and Fock.[72] The equivalence between the resulting steady-state solutions and the Floquet normal modes has been discussed in the recent literature. Young, Deal, and Kestner[44] call the Floquet particular solutions 'quasiperiodic states" and assert that they are the steady-state solutions. Okuniewicz[47] gives a mathematical discussion of the necessary and sufficient conditions for the existence of the quasiperiodic solutions. Young and Deal[45] prove that an adiabatically turned-on periodic perturbation will put a quantum system into a Floquet normal mode. Okuniewicz[46] and Sambe[43] discuss how, after the correspondence between the Floquet solutions and the steady-state solutions has been made, the problem of a quantum system in a periodic time-dependent perturbation may be treated by borrowing some of the techniques used in the time-independent quantum theory. This correspondence is also stressed in papers by Sen Gupta,[40,41,42] Hicks, Hess, and Cooper,[48] as well as Salzman.[49]

VI. THE SALWEN–WINTER–SHIRLEY (SWS)′ PARTITIONING PERTURBATION TECHNIQUE WITH $\delta=0$

When $\delta=0$ and the value of ε approaches a positive integer n_ε, the two-level system approaches a state of resonance in which the Fourier components $|A,0\rangle$ and $|B,-n_\varepsilon\rangle$ become degenerate with respect to the zero-order Floquet Hamiltonian $H_F^{(0)}$ defined by (5.1). In the vicinity of such a resonance, the nondegenerate Rayleigh–Schrödinger perturbation treatment given in the (SG)′ technique of Section V no longer gives useful results. Furthermore, the degenerate Rayleigh–Schrödinger perturbation theory is only applicable when ε is *exactly* equal to n_ε. However, the partitioning perturbation treatments of either Löwdin[73,74] or Certain and Hirschfelder[75,76]* apply to either a degeneracy or near-degeneracy. In the (SWS)′ technique we use the Certain–Hirschfelder formulation without any modifications. Our two-level problem is particularly simple since the

*The Certain–Dion–Hirschfelder[77] paper applies the Certain–Hirschfelder[75,76] formalism to a particular example involving the approach to degeneracy.

degeneracy is resolved in the second order of perturbation and the two degenerate states do not interact with each other until the n_εth order. Indeed, the values of the μ's are not affected by the degeneracy until the $2n_\varepsilon$th order. For $n_\varepsilon = 1$, we obtain a quadratic equation that determines the μ's accurate through the third order in the field strength. Our results are given in both tabular and graphical form. They are to be compared with the results of Autler and Townes[4] (see Eqs. 3.24 through 3.27).

We label this section the (SWS)′ technique because Salwen,[50] Winter,[5] and Shirley[6,25] were the first ones to use a perturbation theory approach to a consideration of resonances. All three of them took $\delta = 0$. Salwen and Shirley also assumed that $\beta = 0$; whereas, Winter allowed β to be nonvanishing. In effect, all three used a static formulation to (2.43). However, their techniques were quite different from ours. Salwen and Shirley use a perturbation procedure such that only approximate solutions to the higher perturbation equations might be obtained. Winter bases his treatment on a theory of Heitler[78] (Chapter 4, Section 14) and his results agree with our (SWS)′.

The reason why we restrict our (SWS)′ formulation to $\delta = 0$ is that the Floquet Hamiltonian for nonvanishing δ is symmetrical but non-Hermitian. A non-Hermitian Hamiltonian does not cause any serious difficulties with respect to a Rayleigh–Schrödinger perturbation procedure such as that which is used in the (SG)′ method of Section V or the (DRS) method of Section VIII because, in those cases, the perturbation equation for each order is solved exactly in terms of the basis set. However, the formulation of the partitioning technique is based upon the Rayleigh–Ritz variational principle for the energy. If the Hamiltonian is non-Hermitian, then such a variation no longer applies.*

*If the Hamiltonian is symmetric, then the matrix of the Hamiltonian can be diagonalized by the use of *simple* scalar products instead of the usual *Hermitian* scalar products. Thus, one can derive in terms of simple scalar products most of the usual relations. For example, one might define an approximate energy in the form

$$E_{\text{app}} = \frac{\int \psi_{\text{app}} H \psi_{\text{app}} \, d\tau}{\int \psi_{\text{app}}^2 \, d\tau}$$

However, it is clear that this could not be a variational principle for E_{app} since, although $\int \psi_{\text{app}}^* \psi_{\text{app}} \, d\tau$ is positive, the integral $\int \psi_{\text{app}}^2 \, d\tau$ can have *any* value: positive, negative, or complex by making a suitable choice for ψ_{app}.

If $\psi = \psi_r + i\psi_i$ is the exact eigenfunction for a complex symmetric Hamiltonian H and E is its exact complex energy, then it is easy to show that

$$\int [\psi_r (H - E) \psi_r + \psi_i (H - E) \psi_i] \, d\tau = 0$$

This equation might be used as the basis for establishing a new variational principle. [See Epstein[79] (p. 4 *et seq.*) for suggestions as to how to proceed.]

If δ does not vanish, it is altogether possible that reasonable results might be obtained by using the partitioning technique with $\delta=0$ and then in the final results replace ε by $(\varepsilon-i\delta)$ every place where it occurs. However, since the justification of the partitioning technique is based upon the Rayleigh–Ritz variational principle which does not apply when $\delta\neq 0$, there would be no guarantee as to the precision of the results.

A. The Formulation of the (SWS)′ Technique

We start with the Floquet Hamiltonian $H=H_F^{(0)}+H_F^{(1)}$ as given by (5.1) through (5.3) with the exception that $\delta=0$. The basic equations that we seek to solve are

$$H_F|\mu_1\rangle=\mu_1|\mu_1\rangle \qquad \text{and} \qquad H_F|\mu_2\rangle=\mu_2|\mu_2\rangle \tag{6.1}$$

where $|\mu_1\rangle$ and $|\mu_2\rangle$ are defined so that

$$\begin{aligned}\lim_{F\to 0}|\mu_1\rangle&=|A,0\rangle=|\mu_1^{(0)}\rangle\\ \lim_{F\to 0}|\mu_2\rangle&=|B,-n_\varepsilon\rangle=|\mu_2^{(0)}\rangle\end{aligned} \tag{6.2}$$

Here n_ε is the integer that minimizes $|\varepsilon-n_\varepsilon|$. Since

$$H_F^{(0)}|A,0\rangle=0|A,0\rangle \qquad \text{and} \qquad H_F^{(0)}|B,-n_\varepsilon\rangle=(-n_\varepsilon+\varepsilon)|B,-n_\varepsilon\rangle \tag{6.3}$$

the two states are degenerate or almost degenerate. The difficulty with the degeneracy or almost degeneracy in the nondegenerate Rayleigh–Schrödinger perturbation theory arises in the n_εth order where the expression for $|\mu_1^{(n_\varepsilon)}\rangle$ is the sum of terms including a term $K_A|B,-n_\varepsilon\rangle/(n_\varepsilon-\varepsilon)$ that becomes infinite as $\varepsilon\to n_\varepsilon$ (for an example see Eq. 5.11). Similarly, $|u_2^{(n_\varepsilon)}\rangle$ contains a term $K_B|A,0\rangle/(n_\varepsilon-\varepsilon)$. All of the other terms in $|u_1^{(n)}\rangle$ and $|u_2^{(n)}\rangle$ are well behaved if $1<n<n_\varepsilon$.

This leads to an interesting proposition: Since the Rayleigh–Schrödinger expansions of $|\mu_1\rangle$ and $|\mu_2\rangle$ are well behaved through the $(n_\varepsilon-1)$st order, we can use Eq. 5.10 to calculate μ_1 and μ_2 through the $(2n_\varepsilon-1)$st order. *The values of μ_1 and μ_2 through the $(2n_\varepsilon-1)$st order show no hint of ε lying in the vicinity of a resonance*. We can prove this assertion in either of two ways: First, suppose that the value of ε is slightly different from n_ε, then there is no question about the applicability of Eq. 5.10. Second, suppose that ε is exactly equal to n_ε, then the Hirschfelder–Certain[68] degenerate Rayleigh–Schrödinger perturbation theory gives the same result. Our conclusion is that as n_ε varies from 1 to 2 to 3 to ..., the resonances become progressively much weaker and less noticeable.

Thus, we shall concentrate on the principal resonance in the vicinity of $\varepsilon=1$. For this case, partitioning perturbation theory is useful. Here the degeneracy is between

$$|\mu_1^{(0)}\rangle=|A,0\rangle \qquad \text{and} \qquad |\mu_2^{(0)}\rangle=|B,-1\rangle \tag{6.4}$$

We know from (5.8) and (5.9) that the first-order Rayleigh–Schrödinger wavefunctions would be, if there were no degeneracy,

$$\begin{aligned} |\mu_1^{(1)}\rangle&=|\chi_1^{(1)}\rangle+\frac{\alpha|B,-1\rangle}{(1-\varepsilon)} \\ |\mu_2^{(1)}\rangle&=|\chi_2^{(1)}\rangle-\frac{\alpha|A,0\rangle}{(1-\varepsilon)} \end{aligned} \tag{6.5}$$

Here the $|\chi_1^{(1)}\rangle$ and $|\chi_2^{(1)}\rangle$ are the parts of $|\mu_1^{(1)}\rangle$ and $|\mu_2^{(1)}\rangle$ that remain well behaved throughout the resonance region, where ε is equal to or almost equal to 1. Thus,

$$\begin{aligned} |\chi_1^{(1)}\rangle&=\frac{-\alpha|B,1\rangle}{(1+\varepsilon)} \\ |\chi_2^{(1)}\rangle&=\beta\big[|B,-2\rangle-|B,0\rangle\big]+\frac{\alpha|A,-2\rangle}{(1+\varepsilon)} \end{aligned} \tag{6.6}$$

Then we form the functions

$$|\chi_1(1)\rangle=|A,0\rangle+|\chi_1^{(1)}\rangle$$

$$|\chi_2(1)\rangle=|B,-1\rangle+|\chi_2^{(1)}\rangle$$

The "best" wavefunctions through the first order of the field strength are those linear combinations

$$|\mu(1)\rangle=C_1|\chi_1(1)\rangle+C_2|\chi_2(1)\rangle \tag{6.7}$$

that satisfy the equations

$$\sum_{k=1}^{2}\langle\chi_j(1)|H_F-\mu(1)|\chi_k(1)\rangle C_k=0 \qquad (j=1 \text{ or } 2) \tag{6.8}$$

The eigenvalues $\mu(1)$ are determined by the secular equation

$$\begin{vmatrix} H_{11}-\mu(1)S_{11} & H_{12}-\mu(1)S_{12} \\ H_{21}-\mu(1)S_{21} & H_{22}-\mu(1)S_{22} \end{vmatrix}=0 \tag{6.9}$$

Here the H's and the S's are the "energy" and overlap integrals, respectively. The values of the μ's that satisfy this secular equation are accurate through the third order of the perturbation and their eigenvectors $|\mu(1)\rangle$ are accurate through the first order.

The matrix elements corresponding to the $\chi(1)$'s are

$$H_{11} = \frac{-\alpha^2}{(\varepsilon+1)}$$

$$H_{12} = \alpha\left[1 + \frac{\beta^2}{\varepsilon+1}\right] = H_{21}$$

$$H_{22} = (\varepsilon-1)(1+2\beta^2) + \frac{2\varepsilon\alpha^2}{(\varepsilon+1)^2}$$

$$S_{11} = 1 + \frac{\alpha^2}{(\varepsilon+1)^2}$$

$$S_{12} = 0 = S_{21}$$

$$S_{22} = S_{11} + 2\beta^2$$

Thus, the two roots of the secular equation are

$$\mu(1)_{\pm} = \frac{1}{2}\left[\left(\frac{H_{22}}{S_{22}} + \frac{H_{11}}{S_{11}}\right)^2 \pm \left\{\left(\frac{H_{22}}{S_{22}} - \frac{H_{11}}{S_{11}}\right)^2 + \frac{4H_{12}^2}{S_{11}S_{22}}\right\}^{1/2}\right] \tag{6.10}$$

If $\beta = 0$, the sum of the two roots is

$$\mu(1)_+ + \mu(1)_- = \varepsilon - 1$$

which is in accord with Theorem 4. However, if $\beta \neq 0$, then the sum of the two roots differs from $(\varepsilon - 1)$ by a term that is fourth order in the field strength. Dropping this term, which is an error due to the truncation in the approximation procedure, and other terms that are fourth order in the field strength, for all values of ε,

$$\mu(1)_{\pm} = \frac{1}{2}\left\{(\varepsilon-1) \pm \left[(\varepsilon-1)^2 + \frac{8\alpha^2\varepsilon(\varepsilon+1)(1-2\beta^2)}{(\varepsilon+1)^2+\alpha^2}\right]^{1/2}\right\} \tag{6.11}$$

If $\varepsilon < 1$, then through terms in the third order of the field strength,

$$\mu(1)_+ = \frac{-2\alpha^2\varepsilon}{(\varepsilon^2-1)} + \cdots \tag{6.12}$$

in agreement with μ_1 as given by (5.11). If $\varepsilon = 1$,

$$\mu(1)_\pm = \pm\alpha\left[\frac{(1-2\beta^2)}{(1+(\alpha/2))}\right]^{1/2} \tag{6.13}$$

Thus, the value of $\mu(1)_\pm$ is independent of β through terms of the third order except in the resonance region $\varepsilon \approx 1$, where β makes a third-order contribution. Note that the argument of the square root in (6.11) is always positive so that $\mu(1)_+$ can never equal $\mu(1)_-$.

Table IV gives the error in the values of the Eq. 6.11 expression for $\mu(1)_+$. For small values of ε and (α/ε), it is very accurate.

TABLE IV. Error in $\mu(1)_+$ as Determined by Eq. (6.11)[a]

	$\mu(1)_+ - \mu_+$(exact)						
α/ε	$\varepsilon=0.6$	$\varepsilon=0.8$	$\varepsilon=0.9$	$\varepsilon=1.0$	$\varepsilon=1.1$	$\varepsilon=1.2$	$\varepsilon=1.4$
0.1	0.0	0.0	0.0	0.0	0.0	0.0	0.0
0.2	0.0	0.0	0.0	0.0	0.0001	0.0001	0.0002
0.3	0.0	0.0001	0.0001	0.0003	0.0005	0.0007	0.0016
0.4	0.0001	0.0004	0.0008	0.0014	0.0023	0.0035	0.0080
0.5	0.0003	0.0015	0.0026	0.0046	0.0076	0.0122	0.0315
0.6	0.0007	0.0038	0.0070	0.0124	0.0212	0.0359	0.1049
0.7	0.0018	0.0085	0.0162	0.0293	0.0522	0.0922	0.1389

[a]The values of μ_+(exact) are given in Table II. Here $\beta = 0 = \delta$.

From (6.8) it follows that

$$\left(\frac{C_1}{C_2}\right)_{\mu_+} = -\frac{H_{12}}{H_{11}-\mu(1)_+ S_{11}} \quad \text{and} \quad \left(\frac{C_2}{C_1}\right)_{\mu_-} = -\frac{H_{12}}{H_{22}-\mu(1)_- S_{22}} \tag{6.14}$$

When $\beta = 0$, $S_{22} = S_{11}$ and $\mu(1)_+ + \mu(1)_- = (H_{11}+H_{22})/S_{11}$; therefore, $(C_1/C_2)_{\mu_+} = -(C_2/C_1)_{\mu_-}$. In Fig. 1, the values of $(C_1/C_2)_{\mu_+}$ are plotted as a function of ε for various values of α/ε and for $\beta = 0 = \delta$. For small values of ε, $(C_1/C_2)_{\mu_+} > 1$ so that $|\mu_+\rangle$ is principally $|A,0\rangle$ and $-(C_2/C_1)_{\mu_-} > 1$

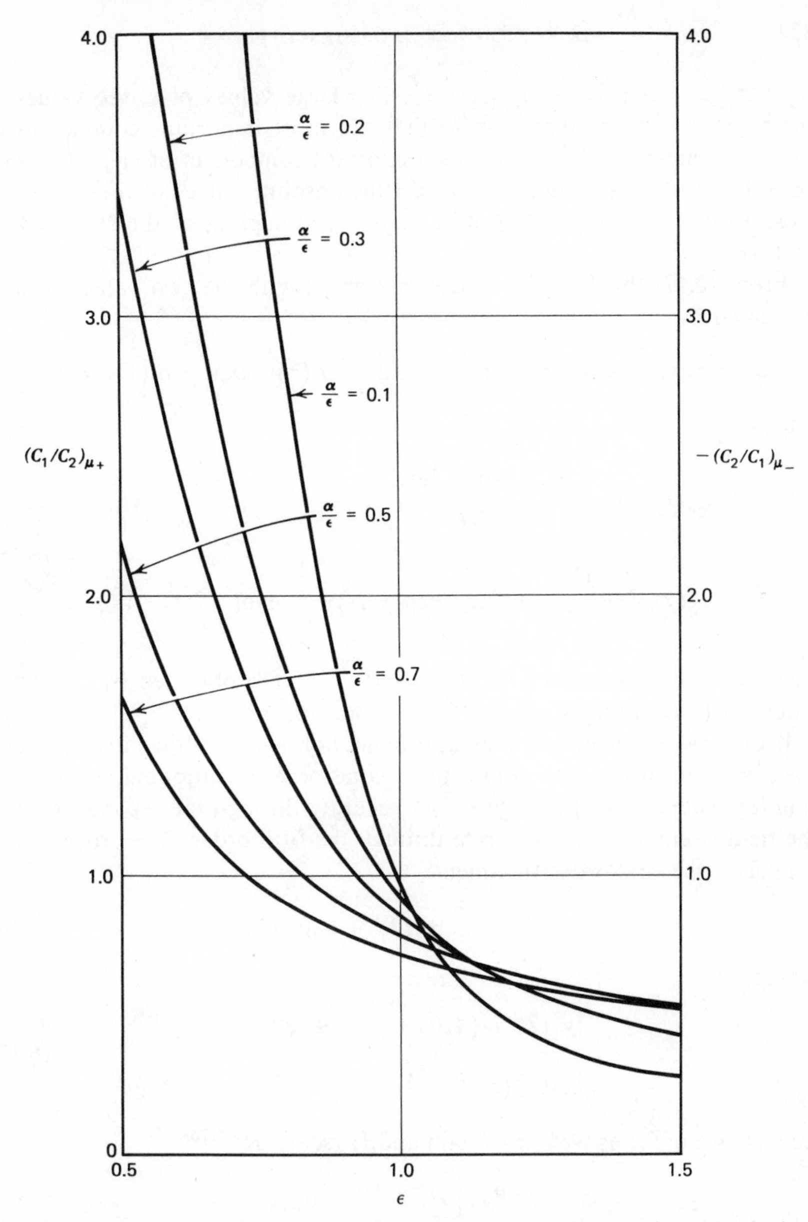

Fig. 1. $(C_1/C_2)_{\mu_+} = (C_2/C_1)_{\mu_-}$ as a function of ϵ for various values of α/ϵ and for $\beta = 0 = \delta$.

so that $|\mu_-\rangle$ is principally $|B,-1\rangle$. For large values of ε, the values of these ratios are less than one so that the most important component of $|\mu_-\rangle$ becomes $|A,0\rangle$ and the most important component of $|\mu_+\rangle$ becomes $|B,-1\rangle$. In the resonance region $\varepsilon\approx 1$ the absolute values of these ratios is close to unity corresponding to a very close coupling of the states $|A,0\rangle$ and $|B,-1\rangle$.

From (2.47) the Floquet modes associated with the two values of $\mu(1)$ are given by

$$a(\tau)=\exp[-i\mu(1)\tau]\phi_a(\tau) \qquad \text{and} \qquad b(\tau)=\exp[-i\mu(1)\tau]\phi_b(\tau)$$

where

$$\begin{aligned} \phi_a(\tau) &= C_1+\frac{C_2\alpha\exp(-2i\tau)}{(1+\varepsilon)} \\ \phi_b(\tau) &= \frac{-C_1\alpha e^{i\tau}}{(1+\varepsilon)}+C_2\{\exp(-i\tau)+\beta[\exp(-2i\tau)-1]\} \end{aligned} \tag{6.15}$$

The modes corresponding to $\mu(1)_+$ and $\mu(1)_-$ are obtained by using the values of $(C_1)_{\mu_\pm}$ and $(C_2)_{\mu_\pm}$ in (6.15).

If this first-order partitioning approximation is not sufficiently accurate, then one can proceed to obtain the second-order partitioning approximation [indicated by (2)] that gives $|\mu\rangle$ accurate through the second order in the field strength and μ accurate through the fifth order. The procedure is similar to the first-order treatment:

$$|\mu(2)\rangle = C_1|\chi_1(2)\rangle + C_2|\chi_2(2)\rangle \tag{6.16}$$

where

$$\begin{aligned} |\chi_1(2)\rangle &= |A,0\rangle + |\chi_1^{(1)}\rangle + |\chi_1^{(2)}\rangle \\ |\chi_2(2)\rangle &= |B,-1\rangle + |\chi_2^{(1)}\rangle + |\chi_2^{(2)}\rangle \end{aligned} \tag{6.17}$$

and using Eq. 5.7 as well as Certain and Hirschfelder[75,76*]

$$\begin{aligned} |\chi_1^{(2)}\rangle &= R_{A,0}\left[H_F^{(1)}|\chi_1^{(1)}\rangle - \alpha|\chi_2^{(1)}\rangle\right] \\ |\chi_2^{(2)}\rangle &= R_{B,-1}\left[H_F^{(1)}|\chi_2^{(1)}\rangle - \alpha|\chi_1^{(1)}\rangle\right] \end{aligned} \tag{6.18}$$

*In using Certain and Hirschfelder,[75,76] it is clear that our $|\chi\rangle$'s are their ϕ's and our μ's are their E's. However, our $R_{A,0}$ should be used in place of their $R^{(0)}$ in determining $|\chi_1^{(n)}\rangle$, and $R_{B,-1}$ should be used in place of $R^{(0)}$ in determining $|\chi_2^{(n)}\rangle$. Note that it is important to satisfy the normalization given by their equation (17).

Then the values of $\mu(2)$ and the values of the C_1 and C_2 are determined by solving the linear equations

$$\sum_{k=1}^{2} \langle\chi_j(2)|H_F-\mu(2)|\chi_k(2)\rangle C_k=0 \qquad (j=1 \text{ or } 2) \tag{6.19}$$

If necessary, the partitioning procedure can be carried out to arbitrarily high orders. Thus, to obtain $|\mu\rangle$ accurate through the nth order and μ accurate through the $(2n+1)$st order, it is necessary to determine $|\chi_1(n)\rangle$ and $|\chi_2(n)\rangle$, which are given by

$$\begin{aligned}|\chi_1(n)\rangle&=|A,0\rangle+\sum_{k=1}^{n}|\chi_1^{(k)}\rangle\\ |\chi_2(n)\rangle&=|B,-1\rangle+\sum_{k=1}^{n}|\chi_2^{(k)}\rangle\end{aligned} \tag{6.20}$$

Here

$$\begin{aligned}|\chi_1^{(n)}\rangle&=R_{A,0}\left[H_F^{(1)}|\chi_1^{(n-1)}\rangle-\sum_{k=1}^{n-1}\{|\varepsilon_{11}^{(k)}\chi_1^{(n-k)}\rangle+|\varepsilon_{21}^{(k)}\chi_2^{(n-k)}\rangle\}\right]\\ &\quad+|N_{11}^{(n)}A,0\rangle+|N_{21}^{(n)}B,-1\rangle\end{aligned} \tag{6.21}$$

$$\begin{aligned}|\chi_2^{(n)}\rangle&=R_{B,-1}\left[H_F^{(1)}|\chi_2^{(n-1)}\rangle-\sum_{k=1}^{n-1}\{|\varepsilon_{12}^{(k)}\chi_1^{(n-k)}\rangle+|\varepsilon_{22}^{(k)}\chi_2^{(n-k)}\rangle\}\right]\\ &\quad+|N_{12}^{(n)}A,0\rangle+|N_{22}^{(n)}B,-1\rangle\end{aligned}$$

where

$$\begin{aligned}N_{ij}^{(2m+1)}&=-\sum_{k=1}^{m}\langle\chi_i^{(k)}|\chi_j^{(2m+1-k)}\rangle\\ N_{ij}^{(2m)}&=-\frac{1}{2}\langle\chi_i^{(m)}|\chi_j^{(m)}\rangle-\sum_{k=1}^{n-1}\langle\chi_i^{(k)}|\chi_j^{(2m-k)}\rangle\end{aligned} \tag{6.22}$$

and

$$\begin{aligned}\varepsilon_{1j}^{(n)}&=\langle A,0|H_F^{(1)}|\chi_j^{(n-1)}\rangle-\sum_{k=1}^{n-1}\left[\langle A,0|\chi_1^{(k)}\rangle\varepsilon_{1j}^{(n-k)}+\langle A,0|\chi_2^{(k)}\rangle\varepsilon_{2j}^{(n-k)}\right]\\ \varepsilon_{2j}^{(n)}&=\langle B,0|H_F^{(1)}|\chi_j^{(n-1)}\rangle-\sum_{k=1}^{n-1}\left[\langle B,-1|\chi_1^{(k)}\rangle\varepsilon_{1j}^{(n-k)}+\langle B,-1|\chi_2^{(k)}\rangle\varepsilon_{2j}^{(n-k)}\right]\end{aligned} \tag{6.23}$$

Once the $|\chi_1(n)\rangle$ and $|\chi_2(n)\rangle$ have been determined, the values of $\mu(n)$ and the corresponding values of C_1 and C_2 are given by solutions to Eq. 6.19 with n replacing 2. Then $|\mu(n)\rangle$ is given by Eq. 6.16 with n replacing 2.

B. Related Approximate Treatments

Resonances in a two-level system are very interesting phenomena. Accordingly, much work has been done on the topic and, surprisingly, some of it is very recent. The recent interest was sparked by Chang and Stehle[9,10] who derived the Bloch-Siegert shift for a two-level system by a quantum electrodynamics calculation. They found disagreement between their results and Shirley's[25] results. Pegg and Series[51] and Pegg[8] checked Shirley's results and found them essentially correct. Stenholm,[11] using techniques he developed for treating the density matrix of a semiclassical two-level system (see, Stenholm's[12,13,14] other work), also confirmed Shirley's results. The matter was finally resolved when Cohen-Tannoudji, DuPont-Roc, and Fabre[52] redid the fully quantized calculation to find that, when correctly done, it and the semiclassical treatment were in full accord.

Although we always recommend (SWS)′ to treat the two-level system's resonances, we will explain and critically review five alternative approaches to treating the problem. All but one of them were formulated for $n_\varepsilon = 1$, and in none of them are both β *and* δ allowed to be nonvanishing. In all cases a lengthier discussion is found in Dion's[56] thesis. The list of five obviously does not exhaust all the work ever done on the problem. It does, however, give the flavor of how others approach the problem.

1. *Textbook Solutions:* $\beta = \delta = 0$; $n_\varepsilon = 1$

The usual textbook (see, for example, Landau and Lifshitz,[80] p. 139) treatment of the main ($\varepsilon \approx 1$) resonance usually requires both β and δ to vanish. It consists of letting $b(\tau)$ in (1.10) and (1.11) be given by

$$b(\tau) = b'(\tau)\exp(-i\tau)$$

The prescription is to neglect all terms in the equations for $a(\tau)$ and $b'(\tau)$ that are multiplied by a time-dependent coefficient to thereby obtain a solvable set of coupled equations with constant coefficients. The solutions obtained in this manner are the *exact* solutions to the Schrödinger equation for a spin $\frac{1}{2}$ particle in a *rotating* magnetic field. These, of course, were first derived by Rabi[1] and this is why these textbook solutions are often called the "Rabi approximation." They are also just the solutions that the zeroth-order partitioning theory gives, that is, the (SWS)′ technique with

$N_F=0$, $n_\varepsilon=1$, and $\beta=\delta=0$. The partitioning theory, therefore, gives a justification for the *ad hoc* textbook solutions as well as a way to systematically improve them.

2. *Bloch–Siegert Solutions: $\beta=\delta=0$; $n_\varepsilon=1$*

The first attempt to improve upon the Rabi rotating field solutions was made by Bloch and Siegert.[2] Their important work led to the realization that the main resonance of a spin $\frac{1}{2}$ system (or equivalently of any two-level system) in an oscillating linearly polarized magnetic (or electric) field would *not* occur at the frequency $\omega=\Delta W/\hbar$ (or $\varepsilon=1$), but would be slightly "shifted," that is, would occur at a frequency ω_0 such that

$$\omega_0=\frac{\Delta W}{\hbar}+\text{terms proportional to the field strength}$$

The Bloch–Siegert treatment consists of changing the independent variable in (1.10) and (1.11) (in which $\beta=\delta=0$) from τ to

$$x=2\tau+2\pi$$

They then derive the equation for the quotient $u(x)=a(x)/b(x)$ and solve it by assuming that $u(x)$ is a certain specified function of the new function $z(x)$. A nonlinear equation for $z(x)$ is obtained that is solved by perturbation theory. The equation for $z(x)$ is such that it has the zeroth-order solution $z^{(0)}(x)=0$. $u(0)$ is such that if $z(0)=0$, then $u(0)$ equals ∞, where this condition corresponds to the two-level system being in state $\psi_a(\mathbf{r})$ at $x=0$. Furthermore, letting $z(x)=0$, the transition amplitude $a^*(x)a(x)$ is identical to the one obtained by the Rabi rotating field approximation. Finding the higher order corrections to $z(x)$, therefore, corresponds to finding the corrections to the Rabi rotating field approximation.

We do not recommend the use of the Bloch–Siegert technique since it is cumbersome to work with, makes no reference to the known Floquet form of solution, and lacks ease of extension to quantum systems with more than two energy levels.

3. *Stevenson–Moulton Approach: $\beta=\delta=0$; $n_\varepsilon=1$*

Shortly after Bloch and Siegert's work was published, Stevenson[3] rederived their result for the "resonant shift" using a less cumbersome technique due to Moulton et al.[35]

Starting with equations equivalent to (1.10) and (1.11) in which he lets $\beta=\delta=0$, Stevenson makes use of Floquet theory by assuming

$$a(\tau)=\phi_a(\tau)\exp(-i\mu\tau); \qquad b(\tau)=\phi_b(\tau)\exp(-i\mu\tau)\exp(-i\tau) \quad (6.24)$$

Therefore,

$$\begin{aligned} \dot{\phi}_a(\tau) &= i\mu\phi_a(\tau) - i\alpha\left[1+\lambda\exp(-2i\tau)\right]\phi_b(\tau) \\ \dot{\phi}_b(\tau) &= i\mu\phi_b(\tau) - i(\varepsilon-1)\phi_b(\tau) - i\alpha\left[1+\lambda\exp(2i\tau)\right]\phi_a(\tau) \end{aligned} \tag{6.25}$$

where the ordering parameter λ multiplies only the time-dependent coefficients in (6.25). His procedure is to now expand μ, $\phi_a(\tau)$, and $\phi_b(\tau)$ in powers of λ to derive a set of perturbation equations that may be solved under the stipulations:

(1) $\mu^{(n)}$ is constant for all $n \geqslant 0$.

(2) $\phi_j^{(n)}(\tau) = \phi_j^{(n)}(\tau+\pi)$ for all $n \geqslant 0; j = a, b$.*

The zeroth-order solutions obtained in the Stevenson treatment are just the Rabi rotating field solutions and the higher order correction terms therefore improve the Rabi solutions. The Stevenson–Moulton approach differs from the Sen Gupta approach in that in the former the equations for the correction functions are *inhomogeneous coupled* (rather than homogeneous coupled) differential equations. The two are similar in that the $\mu^{(n)}$'s and the constants of integration (aside from normalization) are determined by the periodicity requirements on the ϕ-parts of the Floquet normal modes.

There is one difficulty with the Stevenson–Moulton approach: The N_Fth order correction can contain a term proportional to $(N_F - 1)$th order in the field strength ($N_F \geqslant 1$). The (SWS)′ approach is, therefore, the preferred way to treat $n_\varepsilon = 1$ since the N_Fth order correction term in it is proportional to the N_Fth order of field strength ($N_F \geqslant 0$). Furthermore, the (SWS)′ treatment applies to arbitrary n_ε.

4. *Pegg–Series Technique:* $\beta = \delta = 0$; n_ε *Arbitrary*

Pegg and Series[51,81] have developed techniques to handle the problem of quantum mechanical spin systems in periodic classical fields. Pegg[8] applies these techniques to the two-level system at its resonance frequencies. It is this latter paper that is discussed here.

The Pegg–Series technique is one of transforming the reference frame to one in which the transformed Hamiltonian is static. The transformed

*Note that because of transformations (6.24), $\phi_a(\tau)$ and $\phi_b(\tau)$ in (6.25) have periodicity π rather than periodicity 2π.

Hamiltonian thereby gives rise to an exactly solvable problem.

In detail, rewrite Eqs. 1.10 and 1.11 in matrix notation*:

$$i\dot{\mathsf{A}}(\tau)=\mathsf{H}(\tau)\mathsf{A}(\tau) \tag{6.26}$$

where

$$\mathsf{H}(\tau)=\begin{pmatrix} 0 & 2\alpha\cos\tau \\ 2\alpha\cos\tau & 2\beta\cos\tau+(\varepsilon-i\delta) \end{pmatrix} \tag{6.27}$$

The two-by-two solution matrix $\mathsf{A}(\tau)$ is given by

$$\mathsf{A}(\tau)=\begin{pmatrix} a_1(\tau) & a_2(\tau) \\ b_1(\tau) & b_2(\tau) \end{pmatrix} \tag{6.28}$$

where the solution pairs $\{a_1(\tau);\ b_1(\tau)\}$ and $\{a_2(\tau);\ b_2(\tau)\}$ form linearly independent solutions to (1.10) and (1.11).

The approach to solving Eq. (6.26) is this: let $\mathsf{S}(\tau)$ be a two-by-two time-dependent matrix that has an inverse and that transforms $\mathsf{A}(\tau)$ according to: $\mathsf{A}'(\tau)=\mathsf{S}(\tau)\mathsf{A}(\tau)$. The equation for $\mathsf{A}'(\tau)$ is

$$\begin{aligned} i\dot{\mathsf{A}}'(\tau)&=\left[i\dot{\mathsf{S}}(\tau)\mathsf{S}^{-1}(\tau)+\mathsf{S}(\tau)\mathsf{H}(\tau)\mathsf{S}^{-1}(\tau)\right]\mathsf{A}'(\tau) \\ &=\overline{\mathsf{H}}(\tau)\mathsf{A}'(\tau) \end{aligned} \tag{6.29}$$

If $\mathsf{S}(\tau)$ is chosen so as to make $\overline{\mathsf{H}}(\tau)$ *time independent*, Eq. 6.29 is easily solved by elementary methods. Furthermore, if $\mathsf{S}(\tau)$ is periodic *and* makes $\overline{\mathsf{H}}(\tau)$ static, then the Floquet solution for $\mathsf{A}(\tau)$ is obtained by this procedure.

The fundamental idea in the Pegg–Series approximation is to choose $\mathsf{S}(\tau)$ so that the time-dependent terms in $\overline{\mathsf{H}}(\tau)$ are small. Ignoring these small terms, an approximation to $\mathsf{A}(\tau)$ is obtained that is of the Floquet form as long as $\mathsf{S}(\tau)$ is chosen to be properly periodic. For example,

*Pegg formulates the problem in terms of spin operators. Our matrix formulation is, of course, equivalent. In recent papers Ansbacher[82,83] also uses a matrix formulation. Letting $\beta=\delta=0$, he replaced the Hamiltonian in our Eq. 1.1 by an approximate Hamiltonian H_a that makes the Schrödinger equation solvable. H_a, in general, contains adjustable parameters that are chosen to minimize $(H-H_a)^2$. A problem with such a treatment is that the choice of H_a is *ad hoc* rather than systematic.

following this procedure with the following specific choice of $\mathsf{S}(\tau)$ and with $\beta=0=\delta$,

$$(\mathsf{S})_{11}=\frac{1}{(\mathsf{S})_{22}}=\exp(-i\tau/2); \qquad (\mathsf{S})_{12}=(\mathsf{S})_{21}=0$$

gives the Rabi rotating field approximate solutions.

Pegg* gives two *ad hoc* transformations for the case of $n_\varepsilon=1$. Both are periodic and both, therefore, give rise to Floquet solutions. One gives characteristic exponents correct through F^3 and† "ϕ-parts" correct through F^1. The other gives characteristic exponents correct through F^5 and "ϕ-parts" correct through F^2. Neither one is extendable to the case of nonvanishing β.

For the case of $n_\varepsilon \geqslant 2$, he gives a general *ad hoc* prescription for $\mathsf{S}(\tau)$ that gives characteristic exponents correct through F^1 and "ϕ-parts" correct through F^0.

For all n_ε, (SWS)′ is preferred to the Pegg–Series approach because the (SWS)′ technique is easily extendable to $\beta\neq 0$, it is not *ad hoc*, and it gives a systematic way to improve on a solution of some given degree of accuracy to an even greater degree of accuracy.

In fact, the (SWS)′ technique may be viewed as a systematic way of determining a periodic transformation, $\mathsf{S}(\tau)$, which makes $\bar{\mathsf{H}}$ static. To see this, let $\delta=0$ and write the exact Floquet solutions to (6.26) as

$$\mathsf{A}(\tau)=\underline{\Phi}_F(\tau)\exp(-i\,\underline{\mu}_F\tau) \tag{6.30}$$

where

$$\underline{\Phi}_F(\tau)=\begin{pmatrix}\phi_{a1}(\tau) & \phi_{a2}(\tau)\\ \phi_{b1}(\tau) & \phi_{b2}(\tau)\end{pmatrix}; \qquad \underline{\mu}_F=\begin{pmatrix}\mu_1 & 0\\ 0 & \mu_2\end{pmatrix} \tag{6.31}$$

$\underline{\mu}_F$ is, therefore, a matrix of characteristic exponents and $\underline{\Phi}_F(\tau)$ is the matrix of "ϕ-parts." Now if $\mathsf{S}(\tau)=\mathsf{C}\underline{\Phi}_F^{-1}(\tau)$ (where C is a two-by-two nonsingular matrix of constants) and $\underline{\Phi}_F^{-1}(\tau)$ is the inverse of $\underline{\Phi}_F(\tau)$, then

$$\bar{\mathsf{H}}(\tau)=\mathsf{C}\,\underline{\mu}_F\mathsf{C}^{-1}$$

which is clearly a (in general nondiagonal) matrix of constants.

*There is a typographical error in Pegg's[8] paper. His equation (5) should read:

$$\hat{S}(t)=\exp\{i\hat{J}_Z(\alpha\sin 2\omega t+(p+1)\omega t)\}\hat{R}^{-1}(\theta)\exp(-i\hat{J}_Z\omega t).$$

A similar correction must be made in his equation (6).

†By "F^n" we, of course, mean nth order in the field strength.

$\mathsf{S}(\tau)$ was chosen so that the matrix C merely scrambles the elements of $\underline{\Phi}_F^{-1}(\tau)$. Since the (SWS)′ technique is equivalent to finding linear combinations of the "ϕ-parts" of the Floquet solutions, (SWS)′ is a way of *systematically* finding a properly periodic matrix $\mathsf{S}(\tau)$ which makes $\overline{\mathsf{H}}(\tau)$ static.

5. *Silverman–Pipkin Technique:* $\beta=0, \delta\neq 0$; $n_\varepsilon=1$

Silverman and Pipkin[7]* have studied the two-level system letting δ assume nonvanishing values and considering $n_\varepsilon=1$.

They, too, give an *ad hoc* prescription for finding the transformation $\mathsf{S}(\tau)$ in (6.29). Their choice of $\mathsf{S}(\tau)$ leads to Floquet solutions correct through F^1 in the characteristic exponents and correct through F^0 in the "ϕ-parts."

This technique is not recommended because it lacks ease of extension to obtain results of arbitrary accuracy. Furthermore, it cannot take into account nonvanishing values of β and values of n_ε such that $n_\varepsilon>1$.

VII. THE SHIRLEY AND SERIES (SSE)′ DEGENERATE RAYLEIGH–SCHRÖDINGER EXPANSION IN POWERS OF $(\varepsilon - i\delta)$

Shirley[6] and Series[31] developed a very interesting technique for approximating the two-level problem solutions when the frequency of the radiation field is large compared to the energy of excitation. They assumed that $\beta=\delta=0$. Our contribution is to extend their technique to nonvanishing values of β and δ. In this modified (SSe)′ form, the method corresponds to expanding the μ's, $\phi_a(\tau)$'s, and $\phi_b(\tau)$'s in powers of $(\varepsilon - i\delta)$. These power series converge rapidly for arbitrary values of α and β as long as *both* ε and δ are much smaller than unity.

The elementary "variations of constants" method of solving differential equations is applied to the solution of (1.10) and (1.11). We express $a(\tau)$ and $b(\tau)$ in the form

$$a(\tau)=F(\tau)\exp[-i(q_+)\sin\tau]+G(\tau)\exp[-i(q_-)\sin\tau]$$

$$2\alpha b(\tau)=(q_+)F(\tau)\exp[-i(q_+)\sin\tau]+(q_-)G(\tau)\exp[-i(q_-)\sin\tau] \tag{7.1}$$

where

$$q_+=\beta+p, \qquad q_-=\beta-p, \qquad \text{and} \qquad p^2=\beta^2+4\alpha^2 \tag{7.2}$$

*See the erratum: *J. Phys. B: Atom. Mol. Phys.*, **7**, 768 (1974).

In the limit that $(\varepsilon - i\delta)$ approaches zero, the $a(\tau)$ and $b(\tau)$ of (7.1) satisfy Eqs. 1.10 and 1.11 where $F(\tau)$ and $G(\tau)$ become arbitrary constants. For nonvanishing values of $(\varepsilon - i\delta)$, the functions $F(\tau)$ and $G(\tau)$ can be determined by substituting the $a(\tau)$ and $b(\tau)$ of (7.1) into Eqs. 1.10 and 1.11. This leads to the coupled differential equations

$$\dot{F} = -\frac{(i\varepsilon + \delta)}{2p}\left[(q_+)F(\tau) + (q_-)\exp[2ip\sin\tau]G(\tau)\right]$$
$$\dot{G} = \frac{(i\varepsilon + \delta)}{2p}\left[(q_+)F(\tau)\exp[-2ip\sin\tau] + (q_-)G(\tau)\right] \tag{7.3}$$

These dynamical equations for $F(\tau)$ and $G(\tau)$ are then expressed in terms of an analogous static eigenvalue–eigenvector problem by making use of Floquet theory and a Fourier expansion. The static problem is then solved by the use of a degenerate Rayleigh–Schrödinger procedure.

A. Formulation of the (SSe)′ Technique

According to Floquet's theorem and Fourier's theorem, there exists a solution to Eqs. (7.3) of the form

$$F(\tau) = \exp(-iu\tau)\exp[-i(\varepsilon - i\delta)\tau/2]\sum_{j=-\infty}^{\infty} F_j\exp(ij\tau)$$
$$G(\tau) = \exp(-iu\tau)\exp[-i(\varepsilon - i\delta)\tau/2]\sum_{j=-\infty}^{\infty} G_j\exp(ij\tau) \tag{7.4}$$

where the F_j's and G_j's are the Fourier expansion coefficients and

$$\mu = u + \frac{(\varepsilon - i\delta)}{2} \tag{7.5}$$

Coupled equations can be obtained for the Fourier coefficients if use is made of the expansion (see Olver[62]),

$$\exp[\pm 2ip\sin\tau] = \sum_{k=-\infty}^{\infty} J_k(2p)\exp[\pm ik\tau] \tag{7.6}$$

where $J_k(2p)$ are the usual Bessel functions of integer order k and argument $2p$. Equations 7.4 and 7.6 are then substituted into Eq. 7.3. Since Eq. 7.3 must be valid for all values of τ, the coefficients of $\exp(in\tau)$ on the two sides of the equation must be equal to each other for every value of n.

Thus, we obtain

$$\begin{aligned}
&[2p(j-u)+\beta(\varepsilon-i\delta)]F_j+(\varepsilon-i\delta)(q_-)\sum_{k=-\infty}^{\infty}J_{j-k}(2p)G_k=0\\
&[2p(j-u)-\beta(\varepsilon-i\delta)]G_j-(\varepsilon-i\delta)(q_+)\sum_{k=-\infty}^{\infty}J_{k-j}(2p)F_k=0
\end{aligned} \tag{7.7}$$

where the value of j ranges from $-\infty$ to ∞.

It is convenient to think of (7.7) as being equivalent to the following Schrödinger-type equation:

$$h_F|u\rangle=u|u\rangle \tag{7.8}$$

Here, the characteristic exponent is an eigenvalue and the eigenvector $|u\rangle$ is assumed to be expandable in an orthonormal basis composed of kets $|F,j\rangle$ and $|G,j\rangle$ where j ranges in integer steps from $-\infty$ to ∞. Thus,

$$\begin{aligned}
&\langle F,j'|F,j\rangle=\delta_{j',j}=\langle G,j'|G,j\rangle\\
&\langle F,j'|G,j\rangle=0
\end{aligned} \tag{7.9}$$

The Fourier expansion coefficients in (7.4) are related to the eigenvector $|u\rangle$ by

$$\begin{aligned}
&\langle F,j|u\rangle=F_j\\
&\langle G,j|u\rangle=G_j
\end{aligned} \tag{7.10}$$

or

$$|u\rangle=\sum_{j=-\infty}^{\infty}[F_j|F,j\rangle+G_j|G,j\rangle] \tag{7.11}$$

Since we wish to solve Eq. 7.8 by a Rayleigh–Schrödinger perturbation expansion in powers of $(\varepsilon-i\delta)$, the Floquet Hamiltonian h_F is expressed in the form

$$h_F=h_F^{(0)}+(\varepsilon-i\delta)h_F^{(1)} \tag{7.12}$$

where

$$h_F^{(0)}|F,j\rangle=j|F,j\rangle \qquad \text{and} \qquad h_F^{(0)}|G,j\rangle=j|G,j\rangle \tag{7.13}$$

and

$$
\begin{aligned}
h_F^{(1)}|F,j\rangle &= \left(\frac{1}{2p}\right)\left[\beta|F,j\rangle - (q_+)\sum_{k=-\infty}^{\infty} J_{j-k}(2p)|G,k\rangle\right] \\
h_F^{(1)}|G,j\rangle &= \left(\frac{1}{2p}\right)\left[-\beta|G,j\rangle + (q_-)\sum_{k=-\infty}^{\infty} J_{k-j}(2p)|F,k\rangle\right]
\end{aligned}
\tag{7.14}
$$

Substituting Eqs. 7.11 through 7.14 into Eq. 7.8, it is easy to verify that Eq. 7.7 is equivalent to

$$\langle F,j|h_F - u|u\rangle = 0 = \langle G,j|h_F - u|u\rangle \tag{7.15}$$

When $\beta = 0 = \delta$, h_F is Hermitian and all of the usual degenerate Rayleigh–Schrödinger procedures can be used. Otherwise, h_F is not Hermitian but we can solve each of the perturbation equations exactly in a step-by-step fashion. Because of the degeneracy of the $|F,0\rangle$ and $|G,0\rangle$, each of the perturbation equations must be mathematically consistent with respect to multiplication from the left by $\langle F,0|$ and by $\langle G,0|$.

First, we expand u and $|u\rangle$ in powers of $(\varepsilon - i\delta)$

$$u = \sum_{n=0}^{\infty} u^{(n)}(\varepsilon - i\delta)^n \qquad \text{and} \qquad |u\rangle = \sum_{n=0}^{\infty} (\varepsilon - i\delta)^n |u^{(n)}\rangle \tag{7.16}$$

where according to (7.11)

$$|u^{(n)}\rangle = \sum_{j=-\infty}^{\infty} \left[F_j^{(n)}|F,j\rangle + G_j^{(n)}|G,j\rangle\right] \tag{7.17}$$

The Rayleigh–Schrödinger perturbation equations are then similar to (5.5). Also, we assume intermediate normalization so that Eq. 5.6 also applies. We therefore have the perturbation equations multiplied by either $\langle F,j|$ or $\langle G,j|$,

$$
\begin{aligned}
(j - u^{(0)})F_j^{(n)} + \left(\frac{\beta}{2p} - u^{(1)}\right)F_j^{(n-1)} - \sum_{r=2}^{n} u^{(r)} F_j^{(n-r)} \\
+ \frac{(q_-)}{2p}\sum_{k=-\infty}^{\infty} J_{j-k}(2p) G_k^{(n-1)} = 0
\end{aligned}
\tag{7.18}
$$

$$
\begin{aligned}
(j - u^{(0)})G_j^{(n)} - \left(\frac{\beta}{2p} + u^{(1)}\right)G_j^{(n-1)} - \sum_{r=2}^{n} u^{(r)} G_j^{(n-r)} \\
- \frac{(q_+)}{2p}\sum_{k=-\infty}^{\infty} J_{k-j}(2p) F_k^{(n-1)} = 0
\end{aligned}
\tag{7.19}
$$

and the normalization conditions

$$\sum_{j=-\infty}^{\infty} \left[F_j^{(0)*} F_j^{(n)} + G_j^{(0)*} G_j^{(n)} \right] = \delta_{0,n} \tag{7.20}$$

The zeroth-order energies $u^{(0)} = k$ are doubly degenerate. The $a(\tau)$ and $b(\tau)$ do not depend upon the value of k (since adding k to $u^{(0)}$ would only have the effect of multiplying the ϕ's by $\exp(ik\tau)$). Therefore, we take $k=0$ so that $u^{(0)}=0$ and

$$|u^{(0)}\rangle = F_0^{(0)}|F,0\rangle + G_0^{(0)}|G,0\rangle \tag{7.21}$$

Considering Eqs. 7.18 and 7.19 for $n=1$ with $j=0$, we obtain two values of $u^{(1)}$,

$$u_1^{(1)} = \frac{s}{2p} \quad \text{and} \quad u_2^{(1)} = -\frac{s}{2p} \tag{7.22}$$

where

$$s^2 = \beta^2 + 4\alpha^2 \left[J_0(2p) \right]^2 \tag{7.23}$$

The corresponding zeroth-order wavefunctions are*

$$\begin{aligned} |u_1^{(0)}\rangle &= N_1^{(0)} \left[|F,0\rangle + \frac{(s-\beta)}{(q_-)J_0(2p)} |G,0\rangle \right] \\ |u_2^{(0)}\rangle &= N_2^{(0)} \left[-\frac{(q_-)J_0(2p)}{(s+\beta)} |F,0\rangle + |G,0\rangle \right] \end{aligned} \tag{7.24}$$

where $N_1^{(0)}$ and $N_2^{(0)}$ are normalization constants determined by (7.20) with $n=0$. The solution associated with $u_1^{(1)}$ corresponds to one Floquet mode and the solution with $u_2^{(1)}$ corresponds to the linearly independent second Floquet mode.

Also from (7.18) and (7.19) for $n=1$, but taking $j \neq 0$, we obtain

$$\begin{aligned} F_j^{(1)} &= -\frac{(q_-)}{2pj} J_j(2p) G_0^{(0)} \\ G_j^{(1)} &= \frac{(q_+)}{2pj} J_{-j}(2p) F_0^{(0)} \end{aligned} \tag{7.25}$$

Then considering Eqs. 7.18 and 7.19 for $n=2$ with $j=0$ [making use of the

*For the special case where $J_0(2p)=0$, $|\mu_1^{(0)}\rangle = |F,0\rangle$, and $|\mu_2^{(0)}\rangle = |G,0\rangle$.

fact that $J_{-k}=(-1)^k J_k$ and Eq. 7.25], we obtain the two relations

$$\left[\frac{\beta}{2p}-u^{(1)}\right]F_0^{(1)}+\frac{(q_-)}{2p}J_0(2p)G_0^{(1)}-u^{(2)}F_0^{(0)}=0 \tag{7.26}$$

$$\frac{(q_+)}{2p}J_0(2p)F_0^{(1)}+\left[\frac{\beta}{2p}+u^{(1)}\right]G_0^{(1)}+u^{(2)}G_0^{(0)}=0 \tag{7.27}$$

From the three relations Eqs. 7.26, 7.27, and the normalization condition (7.20) with $n=1$, we find that $u^{(2)}$, $F_0^{(1)}$, and $G_0^{(1)}$ all vanish. Using this result with (7.25) we obtain

$$\begin{aligned}|u_1^{(1)}\rangle=&\frac{-N_1^{(0)}(s-\beta)}{2pJ_0(2p)}\sum_{j=-\infty}^{\infty}{}'\frac{J_j(2p)}{j}|F,j\rangle\\&+\frac{N_1^{(0)}(q_+)}{2p}\sum_{j=-\infty}^{\infty}{}'\frac{J_{-j}(2p)}{j}|G,j\rangle\end{aligned} \tag{7.28}$$

$$\begin{aligned}|u_2^{(1)}\rangle=&\frac{-(q_-)N_2^{(0)}}{2p}\sum_{j=-\infty}^{\infty}{}'\frac{J_j(2p)}{j}|F,j\rangle\\&+\frac{2\alpha^2N_2^{(0)}J_0(2p)}{p(s+\beta)}\sum_{j=-\infty}^{\infty}{}'\frac{J_{-j}(2p)}{j}|G,j\rangle\end{aligned} \tag{7.29}$$

where the primes are used to indicate that $j=0$ is excluded from the sum.

In this manner we can proceed step-by-step to determine the Floquet normal modes. The procedure is doable although tedious, since finding $u^{(N+1)}$, $F_0^{(N)}$, and $G_0^{(N)}$ involves solving sets of three (in general) homogeneous linear equations in three unknowns. Using Eqs. 7.1, 7.4, and 7.10, we obtain the time-dependent Floquet solutions correct through $(\varepsilon-i\delta)^2$ in the characteristic exponents and correct through $(\varepsilon-i\delta)$ in the periodic parts. Both modes $(k=1,2)$ are functions of the usual Floquet form

$$a_k(\tau)=\exp(-i\mu_k\tau)\phi_{ak}(\tau) \qquad \text{and} \qquad b_k(\tau)=\exp(-i\mu_k\tau)\phi_{bk}(\tau)$$

where

$$\begin{aligned}\phi_{ak}(\tau)&=\phi_{Fk}(\tau)\exp[-i(q_+)\sin\tau]+\phi_{Gk}(\tau)\exp[-i(q_-)\sin\tau]\\2\alpha\phi_{bk}(\tau)&=(q_+)\phi_{Fk}(\tau)\exp[-i(q_+)\sin\tau]+(q_-)\phi_{Gk}(\tau)\exp[-i(q_-)\sin\tau]\end{aligned} \tag{7.30}$$

Thus, we obtain the following modes.

1. *First Floquet Mode*

$$\mu_1 = \frac{(\varepsilon - i\delta)}{2p}(p+s) + (\varepsilon - i\delta)^3(\cdots) + \cdots$$

$$\phi_{F1}(\tau) = N_1^{(0)}\left[1 - \frac{(\varepsilon - i\delta)(s-\beta)}{2pJ_0(2p)} \sum_{j=-\infty}^{\infty}{}' \frac{J_j(2p)\exp(ij\tau)}{j} + (\varepsilon - i\delta)^2(\cdots) + \cdots\right]$$

$$\phi_{G1}(\tau) = N_1^{(0)}\left[\frac{(s-\beta)}{(q_-)(s+\beta)} + \frac{(\varepsilon - i\delta)(q_+)}{2p} \sum_{j=-\infty}^{\infty}{}' \frac{J_{-j}(2p)\exp(ij\tau)}{j} + (\varepsilon - i\delta)^2(\cdots) + \cdots\right] \tag{7.31}$$

where

$$N_1^{(0)} = \left[1 + \left(\frac{s-\beta}{(q_-)J_0(2p)}\right)^2\right]^{-1}$$

2. *Second Floquet Mode*

$$\mu_2 = \frac{(\varepsilon - i\delta)}{2p}(p-s) + (\varepsilon - i\delta)^3(\cdots) + \cdots$$

$$\phi_{F2}(\tau) = -N_2^{(0)}\left[\frac{(q_-)J_0(2p)}{(s+\beta)} + \frac{(\varepsilon - i\delta)(q_-)}{2p} \sum_{j=-\infty}^{\infty}{}' \frac{J_j(2p)\exp(ij\tau)}{j} + (\varepsilon - i\delta)^2(\cdots) + \cdots\right]$$

$$\phi_{G2}(\tau) = N_2^{(0)}\left[1 + \frac{2(\varepsilon - i\delta)\alpha^2 J_0(2p)}{p(s+\beta)} \sum_{j=-\infty}^{\infty}{}' \frac{J_{-j}(2p)\exp(ij\tau)}{j} + (\varepsilon - i\delta)^2(\cdots) + \cdots\right] \tag{7.32}$$

where

$$N_2^{(0)} = \left[1 + \left(\frac{(q_-)J_0(2p)}{s+\beta} \right)^2 \right]^{-1}$$

B. Convergence of the (SG)′ Technique

Each of the coefficients $F_j^{(n)}$ and $G_j^{(n)}$ in (7.17) can be expressed as the sum of terms each one of which is the product of n factors of the form

$$\mathcal{G} = \frac{\langle k,n|h_F^{(1)}|k',n'\rangle}{I} \tag{7.33}$$

Here I is any positive or negative integer except zero. The k's can be either F or G, and the n's can be any integer. Because of the form of the perturbation in Eq. (7.14), there are only two general forms which $|\mathcal{G}|$ can take

$$\left| \frac{\beta}{2pI} \right|, \qquad \left| \frac{(q_\pm)J_m(2p)}{2pI} \right| \tag{7.34}$$

Regardless of the value of the integer m and of the argument $(2p)$, $J_m(2p)$ never exceeds unity. By definition $|\beta/2p|$ and $|(q_-)/2p|$ are always less than or equal to one half, and $|(q_+)/2p|$ is always less than or equal to unity. $|\mathcal{G}|$, therefore, never exceeds unity. Since in the perturbation series results the coefficients $F_j^{(n)}$ and $G_j^{(n)}$ always get multiplied by $(\varepsilon - i\delta)^n$, the (SSe)′ technique will be quickly convergent as long as both ε and δ are much less than unity.

The expansion of the Floquet μ's and ϕ's is in general asympotically convergent. These approximate solutions are not, however, numerically convenient. Infinite sums over Bessel functions occur in the higher order perturbation corrections. Since $|J_n(2p)|$ gets small only very slowly as a function of increasing $|n|$, many terms must be retained in the sums before numerical convergence is obtained.

Because of this, the use of (SSe)′ is only recommended when either α or β exceeds unity and both δ and ε are smaller than unity.

VIII. THE DEGENERATE RAYLEIGH–SCHRÖDINGER (DRS) TECHNIQUE FOR CONSIDERING HIGH-FREQUENCY FIELDS

We developed the (DRS) technique to treat problems involving high-frequency fields where α, β, δ, and ε are *all* much smaller than unity and where δ and ε are both much smaller than either α or β. We do not know of any other technique that has been used to consider this set of conditions. The (DRS) is a simple application of degenerate Rayleigh–Schrödinger perturbation theory to a system having the Floquet Hamiltonain H_F that was used in both the (SG)′ and the (SWS)′ techniques. However, we use a different resolution of H_F into a zeroth- and first-order operator. If $\delta \neq 0$, then special care is required since H_F is non-Hermitian. However, this is not a serious difficulty since, as in the (SSe)′ treatment, we can proceed step-by-step to determine *exact* solutions to the perturbation equation of each order. Indeed, our present problem is so similar to the (SSe)′ problem that we can parallel the development given in Section VII with very little additional explanation.

A. Formulation of the (DRS) Technique

For present purposes the Hamiltonian H_F is resolved in the following manner:

$$H_F = \overline{H}_F^{(0)} + \overline{H}_F^{(1)} \tag{8.1}$$

where the bars have been used to distinguish our present zeroth- and first-order operators from the corresponding operators used in the (SG)′ and (SWS)′ techniques. Here

$$\overline{H}_F^{(0)}|A,j\rangle = j|A,j\rangle \qquad \text{and} \qquad \overline{H}_F^{(0)}|B,j\rangle = j|B,j\rangle \tag{8.2}$$

and

$$\begin{aligned} \overline{H}_F^{(1)}|A,j\rangle &= \alpha\big[|B,j-1\rangle + |B,j+1\rangle\big] \\ \overline{H}_F^{(1)}|B,j\rangle &= \beta\big[|B,j-1\rangle + |B,j+1\rangle\big] + (\varepsilon - i\delta)|B,j\rangle \\ &\quad + \alpha\big[|A,j-1\rangle + |A,j+1\rangle\big] \end{aligned} \tag{8.3}$$

We seek to obtain the eigenvalues and eigenfunctions corresponding to the Schrödinger-type equation

$$H_F|\mu\rangle = \mu|\mu\rangle \tag{2.43}$$

The eigenvalue μ and the eigenfunction $|\mu\rangle$ are expanded in the usual Rayleigh–Schrödinger perturbation series.

There is great similarity between the present problem and the determination of the u and $|u\rangle$ in the (SSe)′ technique of Section VII. In the present case our basis functions $|A,j\rangle$ and $|B,j\rangle$ correspond to the $|F,j\rangle$ and $|G,j\rangle$;

our $\overline{H}_F^{(0)}$ is then equivalent to $h_F^{(0)}$; there is, however, a difference between $\overline{H}_F^{(1)}$ and $h_F^{(1)}$. Both problems involve a double degeneracy that is split in the first order. Thus, we start with

$$|\mu^{(0)}\rangle = C_{A,0}^{(0)}|A,0\rangle + C_{B,0}^{(0)}|B,0\rangle \tag{8.4}$$

where the constants $C_{A,0}^{(0)}$ and $C_{B,0}^{(0)}$ are determined by the eigenvectors of $\mu^{(1)}$ together with the normalization condition. Indeed, we completely parallel the development and procedures given in connection with Eqs. (7.17) through (7.27). Thus, we determined the μ's through the third order and the $|\mu\rangle$ functions through the second order.

1. First Floquet Mode

$$\mu_1 = 2\alpha^2(\varepsilon - i\delta) + \cdots$$

$$\begin{aligned} |\mu_1\rangle = {} & |A,0\rangle + \alpha[|B,-1\rangle - |B,1\rangle] \\ & + \frac{\alpha^2}{2}[|A,2\rangle + |A,-2\rangle] + \frac{\alpha\beta}{2}[|B,2\rangle + |B,-2\rangle] \\ & + \alpha(\varepsilon - i\delta)[|B,1\rangle + |B,-1\rangle] - 2\alpha\beta|B,0\rangle + \cdots \end{aligned}$$

Thus,

$$a_1(\tau) = \exp(-i\mu_1\tau)\phi_{a1}(\tau) \qquad \text{and} \qquad b_1(\tau) = \exp(-i\mu_1\tau)\phi_{b1}(\tau)$$

where

$$\phi_{a1}(\tau) = 1 + \alpha^2\cos(2\tau) + \cdots$$

$$\phi_{b1}(\tau) = -2i\alpha\sin\tau - 2\alpha\beta + 2\alpha(\varepsilon - i\delta)\cos\tau + \alpha\beta\cos(2\tau) + \cdots$$

2. Second Floquet Mode

$$\mu_2 = (\varepsilon - i\delta) - 2\alpha^2(\varepsilon - i\delta) + \cdots$$

$$\begin{aligned} |\mu_2\rangle = {} & |B,0\rangle + \alpha[|A,-1\rangle - |A,1\rangle] + \beta[|B,-1\rangle - |B,1\rangle] \\ & - \alpha(\varepsilon - i\delta)[|A,1\rangle + |A,-1\rangle] + \frac{\alpha\beta}{2}[|A,2\rangle + |A,-2\rangle] \\ & + \frac{1}{2}(\alpha^2 + \beta^2)[|B,2\rangle + |B,-2\rangle] + \cdots \end{aligned}$$

Thus,

$$a_2(\tau)=\exp(-i\mu_2\tau)\phi_{a2}(\tau)$$
$$b_2(\tau)=\exp(-i\mu_2\tau)\phi_{b2}(\tau)$$

where

$$\phi_{a2}(\tau)=-2i\alpha\sin\tau-2\alpha(\varepsilon-i\delta)\cos\tau+\alpha\beta\cos(2\tau)+\cdots$$
$$\phi_{b2}(\tau)=1-2i\beta\sin\tau+(\alpha^2+\beta^2)\cos(2\tau)+\cdots$$

IX. THE $(1/\varepsilon)$ SINGULAR PERTURBATION EXPANSION (SPT) TECHNIQUE

The (SPT) technique involves finding the Floquet normal modes as expansions in powers of $(1/\varepsilon)$. Rather than solving the equations for $a(\tau)$ and $b(\tau)$, the equations for the quotients

$$\Phi_1(\tau)=\frac{b_1(\tau)}{a_1(\tau)},\qquad \Phi_2(\tau)=\frac{a_2(\tau)}{b_2(\tau)}$$

are solved. Finding a $(1/\varepsilon)$-expansion for $\Phi_1(\tau)$ gives rise to one of the Floquet normal modes, and finding a $(1/\varepsilon)$-expansion for $\Phi_2(\tau)$ gives rise to the other Floquet normal mode.

After multiplying both sides of (2.26) by (i/ε), we have

$$\frac{i}{\varepsilon}\dot{\Phi}_1=\Phi_1-\frac{i\delta}{\varepsilon}\Phi_1+\frac{2\beta}{\varepsilon}\cos\tau\Phi_1+\frac{2\alpha}{\varepsilon}\cos\tau-\frac{2\alpha}{\varepsilon}\cos\tau(\Phi_1)^2 \tag{9.1}$$

and after multiplying Eq. 2.29 by $-(i/\varepsilon)$,

$$-\frac{i}{\varepsilon}\dot{\Phi}_2=\Phi_2-\frac{i\delta}{\varepsilon}\Phi_2+\frac{2\beta}{\varepsilon}\cos\tau\Phi_2-\frac{2\alpha}{\varepsilon}\cos\tau+\frac{2\alpha}{\varepsilon}\cos\tau(\Phi_2)^2 \tag{9.2}$$

If $(1/\varepsilon)$, (δ/ε), $(2\beta/\varepsilon)$, and $(2\alpha/\varepsilon)$ are all much smaller than unity, then every term in (9.1) and (9.2) may be taken to be a perturbation on the terms Φ_1 and Φ_2, respectively. Since the highest order derivative term is included in the perturbation, the perturbation is called "singular"; or Eq. 9.1 is called a "stiff equation."

There are three novel aspects to the (SPT) method of solution:

(1) It gives solutions that quickly converge when $(1/\varepsilon)$, (δ/ε), $(2\alpha/\varepsilon)$, and $(2\beta/\varepsilon)$ are all much less than unity thereby giving solutions for the

two-level system in a parameter range for which no previous work applies. It is especially useful when the field strength is large enough to make α and/or β larger than unity.

(2) It represents the first example of applying singular perturbation theory to the two-level system.

(3) The method exhibits a behavior atypical of ordinary singular perturbation theory: Separate computation of inner solutions is never needed. $\Phi_1(\tau)$ itself is the inner solution for $\Phi_2(\tau)$ and vice versa.

A. Description of the (SPT) Technique

We assume that $\Phi_1(\tau)$ and $\Phi_2(\tau)$ may be expanded in power series in $(1/\varepsilon)$:

$$\Phi_1(\tau)=\sum_{n=n_1}^{\infty}\left(\frac{1}{\varepsilon}\right)^n\Phi_1^{(n)}(\tau) \tag{9.3}$$

and

$$\Phi_2(\tau)=\sum_{n=n_2}^{\infty}\left(\frac{1}{\varepsilon}\right)^n\Phi_2^{(n)}(\tau) \tag{9.4}$$

Substituting Eqs. 9.3 and 9.4 into (9.1) and (9.2), respectively, and setting the coefficients of each power of $(1/\varepsilon)$ equal to zero, we obtain the general nth order equations:

$$\Phi_1^{(n)}(\tau)=i\dot{\Phi}_1^{(n-1)}+\left[i\delta-2\beta\cos\tau\right]\Phi_1^{(n-1)} + 2\alpha\cos\tau\left[-\delta_{n1}+\sum_{j=n_1}^{n-1-n_1}\Phi_1^{(j)}\Phi_1^{(n-1-j)}\right] \tag{9.5}$$

and

$$\Phi_2^{(n)}(\tau)=-i\dot{\Phi}_2^{(n-1)}+\left[i\delta-2\beta\cos\tau\right]\Phi_2^{(n-1)} + 2\alpha\cos\tau\left[\delta_{n,1}-\sum_{j=n_2}^{n-1-n_2}\Phi_2^{(j)}\Phi_2^{(n-1-j)}\right] \tag{9.6}$$

There are only two values of n_1 and two values of n_2 that satisfy the indicial equations:

$$n_1=1=n_2 \qquad \text{and} \qquad n_1=-1=n_2$$

The expansions with $n_1 = 1 = n_2$ generate asymptotic expansions corresponding to the two Floquet modes. The expansions with $n_1 = -1 = n_2$ are not useful because they contain terms which become infinite for certain values of τ. For example,

$$\Phi_1^{(-1)}(\tau) = [2\alpha\cos\tau]^{-1} = -\Phi_2^{(-1)}(\tau) \tag{9.7}$$

Indeed, the $n_1 = -1$ expansion of $\Phi_1(\tau)$ turns out to correspond to the *reciprocal* of the $n_2 = 1$ expansion of $\Phi_2(\tau)$. Thus we ignore the $n_1 = -1 = n_2$ expansions and take $n_1 = 1 = n_2$.

1. *First Floquet Mode*

The first two terms of $\Phi_1(\tau)$ are

$$\begin{aligned} \Phi_1^{(1)} &= -2\alpha\cos\tau \\ \Phi_1^{(2)} &= 2i\alpha\sin\tau - 2\alpha(i\delta - 2\beta\cos\tau)\cos\tau \end{aligned} \tag{9.8}$$

Once $\Phi_1(\tau)$ is known, $a_1(\tau)$ is recovered by using Eq. 2.27. Then $b_1(\tau)$ is found through the definition of $\Phi_1(\tau)$ or Eq. 2.25. Since $\Phi_1(\tau)$ contains *only* periodic terms and since the indefinite integral of $\Phi_1(\tau)$ contains only terms linear in τ in addition to periodic terms, we find that the expressions for $a_1(\tau)$ and $b_1(\tau)$ are Floquet normal mode solutions. The explicit expression for this Floquet solution correct through second order in $(1/\varepsilon)$ is

$$a_1(\tau) = \exp(-i\mu_1\tau)\phi_{a1}(\tau) \qquad \text{and} \qquad b_1(\tau) = \exp(-i\mu_1\tau)\phi_{b1}(\tau) \tag{9.9}$$

where

$$\mu_1 = -\frac{2\alpha^2}{\varepsilon} - \frac{2i\alpha^2\delta}{\varepsilon^2} + \cdots$$

$$\phi_{a1} = \exp\left[\frac{i\alpha^2}{\varepsilon}\sin(2\tau) + \frac{\alpha^2}{\varepsilon^2}\left\{2\sin^2\tau - \delta\sin(2\tau) - \frac{2i\beta}{3}\sin(3\tau) - 6i\beta\sin\tau\right\} + \cdots\right]$$

$$\phi_{b1} = \phi_{a1}\left[-\frac{2\alpha}{\varepsilon}\cos\tau + \frac{2\alpha}{\varepsilon^2}[i\sin\tau - (i\delta - 2\beta\cos\tau)\cos\tau] + \cdots\right]$$

2. *Second Floquet Mode*

The first two terms of $\Phi_2(\tau)$ are

$$\begin{aligned}\Phi_2^{(1)} &= 2\alpha\cos\tau \\ \Phi_2^{(2)} &= 2i\alpha\sin\tau + 2\alpha\left[i\delta - 2\beta\cos\tau\right]\cos\tau\end{aligned} \tag{9.10}$$

Thus using Eq. 2.30 we obtain the second linearly independent Floquet mode correct through the second power of $(1/\varepsilon)$,

$$a_2(\tau) = \exp(-i\mu_2\tau)\phi_{a2}(\tau) \qquad \text{and} \qquad b_2(\tau) = \exp(-i\mu_2\tau)\phi_{b2}(\tau) \tag{9.11}$$

where

$$\mu_2 = \varepsilon - i\delta + \frac{2\alpha^2}{\varepsilon} + \frac{2i\alpha^2\delta}{\varepsilon^2} + \cdots$$

$$\phi_{a2} = \phi_{b2}\left[\frac{2\alpha}{\varepsilon}\cos\tau + \frac{2\alpha}{\varepsilon^2}\left[i\sin\tau + \cos\tau(i\delta - 2\beta\cos\tau)\right] + \cdots\right]$$

$$\begin{aligned}\phi_{b2} = \exp\bigg[&-2i\beta\sin\tau - \frac{i\alpha^2}{\varepsilon}\sin(2\tau) \\ &+ \frac{\alpha^2}{\varepsilon^2}\left[2\sin^2\tau + \delta\sin(2\tau) + \frac{2i\beta}{3}\sin(3\tau) + 6i\beta\sin\tau\right] + \cdots\bigg]\end{aligned}$$

In order to show that the two solutions given by (9.9) and (9.11) are indeed linearly independent, it is only necessary by virtue of Theorem 3 to show that

$$D(0) = a_1(0)b_2(0) - b_1(0)a_2(0) \neq 0$$

Indeed, we find that

$$D(0) = a_1(0)b_2(0)\left[1 - \Phi_1(0)\Phi_2(0)\right] = 1 + \frac{4\alpha^2}{\varepsilon^2} + \cdots \tag{9.12}$$

Therefore, the two Floquet modes given by (9.9) and (9.11) are linearly independent.

B. Convergence of the (SPT) Technique

We believe that the (SPT) method gives rapidly converging asymptotic approximations to the Floquet normal modes provided that $(1/\varepsilon)$, (δ/ε),

$(2\alpha/\varepsilon)$, and $(2\beta/\varepsilon)$ are *all* much less than one. Indeed, it is characteristic of singular perturbation expansions that the first few terms give an excellent approximation to the solution or else the method is not appropriate.

The general form of the $\Phi^{(n)}(\tau)$'s is

$$\Phi^{(n)}(\tau) = \sum_{a,b,c} \frac{(2\alpha)^a (2\beta)^b \delta^c}{\varepsilon^n} f_{abc}^{(n)}(\cos\tau, \sin\tau)$$

where $f_{abc}^{(n)}(\cos, \sin)$ is a polynomial in $\cos\tau$ and $\sin\tau$ and it is, therefore, bounded. a, b, and c are positive integers (or zero) such that

$$a + b + c \leqslant n$$

If $(1/\varepsilon)$, (δ/ε), $(2\alpha/\varepsilon)$, and $(2\beta/\varepsilon)$ are *all* much less than unity, the coefficient of the functions $f_{abc}^{(n)}$ are of order n in small quantities. Without further study of the functions $f_{abc}^{(n)}$, we cannot be sure that the $\Phi^{(n+1)}(\tau)$'s will be less than the $\Phi^{(n)}(\tau)$'s for all values of n and τ. Actually, since we expect the series to converge asymptotically, the $\Phi^{(n+1)}(\tau)$'s are probably only smaller than the $\Phi^{(n)}(\tau)$'s for small values of n.

C. Relationship of (SPT) to Usual Singular Perturbation Solutions

When ε is very large, Eqs. 9.1 and 9.2 are "stiff equations" and the perturbation is singular since the coefficient of the highest order derivative is very small.* However, our Floquet solutions for $\Phi_1(\tau)$ and $\Phi_2(\tau)$ as given by (9.5), (9.6), (9.8), and (9.10) are atypical.

The solutions to stiff equations typically† have the following behavior. All solutions, regardless of their boundary conditions, rapidly approach a single function as the time variable moves forward or backward. The approached solutions are called "outer solutions." The singular perturbation series usually gives an asymptotic approximation to such an outer solution. The solution in the immediate vicinity of the boundary is called an "inner solution," and it must be found by some technique other than singular perturbation theory.

The present results differ from the typical singular perturbation problem in two respects:

(1) The first difference is that we never have to determine the inner solutions. From the knowledge of the Floquet Φ_1 and Φ_2, we determine the

*The textbooks by Nayfeh[84] and Cole[85] have excellent discussions of singular perturbations. Also, see Curtiss and Hirschfelder[86].

†See Cole's[85] discussion (in Chapter 2) of an overdamped harmonic oscillator of extremely small mass as an example of a "typical" singular perturbation problem.

two linearly independent Floquet modes $a_1(\tau)$, $b_1(\tau)$ and $a_2(\tau)$, $b_2(\tau)$. A solution obeying *arbitrary* boundary conditions can then be expressed as a linear combination of the Floquet modes.

(2) The second atypical aspect of the (SPT) is that the Floquet Φ_1 and Φ_2 only play the typical role of outer solutions when* $\delta \neq 0$ *and* $-\delta + 2\,\mathrm{Re}(i\mu_1) \neq 0$. This is surprising since the Floquet Φ_1 and Φ_2 are approximated according to the usual prescription for obtaining outer solutions, namely expressing them as asymptotically converging power series in the perturbation parameter. In order to show the conditions under which the Floquet Φ_1 is a typical outer solution, consider the $\Phi_1 = b(\tau)/a(\tau)$ for functions $a(\tau)$ and $b(\tau)$ that satisfy arbitrary boundary conditions

$$\Phi_1(\text{general}) = \frac{c_1\Phi_{b1} + c_2\Phi_{b2}\exp[-i\varepsilon\tau - \delta\tau + 2i\mu_1\tau]}{c_1\Phi_{a1} + c_2\Phi_{a2}\exp[-i\varepsilon\tau - \delta\tau + 2i\mu_1\tau]} \tag{9.13}$$

Here c_1 and c_2 are arbitrary constants, and we have made use of Theorem 4 or Eq. 2.13, which states that $\mu_2 = -\mu_1 + \varepsilon - i\delta$. The two Floquet modes correspond to either c_2 or c_1 being zero. If the Floquet modes can be regarded as outer solutions, then as τ approaches either $-\infty$ or else $+\infty$, Φ_1(general) must approach either ϕ_{b1}/ϕ_{a1} or ϕ_{b2}/ϕ_{a2}. This condition is only satisfied if $\delta \neq 0$ *and* $-\delta + 2\,\mathrm{Re}(i\mu_1) \neq 0$.

APPENDIX A: EXAMPLES OF SIMPLE SYSTEMS THAT HAVE FORM I, II, AND III SOLUTIONS

It is instructive to illustrate each of the three forms of possible solutions of (2.1) by a simple example.

A. Form I

$\theta_{11}(\tau) = \theta_{22}(\tau) = \cos\tau$ and $\theta_{12}(\tau) = \theta_{21}(\tau) = \alpha$(a noninteger).

The differential equations for $a(\tau)$ and $b(\tau)$ are therefore

$$\dot{a}(\tau) = -i\cos\tau\, a(\tau) - i\alpha b(\tau)$$

$$\dot{b}(\tau) = -i\alpha a(\tau) - i\cos\tau\, b(\tau) \tag{A.1}$$

Letting $a(\tau) = A(\tau)\exp(-i\sin\tau)$ and $b(\tau) = B(\tau)\exp(-i\sin\tau)$, the functions $A(\tau)$ and $B(\tau)$ obey

$$\dot{A}(\tau) = -i\alpha B(\tau); \qquad \dot{B}(\tau) = -i\alpha A(\tau) \tag{A.2}$$

*See Dion's[56] thesis (pp. 13.12 through 13.17).

Since $\ddot{A}(\tau)+\alpha^2 A(\tau)=0$, two linearly independent solutions to (A.1) are

$$a_1(\tau)=\exp(-i\alpha\tau)\exp(-i\sin\tau); \qquad b_1(\tau)=\exp(-i\alpha\tau)\exp(-i\sin\tau)$$

$$a_2(\tau)=\exp(i\alpha\tau)\exp(-i\sin\tau); \qquad b_2(\tau)=-\exp(i\alpha\tau)\exp(-i\sin\tau) \quad (A.3)$$

These solutions are clearly Form I Floquet solutions where

$$\mu_1=-\mu_2=\alpha$$

$$\phi_{a1}(\tau)=\phi_{b1}(\tau)=\phi_{a2}(\tau)=-\phi_{b2}(\tau)=\exp(-i\sin\tau) \quad (A.4)$$

B. Form II

$\theta_{11}(\tau)=\theta_{22}(\tau)=\cos\tau$ and $\theta_{12}(\tau)=\theta_{21}(\tau)=\alpha$(an integer).

This system is the same as the first example with the exception that α is now an integer. The solutions to the differential equations are still given by (A.3). However, in the present case, the solutions correspond to Form II. To illustrate the ambiguity in the definition of μ, we can make the identification:

$$\mu=n \text{ (where } n \text{ is an arbitrary integer)}$$

$$\phi_{a1}(\tau)=\phi_{b1}(\tau)=\exp[-i(\alpha-n)\tau-i\sin\tau]$$

$$\phi_{a2}(\tau)=-\phi_{b2}(\tau)=\exp[i(\alpha+n)\tau-i\sin\tau] \quad (A.5)$$

C. Form III

$\theta_{21}(\tau)=-\theta_{12}(\tau)=\frac{1}{2}$; $\theta_{11}(\tau)=\frac{1}{2}-\sin\tau$; $\theta_{22}(\tau)=-(\frac{1}{2}+\sin\tau)$.

The equations for $a(\tau)$ and $b(\tau)$ are

$$\dot{a}(\tau)=-i\left(\frac{1}{2}-\sin\tau\right)a(\tau)+\frac{i}{2}b(\tau)$$

$$\dot{b}(\tau)=-\frac{i}{2}a(\tau)+i\left(\frac{1}{2}+\sin\tau\right)b(\tau) \quad (A.6)$$

Letting

$$a(\tau)=A(\tau)\exp(-i\cos\tau) \quad \text{and} \quad b(\tau)=B(\tau)\exp(-i\cos\tau) \quad (A.7)$$

Equations A.6 become

$$\dot{A}(\tau)=\dot{B}(\tau)=\frac{i}{2}[B(\tau)-A(\tau)] \quad (A.8)$$

Since $\dot{A}(\tau)-\dot{B}(\tau)=0$, it follows that $A(\tau)-B(\tau)=c_1$ where c_1 is an arbitrary constant. Thus, Eq. A.8 can be integrated to give

$$A(\tau)=-i\frac{c_1\tau}{2}+c_2; \qquad B(\tau)=-i\frac{c_1\tau}{2}+(c_2-c_1) \tag{A.9}$$

where c_2 is another arbitrary constant. The general solution of (A.8) therefore is

$$a(\tau)=\left[-i\frac{c_1\tau}{2}+c_2\right]\exp(-i\cos\tau)$$

$$b(\tau)=\left[-i\frac{c_1\tau}{2}+(c_2-c_1)\right]\exp(-i\cos\tau) \tag{A.10}$$

To describe these Form III Floquet solutions in terms of the notation in (2.7) and (2.8), the following identifications are made:

$$\mu_2=\mu_1=\phi_{a2}(\tau)=0$$

$$\phi_{a1}(\tau)=\phi_{b1}(\tau)=\frac{i}{2}\phi_{b2}(\tau)=\exp(-i\cos\tau) \tag{A.11}$$

Acknowledgments

The authors thank L. W. Bruch, P. R. Certain, C. F. Curtiss, J. D. Doll, S. T. Epstein, J. E. Harriman, and P. W. Langhoff for many helpful discussions. They are also grateful to N. Ashby for making J. G. Ross' Doctoral Thesis available to us; and to J. O. Eaves for his help with Section I.D. Thanks are also due to Ms. Patty Spires for her excellent job of typing this manuscript and to Mrs. J. O. H. for proofreading it. One of the authors (D. R. D.) thanks NSF-31-8303A Grant to SUNY at Stony Brook for financial support in the final stages of preparation of this manuscript.

References

1. I. Rabi, *Phys. Rev.*, **51**, 652 (1937).
2. F. Bloch and A. Siegert, *Phys. Rev.*, **57**, 522 (1940).
3. A. F. Stevenson, *Phys. Rev.*, **58**, 1061 (1940).
4. S. H. Autler and C. H. Townes, *Phys. Rev.*, **100**, 703 (1955).
5. J. M. Winter, *Ann. Phys. (Paris)*, **4**, 745 (1959).
6. J. H. Shirley, Thesis, California Institute of Technology (unpublished) (1963).
7. M. P. Silverman and F. M. Pipkin, *J. Phys. B: Atom. Mol. Phys.*, **5**, 1844 (1972).
8. D. T. Pegg, *J. Phys. B: Atom. Mol. Phys.*, **6**, 246 (1973).
9. C. S. Chang and P. Stehle, *Phys. Rev.*, **A4**, 641 (1971).
10. C. S. Chang and P. Stehle, *Phys. Rev.*, **A5**, 1087 (1972).
11. S. Stenholm, *J. Phys. B: Atom. Mol. Phys.*, **6**, 1650 (1973).

12. S. Stenholm, *J. Phys. B: Atom. Mol. Phys.*, **5**, 878 (1972).
13. S. Stenholm, *J. Phys. B: Atom. Mol. Phys.*, **5**, 890 (1972).
14. S. Stenholm, *J. Phys. B: Atom. Mol. Phys.*, **6**, 1097 (1973).
15. J. C. McGurk, T. G. Schmalz, and W. H. Flygare, in *Advances in Chemical Physics*, Vol. XXV, edited by I. Prigogine and S. Rice, Wiley, New York, 1974.
16. P. A. M. Dirac, *Proc. Roy. Soc.* (*London*) *Ser. A*, **114**, 243 (1927).
17. G. Wentzel, *Z. Phys.*, **41**, 828 (1927).
18. O. Klein and Y. Nishina, *Z. Phys.*, **52**, 853 (1929).
19. O. Klein, *Z. Phys.*, **41**, 407 (1927).
20. N. Bloembergen, *Nonlinear Optics*, Benjamin, New York, 1965.
21. M. Scully and M. Sargent III, *Phys. Today*, **25**, 38 (1972).
22. S. Haroche, *Ann. Phys.*, **6**, 189 (1971); **6**, 327 (1971).
23. T. von Foerster, *Am. J. Phys.*, **40**, 854 (1972).
24. S. Stenholm, *Phys. Rept.*, **6**, 1 (1973).
25. J. H. Shirley, *Phys. Rev.*, **138**, **B**979 (1965).
26. R. P. Feynman, F. L. Vernon, and R. W. Hellwarth, *J. Appl. Phys.*, **28**, 49 (1957).
27. C. R. Stroud, Jr., and E. T. Jaynes, *Phys. Rev.*, **A1**, 106 (1970).
28. V. Weisskopf, and E. Wigner, *Z. Phys.*, **63**, 54 (1930).
29. A. Maitland and M. H. Dunn, *Laser Physics*, North-Holland, Amsterdam, 1969.
30. N. Ashby, *Proc. of Boulder Institute of Theoretical Physics*, 599 (1968).
31. G. W. Series, in *Quantum Optics*, edited by S. M. Kay and A. Maitland, Academic, New York, 1970.
32. H. E. Meadows, *Bell System Tech. J.*, **41**, 1275 (1962).
33. G. Floquet, *Ann. de l'Ecole Norm. Sup.* **12**, 47 (1883).
34. J. H. Poincaré, *Les Méthodes Nouvelles de la Méchanique Céleste*, Vols. I, II, III, Paris, 1892, 1893, 1899.
35. F. R. Moulton, D. Buchanan, T. Buck, F. Griffin, W. Longley, and W. MacMillan, *Periodic Orbits*, The Carnegie Institution of Washington, Washington, D. C., 1920.
36. F. R. Moulton, *Differential Equations*, MacMillan, New York, 1930.
37. L. I. Schiff, *Quantum Mechanics*, McGraw-Hill, New York, 1955.
38. P. Langhoff, S. T. Epstein, and M. Karplus, *Rev. Mod. Phys.*, **44**, 602 (1972).
39. J. O. Hirschfelder, C. F. Curtiss, and R. B. Bird, *Molecular Theory of Gases and Liquids*, Wiley, New York, 1964.
40. N. D. Sen Gupta, *J. Phys. A: Gen. Phys.*, **3**, 618 (1920).
41. N. D. Sen Gupta, *J. Phys. A: Gen. Phys.*, **5**, 401 (1972).
42. N. D. Sen Gupta, *Phys. Rev.*, **A7**, 891 (1973).
43. H. Sambe, *Phys. Rev.*, **A7**, 2203 (1973).
44. R. H. Young, W. J. Deal, Jr., and N. R. Kestner, *Mol. Phys.*, **17**, 369 (1969).
45. R. H. Young and W. J. Deal, Jr., *J. Math. Phys.*, **11**, 3298 (1970).
46. J. M. Okuniewicz, Thesis, University of Minnesota (unpublished) (1972).
47. J. M. Okuniewicz, *J. Math. Phys.*, **15**, 1587 (1974).
48. W. W. Hicks, R. A. Hess, and W. S. Cooper, *Phys. Rev.*, **A5**, 490 (1972).
49. W. R. Salzman, *Phys. Rev.*, **A10**, 461 (1974).
50. H. Salwen, *Phys. Rev.*, **99**, 1274 (1955).
51. P. T. Pegg and G. W. Series, *Proc. Roy Soc.* (*London*) *Ser. A,* **A332**, 281 (1973).
52. C. Cohen-Tannoudji, J. DuPont-Roc, and C. Fabre, *J. Phys. B: Atom. Mol. Phys.*, **6**, L214 (1973).
53. J. H. Wilkinson, *The Algebraic Eigenvalue Problem*, Clarendon, Oxford, 1965.
54. S. D. Conté and C. deBoor, *Elementary Numerical Analysis*, McGraw-Hill, New York, 1972.

55. J. H. Wilkinson and C. Reinsch, *Handbook for Automatic Computation*, Springer-Verlag, New York, 1971.
56. D. R. Dion, Ph.D. Thesis, University of Wisconsin—Madison, Theoretical Chemistry Institute Report WIS-TCI-502 (1974).
57. E. L. Ince, *Ordinary Differential Equations*, Dover, New York, 1956.
58. H. Margenau and G. M. Murphy, *The Mathematics of Physics and Chemistry*, Van Nostrand, Princeton, N. J., 1956.
59. L. Brillouin, *J. Appl. Math.*, **6**, 167 (1948).
60. L. Brillouin, *J. Appl. Math.*, **7**, 363 (1950).
61. E. A. Coddington, *Ordinary Differential Equations*, Prentice-Hall, Inc., Englewood Cliffs, N. J., 1961.
62. F. W. J. Olver, in *Handbook of Mathematical Functions*, edited by M. Abramowitz and I. A. Stegun, U. S. Government Printing Office. Washington, D. C., 1964, Chapter 9.
63. J. G. Ross, Thesis, University of Colorado (unpublished) (1969).
64. R. Zucker, in *Handbook of Mathematical Functions*, edited by M. Abramowitz and I. A. Stegun, U. S. Government Printing Office, Washington, D. C., 1964, Chapter 4.
65. St. Bobr, *Math. Z.*, **10**, 1 (1921).
66. L. B. W. Jolley, *Summation of Series*, Chapman and Hall, London, 1925.
67. R. V. Churchill, *Complex Variables and Applications*, McGraw-Hill, New York, 1960.
68. J. O. Hirschfelder and P. R. Certain, *J. Chem. Phys.*, **60**, 1118 (1974).
69. J. O. Hirschfelder, W. Byers Brown, and S. T. Epstein, in *Advances in Quantum Chemistry*, Vol. I, edited by P. O. Löwdin, Academic, New York, 1964.
70. S. T. Epstein, University of Wisconsin—Madison, Theoretical Chemistry Institute Report WIS-TCI-339 (1969).
71. J. Musher, *Ann. Phys.* (*New York*), **27**, 167 (1964).
72. M. Born and V. Fock, *Z. Phys.*, **40**, 165 (1928).
73. P.-O. Löwdin, in *Perturbation Theory and Its Applications in Quantum Mechanics*, edited by C. H. Wilcox, Wiley, New York, 1966.
74. P.-O. Löwdin, *J. Math. Phys.*, **6**, 1341 (1965).
75. P. R. Certain and J. O. Hirschfelder, *J. Chem. Phys*, **52**, 5977 (1970).
76. P. R. Certain and J. O. Hirschfelder, *J. Chem. Phys.*, **52**, 5992 (1970).
77. P. R. Certain, D. R. Dion, and J. O. Hirschfelder, *J. Chem. Phys.*, **52**, 5987 (1970).
78. W. Heitler, *The Quantum Theory of Radiation*, 3rd ed., Clarendon, Oxford, 1960.
79. S. T. Epstein, *Variational Method in Quantum Chemistry*, Wiley, New York, 1974.
80. L. D. Landau and E. M. Lifshitz, *Quantum Mechanics*, Pergamon, London, 1965.
81. D. T. Pegg and G. W. Series, *J. Phys.*, **B3**, L33 (1970).
82. F. Ansbacher, *J. Phys. B: Atom. Mol. Phys*, **6**, 1620 (1973).
83. F. Ansbacher, *J. Phys. B: Atom. Mol. Phys*, **6**, 1633 (1973).
84. A. H. Nayfeh, *Perturbation Methods*, Wiley, New York, 1973.
85. J. D. Cole, *Perturbation Methods in Applied Mathematics*, Blaisdell, Waltham, Mass., 1968.
86. C. F. Curtiss and J. O. Hirschfelder, *Proc. Nat. Acad. Sci.*, **38**, 235 (1952).
87. G. Blanch, in *Handbook of Mathematical Functions*, edited by M. Abramowitz and I. A. Stegun, U. S. Government Printing Office, Washington, D. C., 1964, Chapter 20.
88. P. K. L. Drude, *Theory of Optics*, Longmans Green, London, 1933.
89. H. A. Lorentz, *The Theory of Electrons*, 2nd ed., Dover, New York, 1952.
90. L. Pauling and E. B. Wilson, *Introduction to Quantum Mechanics*, McGraw-Hill, New York, 1935.

AUTHOR INDEX

Numbers in parentheses are reference numbers and show that an author's work is referred to although his name is not mentioned in the text. Numbers in *italics* indicate the pages on which the full references appear.

SUBJECT INDEX